Beyond the Networked City

Cities around the world are undergoing profound changes. In this global era we live in a world of rising knowledge economies, digital technologies and awareness of environmental issues. The so-called "modern infrastructural ideal" of spatially and socially ubiquitous centrally-governed infrastructures providing exclusive, homogeneous services over extensive areas has been the standard of reference for the provision of basic essential services, such as water and energy supply. This book argues that after decades of undisputed domination this ideal is being increasingly questioned, and the network ideology that supports it may be waning.

In order to begin exploring the highly diverse, fluid and unstable landscapes emerging beyond the networked city, this book identifies dynamics through which a 'break' with previous configurations has been operated, and brittle zones of new socio-technical controversy through which urban infrastructure (and its wider meaning) are being negotiated and fought over. It uncovers, across a diverse set of urban contexts, new ways in which processes of urbanization and infrastructure production are being combined with crucial sociopolitical implications: through shifting political economies of infrastructure which rework resource distribution and value creation; through new infrastructural spaces and territorialities which rebundle socio-technical systems for particular interests and claims; and through changing offsets between individual and collective appropriation, experience and mobilization of infrastructure.

With contributions from leading authorities in the field and drawing on theoretical advances and original empirical material, this book is a major contribution to an ongoing infrastructural turn in urban studies, and will be of interest to all those concerned by the diverse forms and contested outcomes of contemporary urban change across the global North and South.

Olivier Coutard is a director of research with the French National Centre for Scientific Research (CNRS), where he has held a full-time research position since 1996.

Jonathan Rutherford holds a research post at LATTS (Laboratoire Techniques, Territoires et Sociétés), Université Paris-Est.

Routledge studies in urbanism and the city

This series offers a forum for original and innovative research that engages with key debates and concepts in the field. Titles within the series range from empirical investigations to theoretical engagements, offering international perspectives and multidisciplinary dialogues across the social sciences and humanities, and encompassing urban studies, planning, geography, geohumanities, sociology, politics, the arts, cultural studies, philosophy and literature.

Published:

The Urban Political Economy and Ecology of Automobility
Driving cities, driving inequality, driving politics
Edited by Alan Walks

Cities and Inequalities in a Global and Neoliberal World
Edited by Faranak Miraftab, David Wilson and Ken E. Salo

Beyond the Networked City
Infrastructure reconfigurations and urban change in the North and South
Edited by Olivier Coutard and Jonathan Rutherford

Forthcoming:

Mega-Urbanization in the Global South
Fast cities and new urban utopias of the postcolonial state
Edited by Ayona Datta and Abdul Shaban

Beyond the Networked City

Infrastructure reconfigurations and urban change in the North and South

Edited by Olivier Coutard and Jonathan Rutherford

Routledge
Taylor & Francis Group
LONDON AND NEW YORK

First published 2016
by Routledge
2 Park Square, Milton Park, Abingdon, Oxon OX14 4RN

and by Routledge
711 Third Avenue, New York, NY 10017

First issued in paperback 2017

Routledge is an imprint of the Taylor & Francis Group, an informa business

British Library Cataloguing in Publication Data
A catalogue record for this book is available from the British Library

Library of Congress Cataloging in Publication Data
Beyond the networked city : infrastructure reconfigurations and urban change in the North and South / edited by Olivier Coutard and Jonathan Rutherford.
pages cm. -- (Routledge studies in urbanism and the city)
Includes bibliographical references and index.
1. Infrastructure (Economics) 2. Urbanization. 3. Information technology. 4. Telecommunication. 5. Municipal services. 6. Regional development. I. Coutard, Olivier, 1965- editor. II. Rutherford, Jonathan, editor.
HC79.C3B49 2016
363--dc23
2015024605

ISBN 13: 978-1-138-30837-4 (pbk)
ISBN 13: 978-1-138-79682-9 (hbk)

Typeset in Times New Roman
by Saxon Graphics Ltd, Derby

Contents

Figures

Tables

Contributors

Sabine Barles is professor of urban planning at Université Paris 1 Panthéon Sorbonne. She is a civil engineer (1988) and holds an MSc in Urban Planning, an MSc in History of Technology and a PhD in Urban Planning (1993). Her main research topics concern urban environmental history and the history of urban technology (eighteenth to twenty-first centuries), urban metabolism and territorial ecology. She has published *La ville délétère, médecins et ingénieurs dans l'espace urbain, XVIIIe-XIXe siècles* (1999), *L'invention des déchets urbains, France, 1790–1970* (2005) and papers in *The Science of the Total Environment, Hydrology and Earth System Sciences, Regional Environmental Change, Journal of Urban History* etc.

Ross Beveridge PhD is a research fellow at the Leibniz Institute for Regional Development and Structural Planning (IRS) in Erkner. His work focuses on environmental and infrastructural politics at the urban and regional levels.

Jérémie Cavé is an independent consultant in urban ecology. After a Master's in political sciences (Sciences Po Paris), he acquired professional experience through assisting urban local agencies with their local public services management. He then conducted a PhD in urban planning at LATTS focusing on the political economy of waste in emerging countries. Since then he has been working both as a researcher and as a consultant in the environmental field for the European Commission, the French Environment and Energy Efficiency Agency (ADEME) and the French Development Agency (AFD).

Olivier Coutard is a director of research with the French National Centre for Scientific Research (CNRS), where he has held a full-time research position since 1996. He was trained as a civil engineer (ENPC 1988) and holds a Master's degree in transportation socio-economics (1988) and a PhD in economics and social sciences (ENPC 1994). His current research addresses urban energy (and "energy transition") policies in Europe, the transformations in the sociotechnical organization of urban services, and the politics of the "post-networked city". He has edited *The Governance of Large Technical Systems* (1999) and *Sustaining Urban Networks* (2005, with Richard Hanley and Rae Zimmerman), both published by Routledge, and *Ecologies Urbaines* (with Jean-Pierre Lévy, 2010, Anthropos/Economica).

Renu Desai is the Coordinator of the Centre for Urban Equity, CEPT University, Ahmedabad. Her research examines urban informality and urban transformation in contemporary Indian cities, with a focus on questions of equitable development and urban citizenship. This work has focused on three sometimes intersecting themes: the first is the articulations of urban redevelopment and Hindu right-wing politics in the reshaping of urban space; the second is informal tenure and the politics of dispossession and contestation; and the third is water and sanitation infrastructures in low-income neighbourhoods. Renu has published several book chapters and journal articles on her research, and is also a co-editor of *Urbanizing Citizenship: Contested Spaces in Indian Cities* (New Delhi: Sage 2012).

Kathryn Furlong is an associate professor in the Department of Geography at the Université de Montréal and holds the Canada Research Chair in Urban, Water and Utility Governance. Her book *Leaky Governance: Alternative Service Delivery and the Myth of Water Utility Independence* will be published by UBC Press in November 2015. Her most recent work looks at the shifting nature of public utility corporations in Colombia and the Netherlands.

Stephen Graham is Professor of Cities and Society at the Global Urban Research Unit in Newcastle University's School of Architecture, Planning and Landscape. Professor Graham has a background in geography, urbanism and the sociology of technology. His research addresses the politics of cities, focusing especially on issues of mobility, infrastructure, technology, security, surveillance and political violence. His books include *Telecommunications and the City* and *Splintering Urbanism* (both with Simon Marvin), *The Cybercities Reader*, and *Cities, War and Terrorism*. His two most recent books are *Disrupted Cities: When Infrastructures Fail* (Routledge, 2009) and *Cities Under Siege: The New Military Urbanism* (Verso, 2010).

André Guillerme is Professor of History of Technology at the Conservatoire National des Arts et Métiers in Paris. His fields of research include the history of waterscape, handcrafts, gesture and industrial ambiance. His publications include *Nuts and Bolts: Culture and technique*, IVth International Congress of Construction History (Picard, 2012) (edited with Carvais, Nègre and Sakarowitch), *La Naissance de l'Industrie à Paris: entre sueurs et vapeurs (1780-1830)* (Champ Vallon, 2007); "Enclosing nature in the city: supplying light and water in Paris, 1770–1840" in the *Journal of Construction History* (2011), "International exhibitions in the late nineteenth century as a showcase for demountable army field huts: a protohistory of prefabrication in building" in *Engineering History and Heritage* (2012), and "Paris 1800–1950" (with S. Barles), in Torje Osticardt, *Water and Urbanization,* published by I.B.Tauris.

Simon Guy is Dean of the Faculty of Arts and Social Sciences at Lancaster University in the UK, and was previously the founding Director of the Manchester Architecture Research Centre at the University of Manchester. A social scientist by background, Professor Guy's research is aimed at critically understanding the co-evolution of design and development strategies and socio-technical-ecological processes shaping cities. His research specialism is in interdisciplinary approaches to sustainable urbanism, engaging with changing forms of architectural knowledge and practice and specifically with debates about buildings, cities and the environment.

Sylvy Jaglin is professor of development and urban studies at Université Paris-Est and currently director of its Doctoral School on Cities, Transport and Territories. She was trained in geography and holds a PhD in urban planning. She is a member of LATTS (Laboratoire Techniques, Territoires et Sociétés), where she co-ordinates research focusing on the interface between cities and technical infrastructure systems that provide basic essential services in the global South, with a special interest in sub-Saharan Africa. She also leads the research network SeRVeD (Services en Réseaux dans les Villes en Développement/Networked Services in Developing Cities), which is a partnership between the Chaire Ville of the École des Ponts ParisTech and the French Agency for Development (AFD).

Andrew Karvonen is Lecturer in Architecture and Urbanism at the University of Manchester and co-director of the Centre for Urban Resilience and Energy. He conducts research on the political and cultural aspects of urban infrastructure with a particular emphasis on sustainable transitions. He has published widely in geography, planning, architecture, and urban studies on water, energy, and the built environment. His 2011 MIT Press monograph, *Politics of Urban Runoff: Nature, Technology, and the Sustainable City* won the John Friedmann Book Award from the Association of Collegiate Schools of Planning for exemplary scholarship in the area of planning for sustainable development.

Simon Marvin is a professor and Director of the Urban Institute at the University of Sheffield. His research interests focus on the changing relationships between cities and infrastructure networks. More recently his work has focused on comparative analysis of low carbon transitions and the interactions between smart technologies and urban contexts.

Colin McFarlane is an urban geographer based at Durham University, UK. His work focuses on urban learning, informality and infrastructure in cities. This has included research on the politics of urban knowledge, urban sanitation, and everyday life in informal neighbourhoods, especially in Mumbai, Cape Town and Kampala. He is author of *Learning the City: Knowledge and Translocal Assemblage* (Wiley-Blackwell), co-editor with Steve Graham of *Infrastructural Lives* (Routledge), co-editor with Jon Anjaria of *Urban Navigations* (Sage), co-editor with Michael Waibel of *Urban Informalities* (Ashgate), and co-editor with Andres Luque and Simon Marvin of *Smart Urbanism: Utopian Vision or False Dawn?* (Routledge).

Jochen Monstadt holds the chair of spatial and infrastructure planning and is the director of the Graduate School for Urban Studies at TU Darmstadt, Germany. His research interests are at the interface between urban/regional studies and social studies of technology. He has conducted and coordinated extensive comparative research on the urban/regional transition of energy, water and wastewater systems, on the political ecology of urban infrastructures, and on infrastructure planning and urban governance in Germany, Switzerland, Canada, the United States and Vietnam. His current research involves urban infrastructure in East Africa, interlinkages between different infrastructure domains, and urban and infrastructure vulnerabilities.

Matthias Naumann PhD is a human geographer at the Leibniz Institute for Regional Development and Structural Planning in Erkner, and a visiting lecturer at Brandenburg University of Technology Cottbus-Senftenberg. His research interests include urban and regional development, infrastructure governance and critical geography.

Jonathan Rutherford holds a research post at LATTS (Laboratoire Techniques, Territoires et Sociétés), Université Paris-Est. His research interests are in the processes and politics of urban socio-technical change through a focus on the shifting relations between infrastructure and cities, and on urban energy transitions.

Sophie Schramm's research focuses on urban infrastructure systems in the global South. She studied Urban and Regional Planning at the HCU in Hamburg. She spent several months at UN-Habitat in Nairobi, Kenya where she studied informal urbanization and slum upgrading initiatives. In 2008, she joined the TU Darmstadt, where she has worked in research projects on urban infrastructures in the global South and as a scholar within the DFG Graduate School Topology of Technology. In 2014, she published her dissertation thesis *City in Flow – Hanoi's wastewater system in light of social and spatial transformations*. Currently she is working on a project on adaptation and creativity in water and sanitation systems in African cities funded by the German Research Foundation (DFG).

Elizabeth Shove is a professor of sociology at Lancaster University and co-Director of the DEMAND research centre (Dynamics of Energy, Mobility and Demand): www.demand.ac.uk. Recent books include *Sustainable Practices: Social Theory and Climate Change* (Routledge, 2013) edited with Nicola Spurling, and *The Dynamics of Social Practice: Everyday life and how it changes* (Sage, 2012) with Mika Pantzar and Matt Watson. Elizabeth has written a fair bit about energy, everyday practices, systems of practice and transitions in comfort, cleanliness and convenience.

Janette Webb is Professor of Sociology at the University of Edinburgh, UK. She has researched and taught the sociology of markets, organizations and consumption for over thirty years. Her current research is about socio-technical innovation in sustainable energy for cities (http://www.heatandthecity.org.uk), in the context of centralized energy markets. The research team collaborates with a network of fifty UK urban authorities, plus government officials and expert practitioners. Her research contributes to policy-making: she is a member of the Scottish Government Expert Commission on District Heating, and she contributed evidence to the UK Government 2013 Heat Strategy. She played a leading role in the development, launch and teaching of the University of Edinburgh's Honours Degree in Sustainable Development.

Acknowledgements

We acknowledge the contributions of the Chaire Ville of the Ecole des Ponts ParisTech, Université Paris-Est Marne-la-Vallée, and the French National Research Agency (ANR) in providing the funding for the July 2012 seminar 'From networked to post-networked urbanism: new infrastructure configurations and urban transitions' in Autun, Burgundy, at which many of the chapters in the book were presented.

We thank Faye Leerink, Emma Chappell and Routledge for their support and help with the publication of the book.

Thanks especially to all the contributors for their prompt responses to our various queries and questions and for great chapters.

We are also both grateful to our colleagues in LATTS over many years for continuing to provide a stimulating environment in which to discuss and work on urban infrastructure change.

Olivier Coutard and Jonathan Rutherford

1 Beyond the networked city

An introduction

Olivier Coutard and Jonathan Rutherford

Beyond the networked city: new infrastructural territorialities

This book examines the diverse and contested processes through which the "networked city" (Tarr and Dupuy 1988) is currently being called into question, both ideally and materially, across the global North and South, and explores emergent new urban landscapes and infrastructural conditions and their socio-political implications.

In a very general sense, the term "networked city" refers to urban configurations that work through the socially and spatially ubiquitous provision of essential services such as transport and telecommunications, water and energy supply, and sanitation and waste management, based on the universal deployment of networked infrastructure systems. These technological networks have been central to urban development and change worldwide over the last one hundred and fifty years. Even in urban contexts where the universal deployment of networks has not been achieved, a "modern infrastructural ideal" (Graham and Marvin 2001) was for a long time, and to a large extent still is, the standard of reference for the provision of basic services for international investors and funding agencies, national and local policy makers, and academic experts alike.

Starting from the now well acknowledged ideas that, after a long period of undisputed domination, the "modern infrastructural ideal" has been increasingly questioned in recent decades (Graham and Marvin 2001) and that the network ideology that supports this ideal may be waning (Coutard 2010), the book explores how this transformation relates to, and is part of, broader processes of urban change. The contributions to this volume spearhead this exploration on three interdependent levels by studying:

- changing socio-technical configurations and political economies of infrastructures;
- changing infrastructural spaces and territorialities;
- changing social, political and urban meanings of infrastructures.

By mapping urban infrastructure landscapes "beyond the networked city" in material, spatial, political and conceptual terms, the volume offers a global perspective on how infrastructural territorialities are being decomposed and reassembled for particular claims and interests through the relational work and situated effects of various combinations of actors, techniques and rationalities. In accounting for the changing forms, meanings and outcomes of urban infrastructural territoriality, we are not seeking, nor indeed promoting, an all-encompassing explanatory logic for 'post-networked' configurations emerging in very different urban contexts. Rather, the book adopts a critical socio-technical perspective in detailing how and why specific configurations are produced/reproduced in certain places, and for and by whom they are put to work. In this regard, it is both analytically heuristic and normatively inescapable to explore the diversity of existing and emerging configurations and how they affect, and are affected by: the political economy of infrastructure industries and services; the more or less extensive socio-spatial deployment of networked infrastructures; the extent of the socioeconomic diversity of, and disparities within, urban societies; and the effectiveness of public control over urbanisation processes.

In this introductory essay, we start by briefly sketching the broad conditions under which the networked city – apprehended as a material/technological, spatial, institutional, ethical, ideological and conceptual assemblage – is being destabilised and potentially disjoined. We then identify three main lines of reconfiguration along which post-'networked city' landscapes are emerging and being shaped – or, as the case may be, are being actively reproduced or reconfigured. In doing so we do not — and indeed we do not aim to — delineate a clear, let alone a general, model of a post-networked city. Rather, we seek to uncover a set of diverse and intensely politicised processes and trends by which infrastructures are used to variously reflect, adapt to and contribute to the shaping of urban spaces with multiple and often crucial socio-spatial, techno-economic and environmental implications.

Building on Graham and Marvin's pioneering critical analysis and the wealth of work on urban infrastructure conducted over the past two decades or so, the book focuses on the socio-technical logics, rationalities, modalities and work (operations) which underpin and feed into practices and politics of producing, managing and using specific hybrid or alternative infrastructure configurations and arrangements. Through this focus and entry point, it uncovers, across a very diverse set of urban contexts, new ways in which urbanisation processes and infrastructure production are being combined with important sociopolitical implications:

- through shifting political economies of infrastructure (resource distribution, value creation, particular calculations and trade-offs between social, political, regulatory and economic management and integration…);
- through new infrastructural spaces and territorialities formed from these configurations (unbundling and rebundling of socio-technical systems for particular interests and claims);
- and through changing offsets between individual and collective appropriation, experience and mobilisation of infrastructure.

Taken together, the chapters in this book demonstrate that infrastructure is more than the simple provision of basic essential urban services, and that how it is continuously made and remade in particular ways is revealing both of how and why it matters (technically, socially, environmentally, politically, economically) in particular contexts, and how these matters are always in flux and open to interpretation in a variety of contingent ways. This notional, material and sociopolitical fluidity to the processes and possibilities of urban infrastructure can open up new theoretical and conceptual avenues for further enquiry into urban socio-technical change.

The networked city revisited/synthesised

We must start, however, by briefly discussing the (notion of the) *networked city* beyond which this book is seeking to proceed empirically, analytically and theoretically.

Over the last thirty years or so, the importance of infrastructure networks in shaping urban environments has gradually become a central object of study in geography, planning, urban history, sociology, political science and associated disciplines. Infrastructure has been shown to be at once a site, an instrument and an outcome of urban economic production and expansion, increasing social inequalities, and environmental transformation. In monographs, edited collections and special issues of journals such as the Journal of Urban History, Technology and Culture, the Journal of Urban Technology and Flux, the pioneering work of Joel Tarr, Gabriel Dupuy and others on 'the rise of networked cities' in history has showed how crucial water, energy, waste, telecommunications and transport infrastructures were to the construction of 'modern' cities in Europe and North America (see for example Tarr 1979; Hughes 1983; Tarr and Konvitz 1987; Tarr and Dupuy 1988; Dupuy 1991; Rose and Tarr 2004).

This body of work showed that during the historical period roughly spanning the mid-nineteenth century to the early twenty-first century, in countries in the North the development of networked infrastructure systems was simultaneously a driver, an instrument and a consequence of the advent of modern forms of relations in territories, organisations, and social practices on and across all spatial scales. This resulted in the socially and spatially ubiquitous provision of, and access to, more or less homogeneous services supplied sometimes on a local and sometimes on a larger scale, by a centrally planned and managed large technological system with exclusive franchise over a given territory which it in turn helped to unify.

The ubiquitous development of networked infrastructures in Western Europe, North America and Japan in particular was associated with the rise of a network ideology which can be summarised by three main elements: the network is the most efficient means of supplying urban services; network performance increases with size (spatial coverage, the number and diversity of connected users); and solutions to problems created by networks reside within the networks themselves (in their expansion, more centralised management and increased technical sophistication) (Coutard 2010). The material development of networks in global

North countries and the network ideology have mutually reinforced each other, and together they have supported the global diffusion of the 'networked city' as a hegemonic socio-technical form sustaining dual areal/relational forms of urban territoriality.

In the North where the universal deployment of networked infrastructure systems was achieved, the advent of the networked city historically, from the mid-nineteenth century, has been shown to be associated with various forms of integration or solidarity:

- socioeconomic integration through the (cross-) subsidised provision of homogeneous essential services across urbanised spaces;
- functional integration, as networked infrastructures have facilitated and sustained spatial practices on the scale of ever broader living areas;
- political integration, through the cooperation between contiguous local governments encouraged by the thrust to provide homogeneous services within entire urban regions;
- and "metabolic integration", as the development of large, monopolistic, centrally-managed networked infrastructure systems resulted in, and rested on, the appropriation of (remote) resources for the benefit of the development of the entire urban region.

It is essential to note, however, that "networked cities" come in very diverse forms, and that the ubiquitous diffusion of networked infrastructure systems has both facilitated and been facilitated by the different types, degrees and combinations of agglomeration/concentration and expansion/dispersion that have marked contemporary urban development in the North (cf. Soja 2000; Castells 2010; Brenner 2013). In other words, infrastructural socio-spatial configurations do not automatically determine urban patterns, but the two are closely intertwined.

Furthermore, if in countries of the global North this ideal figure of the networked city was more or less achieved materially, territorially and politically, in the South, very diverse configurations have prevailed. The networked city model has been actively implemented in some cases (e.g. in Latin America), albeit sometimes with ambivalent aims, while in others it has remained a very broad objective or slogan (cf. the contributions in Coutard 2008; Lorrain 2011). In Santiago, for instance, the connection of all residential units to essential networks has, particularly since the 1980s, facilitated and even legitimated a strong policy of spatial segregation and functional, social, fiscal and political fragmentation at the local authority level within the urban region, as well as sustained processes of urban sprawl (Pflieger and Matthieussent 2008). In Buenos Aires, a policy strictly conforming to the modern infrastructural ideal resulted in the creation of a premium networked space (in the central part of the federal capital) delimited by the perimeter of the water mains and excluding half the population of the urban region (Botton and De Gouvello 2008).

Hence, empirical investigations reveal that more or less ubiquitous infrastructure networks have coexisted and even been related to more or less stark socio-spatial disparities. This fundamentally reflects the fact that networked infrastructures can only be universally deployed and produce dominantly integrative socio-spatial effects under broader socioeconomic, institutional and political conditions, involving in particular increasing revenues and wealth over time, some degree of welfare state or socially progressive revenue redistribution, the elimination of extreme poverty, and a degree of control over urbanisation and urban migration processes. Yet even where it was not materially achieved (in most contexts of the global South), the networked city ideal was politically and ideologically powerful, and it widely informed and orientated the policies promoted by international funding agencies and national and local governments.

Box: Synthesising the characteristics of the networked city

Ubiquitous network infrastructures and services, and associated values and principles (supply-oriented, planned for growth, materialising a linear metabolism, emphasising flows over resources...)

Network ideology (universal reticulation, spatial extension, organisational expansion, sophistication) and supply-side ideology (infrastructure and service provision; limited consideration of demand/use/consumption/practices)

Homogeneous infrastructural spaces (of provision), norms, conventions, expectations

Urban citizenship through "network citizenship" (including ideas of solidarity through use, subscription to collective service...)

Importance of: public providers of essential services, public oversight of (especially privately-owned) providers, welfare state and revenue redistribution, control on urban development

The networked city challenged

Over the last twenty years or so, building on the previous, mainly historical studies of the co-development of cities and infrastructures, more global and more critical perspectives have developed, epitomised by Graham and Marvin's (2001) seminal work on "splintering urbanism". These perspectives analyse the ways in which very particular and selective forms and degrees of infrastructure provision have become bound into and result from wider techno-economic logics and processes of globalisation, liberalisation, privatisation and digitisation, and the regressive social and socio-spatial consequences of unbundled, fragmented network systems around the world. Most recently there has been a focus on 'extreme' splintering tendencies and externalities of reticulation associated with infrastructure crisis, failure and discourses of ecological security (Graham 2009; Hodson and Marvin 2009; 'the purgatory of networks': Dupuy 2011). Through these distinct periods of research, one constant has been an interest in studying the diversity of ways in which infrastructure networks come to matter in particular urban contexts, both as a technology of governing and as a site of contestation of strategies of domination and the reworking of power relations, thus making, unmaking and constantly remaking the urban in particular, highly contested ways (McFarlane and Rutherford 2008).

While an in-depth overview of work on urban infrastructure over the past fifteen to twenty years would mention a number of interconnected strands and perspectives (see for recent overviews: Coutard 2008; McFarlane and Rutherford 2008; Monstadt 2009; Guy *et al.* 2011), two streams of research have been particularly influential and indeed help to situate the argument and chapters in this collection. In recent research on infrastructure in cities in the South, on the one hand, researchers have explored the widespread urban implications of the absence or dysfunction of, but also long-standing alternatives to, large technical systems for access to essential services (mainly water and sanitation), underscoring both how the 'modern infrastructural ideal' has rarely been achieved in cities of the South (Swyngedouw 2004; Jaglin 2005; the contributions to Coutard 2008; Gandy 2008), and, indeed, crucially, how this Northern ideal and construction is problematic in its transposition to a very different set of diverse and fragmented (postcolonial) urban contexts in Africa, Asia and Latin America (Bakker 2003; Simone 2004; McFarlane 2008). In work in the field of urban environmental change and urban political ecology, on the other hand, infrastructure has been shown to be one of the key sites for contestations over access to environmental amenities and resource flows (Melosi 2000; Guy *et al.* 2001; Hodson and Marvin 2009; Karvonen 2011; Bulkeley *et al.* 2014) and the production of infrastructure to be a major urban political process working through unequal socio-natural spaces of environmental transformation, thus utterly blurring artificial distinctions between nature and the city (Kaika and Swyngedouw 2000; Gandy 2002; Keil 2003; Luke 2003; Kaika 2005; Heynen *et al.* 2006).

These bodies of literature have amply demonstrated that in all parts of the world, albeit for different reasons according to national, regional and local contexts, conventional networked infrastructure systems have been increasingly contested as a hegemonic way of delivering essential services. This shift is having tangible implications on infrastructure policies worldwide. Where conventional networks had historically become hegemonic, there is now support for the development of diverse system configurations. Where networked systems were not (fully) rolled out, other existing configurations are perpetuated.

It should be emphasised, however, that shifting forms of infrastructure never consist of big, paradigm-busting transitions from one large technical system to another. Existing systems are not being dismantled and withdrawn to be replaced by brand new alternative systems as straight technological fixes. Incremental changes, small-scale adjustments and repairs, the integration of 'mediating technologies' (Furlong 2011) or new layers of devices offering increased flexibility and feedback into existing (already multilayered) infrastructure systems are all potentially or actually producing smaller or bigger reconfigurations of urban functioning and experience (see, for example, Moss 2000). The same applies to the renaissance of previously existing technological systems which might have disappeared or lain dormant for a time, only to re-emerge as their qualities or components are found to be relevant again (Shove 2012).

These highly differentiated forms and degrees of technical change are obviously also closely articulated in distinct ways with (indeed, they cannot be demarcated from) evolving social relations and processes, such that infrastructure must be seen as constituted by co-configurations of various 'hardwares' and 'softwares' (Walker and Cass 2007). As past and current STS and transitions literatures demonstrate, socio-technological change is always about a contingent mix of different types and degrees of change, in relation to evolutionary dynamics of innovation and more stable components of inertia, path dependency and obduracy (Bijker *et al.* 1987; Hommels 2005; Geels and Schot 2007).

Hence change in infrastructure systems never occurs in one movement; it always involves frictions and tensions between the largely independent dynamics of these various dimensions. Yet some general features of ongoing transformations in material infrastructures and/or the dominant infrastructural ideal may be identified:

- a shift from (effective or targeted) homogeneity to (acknowledged) diversity (and targeted selectivity) of infrastructural spaces, including rebundling of services, practices, technologies, standards, flows, service suppliers, etc. in particular ways;
- reconfigurations and rescalings in the "spaces of solidarity" associated with these systems: large scale solidarities are recombining with forms of local "autonomy";
- socio-spatial solidarity increasingly based on the division (which presupposes the appropriation) of resources, which combine with conventional solidarities based on the provision of homogeneous, shared services;
- a shift from collective to more individualised or diversified practices, norms and expectations in terms of services; and related to this, a shift from "one size fits all" to "tailor-made" customised infrastructure;
- the rise of new forms of individual and collective appropriation of infrastructure; new social and political salience and meanings of infrastructure;
- the international and interurban circulation of models supporting these shifts.

In order to explore the nature, signification and uncertain outcomes of these transformations, we argue that it is necessary from an analytical perspective to change approach and move on from the (implicit or explicit) normative reference to the 'networked city' when attempting to identify, describe, understand and account for ongoing changes. What, then, could an approach to urban infrastructure without this frame of reference imply?

Understanding socio-technical urban change

As many authors have shown (see, for example, Bijker *et al.* 1987; Summerton 1994; Star 1999; Coutard and Guy 2007; Farias and Bender 2010; Guy and Karvonen 2011), and as the chapters that follow will elaborate, speaking across an otherwise artificial divide between "the social" and "the technical" (the 'seamless web' of technology and society) means privileging their mutual constitution, and being cognizant at once of the materiality of the technological systems and infrastructures that are both shaping collective action and shaped by it, and the broader social conditions within which technology becomes meaningful(or not) and is enacted in and through social practices (van Vliet *et al.* 2005).

Infrastructure systems have long been conceived of as "multi-layer" systems comprising morphological, material, functional, regulatory, territorial, imaginary and cognitive dimensions (see Offner 1993). Even though these dimensions are in constant interaction, each of them also obeys its own logics and rhythms. This means that we engage with urban infrastructure by seeing it as more than a hard, fixed technological backdrop or a simple instrument of governance and policymaking, and taking seriously the "leaky, partial and heterogeneous entities" (Graham and Thrift 2007, p.10) that constitute the materiality and technicity of its components and flows, notably by studying the ways in which these are put to work for contrasting often competing social and political projects. In short, we argue that a critical understanding of the politics of urban change demands an analysis of how different social interests and groups frame, engage, implement and contest urban imaginaries, operations and activities *through* mobilisations of the inherent materiality and technicity of the urban fabric.

Infrastructure, then, matters. However, it matters not as a single readily identifiable artifact somehow separated or carved off from the rest of urban development and urban life, but as a 'seamless' (and at the same time messy and contingent, and therefore conflictual and inherently political) relational system or arrangement of technology, actors, skills, knowledges, practices, cultural meanings and values, resources, money and politics. Infrastructure is not just the 'technological sinews' (or 'metabolism') of cities and societies, but this is one key element (as simultaneously presupposition, medium and outcome) of the constant coming into being of the urban fabric. Infrastructure is, in this sense, a relational process before it is a 'result', insofar as the shifting, dynamic 'soft' and 'hard' associations and arrangements are every bit as much a part of an infrastructure as the resulting 'hard' technical network. Indeed, because "it is in the messiness of the city where we can find its vitality" (Guy and Karvonen 2011, p.131), the contributors to this volume delve explicitly into the uncertainties, instabilities, controversies and tensions of urban infrastructural processes to interpret the production, reproduction or transformation of socio-technical configurations, their actual un/making of the urban present, and their possible remaking of alternative urban futures.

The rationales, forms and ramifications of ongoing infrastructure reconfigurations are therefore used in this book as an entry point into urban change. The chapters examine the logics, contradictions and tensions present in 'hybridised' socio-technical infrastructure systems emerging at the margins of, in the interstices of, in combination with, or simply in place of existing (or, indeed, absent) centralised infrastructure networks, in order to critically interrogate the shifting interplay between the 'technical' and the 'urban' across North and South. This analysis captures the myriad of urban-infrastructural configurations in a post-'networked city' era, and highlights how ongoing changes in these configurations fundamentally rework our ideas and understandings of the urban condition (Coutard and Rutherford 2011).

Drawing on theoretical advances and a wealth of original empirical material, the chapters that follow focus in particular on the rationales and processes of production and reproduction of alternative configurations for the supply of basic services in specific urban contexts, and how these are combined into existing infrastructure systems, as a means to explore:

- the urban political economies and ecologies of infrastructure reconfigurations and their associated re-articulations of technical infrastructures and built environments on and across various scales, from individual buildings to urban regions (who initiates, shapes and controls these transformations, how, and for whose benefit?);
- the urban outcomes of these socio-technical changes in terms of patterns of socio-spatial inequalities and fragmentation in access to essential services, and of evolving capacities and capabilities of urban actors and authorities to manage, regulate and govern urban spaces;
- the shifting social and political meaning of infrastructures, related to the changing forms of individual and collective appropriations, experience and mobilisation of infrastructure.

In order to do so, the book develops a socio-technical approach to the study of urban change, paying close attention to:

- the *material* and technical dimensions of processes: components of technical infrastructures; laws and regulations, standards and rules, and other policy instruments; management contracts and regulations; financial arrangements; stocks and flows of resources...
- the broad, gradual, open, contested, inherently political dynamics of socio-technical *change*, rather than (more narrowly defined) technological innovation and diffusion;
- the *situated*, context-dependent reconfigurations of infrastructural territoriality, in terms of control and management of, access to, and use of essential services and resource flows.

With this socio-technical lens, we study the mutual transformations of infrastructures and cities: how and by whom these transformations are actively produced, negotiated, resisted and reshaped; which stakeholders win or lose from them; the extent to which alternative possibilities are opened up or closed down; how instruments, models, values, representations and practices of the various stakeholders contribute to reshape power relations; and so on. It has been long-recognised that "artefacts have politics", but contemporary urban infrastructural change is complex, multifaceted, and does not obey a single clearly-defined pattern, so it remains a ripe area for continuing analysis and conceptualisation.

Advancing infrastructure urbanism: three forward salients[1]

As the above account of a declining paradigm of networked urbanism suggests, reconfigurations in infrastructure systems both encapsulate and contribute to wider societal shifts. These work inherently *through* fundamental reconfigurations of the socio-technical make-up of cities, *within* situated urban contexts, and *towards* (by being productive of) new urban processes, places and politics. By being at once the forbearer, medium and product of the dynamics of globalisation, financialisation, environmentalisation and social differentiation, urban infrastructure becomes a key site and marker of societal meaning and transformation. As 'splintered' socio-technical landscapes and new infrastructural territorialities continue to emerge and settle, further detailed analysis and reflection is required regarding the processes and practices through which these configurations come to be, as well as their wide-ranging outcomes. The chapters in this book contribute to this task.

In order to begin exploring the highly diverse, fluid and unstable landscapes emerging beyond the networked city, we identify and offer cross-cutting reflections on three overlapping lines of analysis present in the chapters. These salients are at once dynamics through which a 'break' with previous configurations has been operated, and brittle zones of new socio-technical controversy through which urban infrastructure (and its wider meaning) will continue to be negotiated and fought over. We focus in turn on:

- i valued 'alternatives' and shifting political economies of infrastructure: the material and idealistic disjoining of the networked city leads to the effective emergence of diverse alternative configurations and the production of new ways of valuing and accounting for infrastructure flows and the political economic distribution of resources;
- ii 'inhabited' sites and spatial ecologies of socio-technical systems: reconfigurations and rescaling in infrastructure systems make visible the ways in which material infrastructure is folded into and reproduces everyday work, lives and environments and the fluid, rebundled ecologies which simultaneously open up and enclose infrastructure space;
- iii appropriation and political transformation of infrastructure: new forms of individual and collective appropriation and experience of infrastructures point to changing socio-political meanings of being connected, of urban citizenship and of claims on and around infrastructure spaces.

1. Valued 'alternatives' and shifting political economies of infrastructure

In none of the chapters do we encounter single, homogeneous, monopolistic networked infrastructure systems. There is always a diverse set of overlapping (sometimes complementary, sometimes competing) socio-technical systems organised around distinct arrangements of technology, management and regulation, cost recovery, uses, practices and customs, "which, despite policy announcements and reforms, and notwithstanding imported models, survive and contribute to the functioning of cities" (Jaglin 2014, p.434). All the chapters are therefore involved in the study of processes and practices of hybridisation, the coexistence and/or coproduction of infrastructure systems constituted by interrelating heterogeneous elements, the study of which takes us beyond analysis of, for example, production-consumption binaries (as production and consumption are intertwined and mutually constitutive), 'systems of provision' and 'battles of systems' (as systems generally overlap and co-exist rather than displace each other).

So it is not just about a binary of or simple hybridisation between ('macro') networked and ('micro') post-networked systems, either. While the networked 'model' exists to varying degrees and in different ways, there are obviously also many forms of 'alternative' or 'decentralised' systems, and thus even greater multiplicity in combinations and overlaps between all these distinct configurations. Monstadt and Schramm develop "a spatial typology of the place-based infrastructure constellations" (p.27) that exist throughout central, suburban and outer areas of Hanoi, placing emphasis on the "hybrid institutional and technological arrangements of networked and non-networked systems, their spatial variation within cities, their co-evolution with urbanisation patterns and their interplay with public planning and service provision" (p.31). There are thus many long-standing socio-technical configurations present in a significant number of cities in the South, which have long been viewed by local populations, groups, providers and some governments as more closely adapted to contexts of great socio-spatial diversity.[2] Indeed, both Furlong and Jaglin show that there is nothing intrinsically superior about a networked system if cultures, customs and practices have embedded other working configurations.

Multiple, co-existing systems of everyday socio-technical arrangements cannot therefore be seen as a precursor or an 'alternative' to a readily identifiable networked model or ideal. The chapter by Furlong illustrates how the network can actually be seen as the 'alternative' to existing arrangements. She studies system coexistence rather than displacement in Quibdó (Colombia), where stable socio-technical practices are constituted by and constitutive of a system in disrepair/dysfunction. Here, it is the piped water network that becomes (partially, selectively) the alternative, supplementing an existing totally decentralised rainwater collection system organised and used by individual households. Coexistence does not emerge through a simple deployment of one system onto or within another, but shift for the co-existence of the two to become "viable" (p.103). While rainwater can continue to be used for non-potable water needs, thus avoiding too much pressure on resources and the network and high costs which would be billed to poor households, extending access to networked water is done for reasons of hygiene and summer water scarcity.

The situation in Quibdó contrasts somewhat with the chapter by Graham *et al.* which shows that there are similarly multiple (juxtaposed, contingent) systems of connection and tactics for obtaining water in Mumbai, but that revanchist urbanism and bourgeois environmentalism have combined to entrench a more splintered infrastructure morphology and experience. Here, the incapacity and unwillingness of public authorities to regulate alternative forms of water supply mean that ubiquitous access and affordability are now even more remote possibilities than in Mumbai's colonial past.

The chapter by Jaglin offers a third, in a way intermediary, perspective on the socio-spatial dynamics and political workings sustaining the emergence of "hybrid delivery configurations" of essential services, or rather their growing acknowledgement in what she calls the "pragmatic turn" of public authorities. This pragmatic turn, she argues, is a response to an "African-style (...) urban transition in its socio-economic (demand for services among emerging middle classes), spatial (periurbanisation, urban sprawl, encroachment into farmland) and political dimensions. The last of these is crucial: (...) it is the middle-class sections of society which today are the main social force demanding public action (p.190)." By "enacting 'differential production' of services within the framework of formalised hybrid delivery configurations", she continues, "this pragmatic turn could (...) constitute the basis for a third generation of reform in network services (after the modernisation-decentralisation and then the privatisation of public utilities)", by importing "the principle of heterogeneity into the operational sphere, by incorporating informal activities" and seeking to coordinate "the distinct socio-technical subsystems of a given delivery configuration (p.198)."

Elsewhere, alternative infrastructure configurations have emerged in recent years within an array of discourses promoting alleged, prospective benefits in terms of financial cost, resource efficiency, and localisation of control and management. Webb analyses the conditions and constraints for the local development of alternatives to centralised energy in two UK cities (using combined heat and power generation in association with district heating). Local authorities and associated actors worked to build social and technical capacity and legitimacy for the deployment of low carbon, decentralised infrastructure, either for the provision of affordable energy or as a contribution to urban regeneration. She highlights "the malleability of the same material CHP/DH technologies to different governance and economic structures … with contrasting primary interests and values being advanced" (p.20). Barles *et al.* study the renaissance of the non-potable water system of the City of Paris under the auspices of resource efficiency (pooling of resources and the use of different water qualities for different uses) and water management for environmental protection ("based primarily on the exploitation (or recovery) and pooling of *local resources*, on flow *(re)circulation* and on *cascade* uses of water", p.67). From a slightly different angle, Beveridge and Naumann's study of attempted remunicipalisations of water and energy systems in Berlin is also a story of the creation of capacity and traction for alternative configurations which do not reproduce the status quo. These are thus locally constructed struggles that are played out against the perceived inadequacies and distanciation of large technical systems which seek to create meaningful socio-technical spaces within which small-scale alternatives focused on local needs can be assembled and incrementally expanded.

However, these reconfigurations are neither easy to do nor guaranteed to succeed. Guy and Karvonen develop a perspective around 'meso-scale infrastructure' as "a new spatial dimension of energy provision for zero carbon housing that resides between the individual house and the universal grid" (p.76). Aiming at least in part to reduce dependence on centralised infrastructures, supply logics, fossil fuels and nuclear energy, "new assemblies of housing, heating, and powering communities are being imagined and designed, and in certain contexts are starting to be constructed" (p.76). By focusing on the complexities and tensions of the processes through which these network-building assemblies are actually achieved, their chapter contrasts the clarity of the overarching objectives with the messiness of the implementation (p.74) when confronted with vested interests, policy routines, and other elements of momentum.

Innovation and novelty in some cases are complemented by a rediscovery of previously existing configurations in others, thus highlighting some degree of reversibility in system pathways. Chapters focus on some renewed infrastructural spaces of intervention/action, whether it is remunicipalisation (Beveridge and Naumann), revaluing of dormant water and heating infrastructure (and 'old' uses thereof) (Barles *et al.*, Guy and Karvonen, Webb), or dynamics of emergence, circulation and disappearance of infrastructures as conjoined materials and practices (Shove).

Particularly at play here is a changing political economy of infrastructure systems, with more attention paid and more value granted to resource flows, services, operation and maintenance and to some extent more diversified expectations and practices, and less to 'hard' infrastructures per se. Whether it is in Quibdó or Paris, coexistence of (competing) infrastructure systems is justified by the differentiated values now accorded to resource flows over pipes and infrastructure (and to circulatory metabolisms as opposed to traditional linear urban metabolisms: see in particular Barles *et al.*, Cavé). There is now greater value (environmental, economic etc.) in what is rendered mobile by/through infrastructure than in systems of pipes and wires simply as fixed assets. Indeed, Marvin conceptualises infrastructure as an inherently mobile atmosphere, as a basic life-support system which becomes selectively differentiated and distributed (as metabolism becomes enveloped by and for particular interests).

Infrastructure overcapacity, duplication and competition ('hyper-reticulation': Barles *et al.*) is tolerated in these contexts by discourses promulgating the environmental benefits of differentiated management of flows and uses of those flows (e.g. potable vs. non-potable water), economic benefits for utilities and users (Quibdó, Vitória and Coimbatore), respect for heritage of infrastructure (Paris) and embedded ways of obtaining and consuming water (Quibdó). Barles *et al.*

summarise this shifting political economy in which "the environmental priorities associated with water resources, and with the quality (or more specifically the different levels of quality) and the uses of water, have taken on new importance, which legitimises the maintenance of a dual system justified by the benefits of managing water differentially in terms of its uses, despite the *de facto* competition between the two water systems in a context of infrastructural overcapacity" (p.51). A wider range of tailored services can be offered over this infrastructure, in some cases allowing companies to extract more value from their differentiated customer base (cf. a focus in the UK and elsewhere on energy services over decentralised systems rather a single network service from the grid), and in other cases justifying local authority/actor interventions in centralised energy systems, which for all the focus on pipes and plants for CHP/DH are very much about promoting and accessing flows of affordable, sustainable heat (Webb; Guy and Karvonen).

Webb delves into the making of a "locally-workable social, environmental and economic calculus of the value of urban CHP and DH infrastructure" (p.212), through which "customised, place-specific socio-technical networks, meshing urban politics and locally-embedded knowledge with non-local expertise" (p.221) aim to bring about material change. She shows that there is "public value" in local authorities working with energy provision, although "the future distribution of costs and benefits" (p.221) will vary between different project configurations (and their evolving trade-offs between short term vs. long term, profit vs. non-profit, least-cost vs. acceptance of risk and so on). In following the resource flows and extraction of value from waste in Vitória (Brazil) and Coimbatore (India), Cavé also queries the future allocation of rights and economic benefits from the waste-as-resource sector, and considers how these might be regulated.

This brings up the question of how to sustain the longer-term reproduction of value within a flow-based political economy. Recent plans for the extension of centralised sanitation systems in Hanoi to reduce the heterogeneity of current solutions and to separate wastewater and stormwater flows are inherently dependent on the social, technical and financial conditions of replacing adequate hybrid systems with a homogeneous network. The reliance on a substantial long-term loan from a Japanese bank to effect this infrastructure transition, and the prospective high costs of maintenance related to an expanded centralised system, cast doubt on the feasibility of such a project (Monstadt and Schramm, p.44–45). This touches on the increasing centrality of urban infrastructure in global finance and investment markets as specialist funds buying up utilities or investing in specific projects, thus rescaling their governance and creating new links and interdependencies between places through global financial flows (see Torrance 2008; Allen and Pryke 2013).

This global circulation leads on to another dimension of this political economy of infrastructural flows discussed in various chapters, which concerns the transfer and circulation of models, ideals, knowledge and expertise about infrastructure configurations between different urban contexts (see, for example, McCann and Ward 2011). Here, we come back to issues of how urban technological modernities forged in the North have been exported and problematically translated into very different contexts in the South (see Furlong and Monstadt and Schramm). Infrastructure configurations beyond the networked city suggest some kind of unsettling or partial obstruction of "this circulating ideal of urban modernity" (Monstadt and Schramm, p.26) for a variety of reasons (see above). However, this does not mean that new ideas and ideals (and investments) are not now circulating instead, and many of the chapters touch on the processes and practices through which 'travelling' models or components of infrastructure configuration come to be grounded and deployed in particular cities. Interest circulates in how and why, for example, many German cities have initiated remunicipalisation processes for water and energy (Beveridge and Naumann), or the Paris non-potable water network has been 'rediscovered' and revalued (Barles *et al.*), or UK cities have suddenly rediscovered district heating (Guy and Karvonen, Webb). Infrastructure configurations and 'innovation' are a key domain of inter-city learning, but the politics of this circulation of knowledge and expertise requires more consideration.

2. 'Inhabited' sites and spatial ecologies of socio-technical systems

While new political economies of infrastructure stress flows over absolute materiality (see above), the chapters also illustrate that infrastructure systems retain important material dimensions, but they evolve to reflect shifting relations between, and situated experiences of, the different heterogeneous components of systems. Rather than conceptualising infrastructure as (just) big technology transforming 'nature' (natural resources) to meet social 'needs', and thus emphasising solely dimensions and logics of supply, provision, technological innovation and the like, it may be more appropriate to view infrastructure 'in use' (Shove), from within people's lives, as a 'life support system' (Marvin).

In this way, infrastructure becomes, in Graham *et al.*'s term, "lived materiality" (p.117). It is practiced, used, operated, deployed, engineered and experienced (in short, embodied) differently, and it is this 'lived' dimension which translates or represents what infrastructure actually is (see Humphrey 2003). Infrastructure does not exist outside of the diverse and quite malleable ways in which it is done, used and appropriated, or how it sustains and reproduces social practices and conventions (Shove). It is not deployed and then taken up, but deployed through being 'inhabited' by engineers and technicians as well as users (see Mitchell 2003). In the chapters, this involves people learning rainwater harvesting techniques involving rooftop storage tanks and floaters (and climate patterns) in Quibdó (Furlong), negotiating "complex circuits of predation, corruption, and patronage" in Mumbai (Graham *et al.*, p.115), or being dependent on networks of electrical power and cooled or frozen space to support shopping and cooking practices (Shove). Technicians inhabit infrastructure in the various contexts studied by making 'density equations' and 'development calculations' for heat systems partly based on things like thermodynamics, angles and obstructions in space (Guy and Karvonen, p.84), forging relations between septic tanks, cleaning filters and building densities in Hanoi (Monstadt and Schramm), deciding on socio-economic calculi for collective reframing of rationales and interests in infrastructure projects (Webb), or blurring the content of contracts between local governments and multinational utilities (Beveridge and Naumann).

So whether in terms of the everyday work of wastepickers in India and Brazil as they negotiate urban circuits of value creation to make a basic living (Cavé), the work done by technicians involving calculations, measurements and constant maintenance (Guy and Karvonen, Webb, Furlong), or the 'self-organisation' arrangements in neighbourhoods in Hanoi for ensuring and maintaining functioning sanitation systems (Monstadt and Schramm), there emerges a multiply embodied/inhabited metabolic process mixing 'natural' resources, technological networks, money and bodies to make the city collectively function (see Gandy 2005).

Furlong shows how apparently 'dysfunctional' infrastructure is 'lived' normally in Quibdó, thus forcing a rethinking of what a network 'ideal' might be: "habits and customs are resistant to change, as they are bound up in the wider social order" (p.98), and therefore they struggle with "the transition from dysfunction to function" (p.104) as a networked system is extended to complement existing arrangements. Water storage practices adapted to intermittent service become problematised as a service becomes gradually more continuous, to say nothing of issues of affordability and cost recovery for utility companies.

This variety of inhabited infrastructural sites reworks the scales and boundaries of techno-ecological systems in distinct ways. On a basic level, growing concern for environmental resources and resource efficiency demands more systematic inclusion in studies of the infrastructural hinterlands from which urban systems extract resources and into which they discharge pollutants and waste (Barles *et al.*). Likewise, the circulation of infrastructure-related practices creates interdependencies between places and across scales that take infrastructure well beyond the perimeter of urban pipes and wires (Shove). Cavé makes connections between the situated, everyday work of wastepickers, merchants and brokers in obtaining and exchanging waste and the global circuits and markets of recovered raw materials into which Brazil and India's industrialising economies are linked.

However, at the same time as this infrastructural extension pushes analysis into new spaces and scales (see also Young *et al.* 2011), there is a concomitant process of rebundling or reconcentrating techno-ecological flows and assemblies in particular places for specific purposes (bringing 'inside' infrastructure 'outsides': Marvin). Some examples of this may be (at least potentially) socially progressive in relocalising control and setting agendas according to local matters (Beveridge and Naumann, Webb). Sometimes, however, there are enormous challenges in creating new infrastructure assemblies. Guy and Karvonen consider the substantive design, planning and development stakes involved in community-level heat systems in the UK: "the necessity of hybrid systems, the risk involved in abandoning gas supplies, the logistical challenges of fuel pellet delivery, the complexities of financial charges, the reshaping of design priorities and the feeling of risk and uncertainty that accompanies the new system..." (p.82). These stakes and tensions undercut the sense of many practitioners that this new spatial configuration of infrastructure "somehow symbolises an ecological turn in and of itself" (p.82).

Meanwhile, other examples risk reproducing and supplementing long-standing social and economic processes of splintering urbanism with new ecological urban fragmentation. Marvin shifts the focus from the flat, horizontal planes of traditional infrastructure grids to a fully 3D volumetric urbanism capable of analysing new rebundled, eco-securitised encapsulations which artificially enclose and reproduce specific urban ecologies. His chapter offers a new view of infrastructure as particular socio-technical configurations of atmosphere and life support envelopes that "disrupt and reconfigure the boundaries between the outside and inside environment", which further splinter the urban fabric and have major implications for urban planning and research.

So in a full conflation of 'nature' and 'the urban', post-networked 'ecological' urbanisation processes capture more and more 'natural' resources and circulate them round (particular bits of) cities to reproduce specific (eco-)urbanisms. Specific notions and understandings of 'ecology' are at play in the different chapters, and the question of the balance between who accesses/uses and who pays is central, both directly in terms of where the finance comes from, and indirectly in terms of externalities from shifting solidarities, cross-subsidisations, secessions and enclosures, and new boundaries between 'public' and 'private'.

Again, there are wider notions of infrastructure, technology, ecology and urban metabolism (than water and energy provision and waste management) bound up here in very particular urban/spatial configurations, demonstrating that beyond traditional views of 'the networked city' and 'networked urbanism' there are multiple, alternative infrastructures and 'networks' that recombine bits of the urban, technological artifacts and specific ecologies in new forms and ways which have significant implications for urban meanings, practices and outcomes.

3. Appropriation and political transformation of infrastructure

From the preceding points, it is clear that there is a fundamental, inerasable, and always changing political context to urban infrastructure. In opposition to nearly half the private sector respondents in a 2009 KPMG survey who see the politicisation of infrastructure as an "impediment to more investment [and] … governmental effectiveness" (KPMG International 2009, p.8), the chapters here underscore a need both to see it as always political, and as a site and a means of the further politicisation of wider urban debate and development. Beveridge and Naumann's chapter on the "awakening of public and political interest in infrastructure networks" (p.138) in Berlin highlights that, if ever it had disappeared from urban agendas, the diversity of ways in which infrastructure is now being reconfigured, and therefore coming to matter again in specific contexts, are pushing it once again to the very forefront of urban politics and competing ideas about possible urban futures.

The examples of technical spaces/sites focused on in the chapters highlight the very different ways in which infrastructure is made or becomes political, and how this politicisation concerns particular "selected components" (Beveridge and Naumann, p.154) or selective enclosures (Marvin) of socio-technical systems which reveal the wider issues at play within each context. There is a constant ambivalence and contingency to infrastructure which means that it can be read in different ways in different contexts as diverse interests mobilise hybridisation and coexistence both through and for "assemblies of housing, heating and community" (Guy and Karvonen), low carbon discourse and policy (Webb, Guy and Karvonen), hygiene and improving access to resources (Furlong), or revanchist urbanism and bourgeois environmentalism (Graham *et al.*).

In turn, being aware of this ambivalence and contingency opens up the spectrum of individuals and interests actively involved in framing, formulating, deploying, calculating and contesting alternatives to dominant infrastructure systems to well beyond the official policymakers and authorities. Beveridge and Naumann see "an ambitious, vibrant and often progressive politics of infrastructure in Berlin, with discourses of both radical and reformist change apparent, and the current and future roles of the state, civil society and private sector heavily contested" (p.138).

Reflecting on the political drivers and significance of novel delivery configurations in African urban contexts, Jaglin argues that the middle classes have a crucial, albeit indirect, influence on current evolutions and that although they are rarely involved in collective mobilisations, "this low level of political engagement should not be interpreted as a lack of interest in the state". Rather, "the potential anti-establishment political force of the emerging middle classes, slated to rise with the urban transition and as a result of electoral factors that reinforce the influence of urban social constituencies, is today neutralised and in a sense diverted into a massive and astonishingly creative engagement with all sorts of urban activities, in particular informal ones." Hence, "formalising hybrid delivery configurations appears as a way for African state authorities to negotiate the urban transition: a partial regularisation of the informal sector facilitates the provision of essential services, in exchange for a civil order that also benefits the urban elites and dominant classes (p.193–194)."

New forms of individual and collective appropriation of infrastructures further point to changing sociopolitical meanings of being connected, and of the forms of urban citizenship permitted and reproduced by socio-technical systems. Some chapters posit a 'right to infrastructure' as a crucial component in any 'right to the city'. In Mumbai, in a context of sobriety and conservation, this means that a right to water is increasingly reserved for 'more legitimate' paying citizens rather than non-paying slum dwellers (Graham *et al.*). In Quibdó (Colombia), a dysfunctional socio-technical system has been "stabilised through custom derived from adaptation to disrepair" (Furlong, p.95), showing that citizens reproduce their everyday lives by learning to cope with and adjust to less than 'model' infrastructure configurations. Cavé shows how waste management practices involve a chain of merchants, brokers, collectors and other individuals making a living by competing for and appropriating recyclable objects.

In Berlin, infrastructure citizenship has been developed through the active support (and, in part, finance) of residents for remunicipalisation projects for water and energy provision (Beveridge and Naumann). By participating in these projects, Berlin residents assert their collective attachment to local, public, not-for-profit basic services, thus reinserting an urban identity into what had become a delocalised multinational infrastructure system.

At the same time, socio-technical change cannot be disarticulated from wider (shifting) urban political economies, socio-political developments and requirements, and vice versa as politics is done through materiality constructed in ever more particular yet transcalar ways. The chapters by Monstadt and Schramm, Barles *et al.* and Shove highlight the influence of successive technical and political regimes over many decades (or longer) in shaping a complex, hybridised socio-technical palimpsest in the city which still affects how people access and experience services today. Webb highlights that "urban political processes are neither separate from, nor completely conditioned by, national and supra-national political dynamics, but are both shaped by, and contribute to reshaping, these" (p.205). Her chapter focuses on the politics against and through which local actors "construct new capacities in socio-technical knowledge and resources in the complex political-economic and technical sphere of energy systems" (p.215). Infrastructure means coordinating across the vast array of functions, services and departments which intervene in the urban ("specialisms including planning, environment, roads, community services, housing, finance and economic development", p.215), which has the potential to join up policy but also to entrench tensions and conflicts between different (local and non-local) knowledge and resource holders. Guy and Karvonen add to this debate by focusing on the contested nature of the transversal factors behind infrastructure project advancement, "involv[ing] protracted debates and deliberations over energy provision, balancing of low carbon design strategies, scalar characteristics of the particular development, economic modelling and financial viability, and long-term maintenance and upkeep" (p.86).

It is striking that many chapters foreground the emerging and re-emerging roles of public actors and authorities (and collective users) in new infrastructure configurations beyond the networked city. Whether it is remunicipalisation in Berlin, renaissance of a dormant network under centralised control in Paris, local authority interventions in otherwise national energy systems in the UK, or the hybrid roles of local actors in Quibdó, Hanoi and elsewhere, infrastructure is once again becoming a public matter and mandate. As Barles *et al.* conclude: "the non-potable water system [in Paris] appears to be an important instrument of the policy of promoting the public – rather than private and commercial – management and use of raw water" (p.68). This goes against ideas of relocalisation as implying or even necessitating decentralisation in a narrow sense. Marvin's trenchant analysis of emerging 'privatised' logics and configurations of domic ecologies is a counter-example here, suggesting that we need to remain attentive to the varying ways in which infrastructure is appropriated and rebundled.

Many of the chapters (e.g. Graham *et al.*, Webb, Beveridge and Naumann, and Guy and Karvonen) conclude by stressing both the need for broader (national) political economic regime change to create the conditions for supporting progressive infrastructure transformations, and the actual and potential role for urban politics in working to bring about such change. This re-emphasises the fact that infrastructure configurations are always both changing in relation to shifts in wider urban political economies, and contributing in part to these shifts. Following Graham *et al.*, what is needed (practically and theoretically) is more understanding of and more challenging of infrastructure politics "at their foundation", recognising that "a democratic infrastructure politics centred on the possibilities of [more inclusive] networking/renetworking..." requires "a sustained and radical change in the nature of infrastructural, urban and national governance" (p.134).

So while there has been a lot of work studying (hierarchical) governing and government of and by networked infrastructure systems, the chapters here, in their analysis of shifting forms and implications of the appropriation and politics of infrastructure (in terms of citizenship and/or the capacities of and conflicts between 'local' actors), point to more horizontal, distributed modes of infrastructure governance and agency (see also Barry 2013; Easterling 2014). This captures the many circulatory and relational dimensions of infrastructure which seep across administrative boundaries and transfer ideas and ideals well beyond rooted contexts: "as grounded, materialised entities, infrastructures are also full of imagination, embodying and carrying ideas and visions of welfare, well-being and profit that circulate and extend far beyond the spots in which they are built" (Shove, p.254). Tracing the transcalar agencies implicated in the making and practicing of infrastructures takes us towards "methods of characterising and describing networks and forms of connection and circulation that link infrastructures and practices, that extend beyond the city but that also make cities what they are" (Shove, p.255).

Conclusion

The forms, significations and implications of infrastructure reconfiguration around the globe suggest that we have moved beyond an era in which an ideal or model of a 'networked city' necessarily dominates imagination, thinking, practice and analysis. This introduction has sketched out some of the ways in which the chapters that follow chart and reflect on the making, experience and contestation of urban infrastructure as a kaleidoscopic process of open-ended, non-linear, flexible socio-technical change. Taken as a whole, the book substantiates three main arguments on the enduring, but constantly evolving, relations between urbanisation and 'infrastructuralisation'.

First, it illustrates how multiple articulations between and reconfigurations of infrastructure systems and the services they provide are one of the key domains where global urban futures are being and will be shaped, implemented and contested for decades to come. From Quibdó to Paris and from Hanoi to Aberdeen, and across potable and non-potable water systems, decentralised energy grids, circuits of waste, and atmospheric life support ecologies, infrastructure transitions are revealed as a socio-technical process constituted in, through and of the urban (and its distinctive 'issues', materialities and politics), and not some 'external' (policy) process impacting a 'separate' urban sphere. A focus on infrastructure is highly relevant for urban studies not just when and where infrastructure is most clearly and visibly changing (through, for example, disruption, crisis and fragmentation), but also more generally because infrastructure is always already coming into being and under constant maintenance (Graham and Thrift 2007), is contingent (in its forms, outcomes and interpretations), and inherently shapes and is shaped by wider socio-political and historical-geographical contexts. In other words, an understanding of the links between what is changing within infrastructure and what is changing in, through and of the urban also requires digging into some of the less visible, perhaps more incremental (but nonetheless potentially consequential) shifts in and controversies over mediations and interrelations between a wide array of actors, technological systems and their wider environments.

Second, and intrinsically bound up with urbanising processes, the book shows that infrastructure is becoming ever more heterogeneous in its constituent elements and meanings, in its topologies, scales and frontiers, and in the multiplicity of ways in which it is engaged with and 'inhabited'. Flows, grids and components of water, energy, waste and other systems are being distended, differentiated, renewed and reshaped into particular, situated arrangements which rework notions of infrastructure and extend the ways in which we understand how it is folded and enveloped into the urban fabric. New infrastructural ecologies emerge at the same time as resource flows and techniques of control and management are rebundled into particular 'local' configurations which rework the boundaries, values and functioning of technological systems. Infrastructure is constantly negotiated, made and remade by technicians and users in their everyday work and lives, and thus customised to and by individual and collective appropriations, experiences and practices. Mapping this vibrant and effervescent sphere demands incessantly extending and renewing analytical perspectives, methods and tools.

Third, the book demonstrates that the socio-political significations of ongoing urban infrastructural transitions are fundamentally ambivalent. In the same way as effective integration in and through 'the networked city' has always depended on broader socio-political conditions, there are no automatic urban effects of 'alternative' socio-technical configurations: they are always shaped by and shaping of the particular characteristics of the urban societies in which they are situated. There is no greater inherent environmental (or other) sustainability in emerging or existing 'decentralised' or distributed infrastructure configurations, which may prove at least as resource-intensive as large-scale, centralised technological systems. New technological arrangements may well perpetuate centralised control and/or be shaped by and for existing dominant social interests reproducing the systems supporting 'business as usual'. Nonetheless, at the same time there is the potential for a diversification and pluralisation of infrastructure configurations to be oriented towards more progressive objectives, shaped by and for collective matters, and opening up, rather than closing down, the diversity of possible urban socio-technical futures.

Understanding and conceptualising socio-technical change needs to be at the heart of current accounts and analyses of contemporary global urbanisation processes and their constant reworkings and re-articulations of areal and relational forms of territoriality. As the above lines or themes suggest, and as the chapters that follow underscore, current infrastructure transitions are contingent, plural and multivocal in form, purpose and repercussion, and in our ways of interpreting and knowing them. This entails mobilising different theories, methods and knowledges to study how, why and for whom current processes and practices of infrastructure reconfiguration connect, disconnect and reconnect in varying ways and across a plurality of places and scales to encompass the multiple actors and socio-technical components of the urban.

New infrastructural territorialities are emerging as discursive ideals, material constructions and situated practices. Tracing their emergence as indissociably social-technical, critically analyzing their rationales, modalities and outcomes, and thus unsettling and repoliticising pre-conceived notions of the changes to which they can contribute is one of the key tasks for urban studies in the years to come.

Notes

1 The idea of a forward salient refers to Thomas Hughes's (1983) concept of a 'reverse salient' as a component in technological system evolution which holds back system development in some way or prevents or delays its advance. We use 'forward salient' here, therefore, to capture areas, sites or themes (across the chapters) which we see as advancing analysis and understanding of infrastructure urbanism as a whole.

2 But which have often been perceived by policymakers and funding agencies as low-grade alternatives to the network.

References

Allen, J. and M. Pryke (2013) "Financialising household water: Thames Water, MEIF, and 'ring-fenced' politics." *Cambridge Journal of Regions, Economy and Society* **6**(3): 419–439.

Bakker, K. (2003) "Archipelagos and networks: urbanisation and water privatisation in the South." *Geographical Journal* **169**(4): 328–341.

Barry, A. (2013) *Material Politics: Disputes along the Pipeline*. Chichester, John Wiley and Sons.

Bijker, W., T. Hughes and T. Pinch, Eds. (1987) *The Social Construction of Technological Systems: New Directions in the Sociology and History of Technology*. Cambridge MA, MIT Press.

Botton, S. and B. De Gouvello (2008) "Water and sanitation in the Buenos Aires metropolitan region: fragmented markets, splintering effects?" *Geoforum* **39**: 1859–1870.

Brenner, N. (2013) "Theses on urbanization." *Public Culture* **25**(1): 85–114.

Bulkeley, H., V. Castan Broto and A. Maassen (2014) "Low carbon transitions and the reconfiguration of urban infrastructure." *Urban Studies* **51**: 1471–1486.

Castells, M. (2010) "Globalisation, networking, urbanisation: reflections on the spatial dynamics of the information age." *Urban Studies* **47**(13): 2737–2745.

Coutard, O. Ed. (2008) *Placing Splintering Urbanism*. Geoforum (special issue) **39**(6).

Coutard, O. (2010) Services urbains: la fin des grands réseaux? In O. Coutard and J. P. Lévy, Eds. *Ecologies Urbaines: État des Savoirs et Perspectives*. Paris, Economica-Anthropos.

Coutard, O. and S. Guy (2007) "STS and the City: Politics and Practices of Hope." *Science, Technology, and Human Values* **32**(6): 713–734.

Coutard, O. and J. Rutherford (2011) The rise of post-networked cities in Europe? Recombining infrastructural, ecological and urban transformations in low carbon transitions. In H. Bulkeley, V. Castan Broto, M. Hodson and S. Marvin, Eds. *Cities and Low Carbon Transitions*. London, Routledge: 107–125.

Dupuy, G. (1991) *L'Urbanisme des Réseaux: Théories et Méthodes*. Paris, Armand Colin.

Dupuy, G. (2011) "Fracture et dépendance : l'enfer des réseaux ?" *Flux* **83**: 6–23.

Easterling, K. (2014) *Extrastatecraft: The Power of Infrastructure Space*. London, Verso.

Farias, I. and T. Bender, Eds. (2010) *Urban Assemblages: How Actor-Network Theory Changes Urban Studies*. London, Routledge.

Furlong, K. (2011) "Small technologies, big change: rethinking infrastructure through STS and geography." *Progress in Human Geography* **34**(4): 460–482.

Gandy, M. (2002) *Concrete and Clay: Reworking Nature in New York City*. Cambridge MA, MIT Press.

Gandy, M. (2005) "Cyborg urbanization: complexity and monstrosity in the contemporary city." *International Journal of Urban and Regional Research* **29**(1): 26–49.

Gandy, M. (2008) "Landscapes of disaster: water, modernity, and urban fragmentation in Mumbai." *Environment and Planning A* **40**: 108–130.

Geels, F. W. and J. Schot (2007) "Typology of sociotechnical transition pathways." *Research Policy* **36**(3): 399–417.

Graham, S., Ed. (2009) *Disrupted Cities: When Infrastructures Fail*. London, Routledge.

Graham, S. and S. Marvin (2001) *Splintering Urbanism: Networked Infrastructures, Technological Mobilities and the Urban Condition*. London, Routledge.

Graham, S. and N. Thrift (2007) "Out of order: understanding repair and maintenance." *Theory, Culture and Society* **24**(3): 1–25.

Guy, S. and A. Karvonen (2011) Using sociotechnical methods: researching human-technological dynamics in the city. In J. Mason and A. Dale, Eds. *Understanding Social Research: Thinking Creatively about Method*. London, Sage: 120–133.

Guy, S., S. Marvin, W. Medd and T. Moss, Eds. (2011) *Shaping Urban Infrastructures: Intermediaries and the Governance of Socio-Technical Networks*. London, Earthscan.

Guy, S., S. Marvin and T. Moss (2001) *Urban Infrastructure in Transition: Networks, Buildings, Plans*. London, Earthscan.

Heynen, N., M. Kaika and E. Swyngedouw, Eds. (2006) *In the Nature of Cities: Urban Political Ecology and the Politics of Urban Metabolism*. London, Routledge.

Hodson, M. and S. Marvin (2009) "Urban ecological security: a new urban paradigm?" *International Journal of Urban and Regional Research* **33**: 193–215.

Hommels, A. (2005) "Studying obduracy in the city: toward a productive fusion between technology studies and urban studies." *Science, Technology, and Human Values* **30**(3): 323–351.

Hughes, T. (1983) *Networks of Power: Electrification in Western Society, 1880–1930*. London, Johns Hopkins University Press.

Humphrey, C. (2003) Rethinking infrastructure: Siberian Cities and the Great Freeze of January 2001. In J. Schneider and I. Susser, Eds. *Wounded Cities: Deconstruction and Reconstruction in a Globalized World*. New York, Berg: 91–110.

Jaglin, S. (2005) *Services d'eau en Afrique subsaharienne : la fragmentation urbaine en question*. Paris, CNRS Editions.

Jaglin, S. (2014) Regulating service delivery in southern cities: rethinking urban heterogeneity. In S. Parnell and S. Oldfield, Eds. *The Routledge Handbook On Cities Of The Global South*. London, Routledge: 434–447.

Kaika, M. (2005) *City of Flows: Modernity, Nature, and the City*. London, Routledge.

Kaika, M. and E. Swyngedouw (2000) "Fetishizing the modern city: the phantasmagoria of urban technological networks." *International Journal of Urban and Regional Research* **24**(1): 120–138.

Karvonen, A. (2011) *Politics of Urban Runoff: Nature, Technology, and the Sustainable City*. Cambridge MA, MIT Press.

Keil, R. (2003) "Urban political ecology." *Urban Geography* **24**(8): 723–738.

KPMG International (2009) *The Changing Face of Infrastructure: Frontline Views from Private Sector Infrastructure Providers*. London, KPMG International in cooperation with The Economist Intelligence Unit.

Lorrain, D., Ed. (2011) *Métropoles XXL en pays émergents*. Paris, Presses de Science Po.

Luke, T. (2003) "Global cities vs. 'global cities': rethinking contemporary urbanism as public ecology." *Studies in Political Economy* **70**: 11–33.

McCann, E. and K. Ward, Eds. (2011) *Mobile Urbanism*. Minneapolis, University of Minnesota Press.

McFarlane, C. (2008) "Governing the contaminated city: infrastructure and sanitation in colonial and post-colonial Bombay." *International Journal of Urban and Regional Research* **32**(2): 415–435.

McFarlane, C. and J. Rutherford (2008) "Political infrastructures: governing and experiencing the fabric of the city." *International Journal of Urban and Regional Research* **32**(2): 363–374.

Melosi, M. (2000) *The Sanitary City: Urban Infrastructure in America from Colonial Times to the Present*. London, Johns Hopkins University Press.

Mitchell, W. (2003) *Me++: The Cyborg Self and the Networked City*. London, MIT Press.

Monstadt, J. (2009) "Conceptualizing the political ecology of urban infrastructures: insights from technology and urban studies." *Environment and Planning A* **41**(8): 1924–1942.

Moss, T. (2000) "Unearthing water flows, uncovering social relations: introducing new waste water technologies in Berlin." *Journal of Urban Technology* **7**(1): 63–84.

Offner, J.-M. (1993) "Le développement des réseaux techniques : un modèle générique." *Flux* **13–14**: 11–18.

Pflieger, G. and S. Matthieussent (2008) "Water and power in Santiago de Chile: socio-spatial segregation through network integration." *Geoforum* **39**(6): 1907–1921.

Rose, M. and J. Tarr (2004) "Introduction: technology, politics, and the structuring of the city." *Journal of Urban History* **30**(5): 643–647.

Shove, E. (2012) "The shadowy side of innovation: unmaking and sustainability." *Technology Analysis and Strategic Management* **24**(4): 363–375.

Simone, A. (2004) "People as infrastructure: intersecting fragments in Johannesburg." *Public Culture* **16**(3): 407–429.

Soja, E. (2000) *Postmetropolis: Critical Studies of Cities and Regions*. Oxford, Blackwell.

Star, S. L. (1999) "The ethnography of infrastructure." *American Behavioral Scientist* **43**(3): 377–391.

Summerton, J., Ed. (1994) *Changing Large Technical Systems*. Boulder, Westview.

Swyngedouw, E. (2004) *Social Power and the Urbanization of Water: Flows of Power*. Oxford, Oxford University Press.

Tarr, J. (1979) "Introduction." *Journal of Urban History* **5**(3): 275–278.

Tarr, J. and G. Dupuy, Eds. (1988) *Technology and the Rise of the Networked City in Europe and America*. Philadephia, Temple University Press.

Tarr, J. and J. Konvitz (1987) Patterns in the Development of the Urban Infrastructure. In H. Gillette and Z. Miller, Eds. *American Urbanism: A Historical Review*. New York, Greenwood Press: 195–226.

Torrance, M. I. (2008) "Forging Glocal Governance? Urban Infrastructures as Networked Financial Products." *International Journal of Urban and Regional Research* **32**(1): 1–21.

van Vliet, B., H. Chappells and E. Shove (2005) *Infrastructures of Consumption: Environmental Innovation in the Utility Industries*. London, Earthscan.

Walker, G. and N. Cass (2007) "Carbon reduction, 'the public' and renewable energy: engaging with socio-technical configurations." *Area* **39**(4): 458–469.

Young, D., P. Burke Wood and R. Keil, Eds. (2011) *In-Between Infrastructure: Urban Connectivity in an Age of Vulnerability*. Kelwona (BC, Canada), Praxis (e)Press.

2 Changing sanitation infrastructure in Hanoi

Hybrid topologies and the networked city

Jochen Monstadt and Sophie Schramm

Introduction

Technical infrastructures have been on the urban research agenda for almost two decades now. Although it has been pointed out that these socio-technical systems display distinct urban technological styles, the spatial variability and diversity of urban infrastructures have, however, been understudied (for this argument see Coutard 2008). This particularly applies to the context-shaped, and context-shaping, interactions between infrastructure provision and urban development in the global South. Until today, planning concepts for, and academic debates on, "modern" cities and infrastructures reflect the conditions and ideals of a "networked city". These include: (1) the concept of a centralised topology of standardised and universal technical networks covering the entire urban area; (2) the notion that urban utility companies provide for ubiquitous infrastructure services to all users; (3) the notion of passive customers who are not actively engaged in the production of infrastructure services; and finally (4) the assumption that the provision of infrastructure services is closely attached to or highly regulated by the state. Engineers, planners and public health officials worldwide have aspired to align with this circulating ideal of urban modernity, hygiene, universalisation and rationalisation of space in (re)producing cities (cf. e.g. Dupuy 2008; Melosi 2008). While the current trends of commercialisation and socio-technical diversification and decentralisation challenge this modern infrastructure ideal worldwide, the urban realities in the global South, which differ greatly from those of the industrial countries, have always challenged the transfer of the networked city model. Since the 1970s engineers and planners have thus promoted a turn toward decentralised, on-site sanitation solutions (cf. van Vliet *et al.* 2010).

The paper addresses the rapid urban transition of the city of Hanoi and its spatially distinct infrastructure configurations in urban sanitation systems. We argue that the transfer of dominant planning models and infrastructure ideals and the attempt to impose the ideal of a unitary, ubiquitously networked city on Hanoi's diverse urban contexts has contributed to the city's severe infrastructure problems. While Hanoi's official urban and infrastructure policies aim at ubiquitous sanitation networks covering the entire urban area, the notion of a networked city has – if at all – been realised only in parts of the city. The severe urban poverty following the American War and rapid urbanisation in the aftermath of economic reforms since the 1980s has impeded the construction of a unitary network. Although labeled non-modern since French colonial times, decentralised socio-technical configurations of wastewater treatment and disposal shape large parts of the city. While these configurations have changed over time along with broader technological and urban dynamics, they have continuously operated beyond the scope of, in the space in between, or complementary to sanitation networks and the activities of the urban government and public utilities. Despite the apparent persistence of infrastructure configurations beyond networked service provision, the state and transnational organisations are currently investing heavily in the universalisation and centralisation of the diverse and fragmented urban sanitation systems.

The objective of this paper is to explore the dynamics and outcomes of attempts to transfer hegemonic infrastructure ideals and models to Hanoi, Vietnam. We argue that although the networked city ideal has never conformed to the urban realities in Hanoi, and is increasingly contested by notions of decentralised or "appropriate" on-site solutions, its legacies remain persistent in urban and infrastructure planning. Until today, state officials, urban planners and public utilities consider the implementation of that ideal as a goal of planning and investment efforts. However, we demonstrate that networked systems in Hanoi are confined to certain urban areas and have been hybridised with alternative socio-technical arrangements. In our paper, we develop a spatial typology of the place-based infrastructure constellations in the former colonial centre, its extensions urbanised in the era of the planned economy, recent "new urban areas", and at the edge of the former city centre. We demonstrate that self-organised and entrepreneurial forms of service production beyond networked sanitation systems are an integral part in each of these urban constellations. Instead of adhering to prevailing dichotomies between centralised, networked and public infrastructure provision and decentralised, on-site and often private solutions, we argue that urban planning and research should address their persistent borderlands, coexistence and hybridities more systematically.

The paper is based on empirical and conceptual research conducted between 2008 and 2013 within a PhD study (Schramm 2014), as well as recent empirical and conceptual work on suburban infrastructure (Monstadt and Schramm 2013).[1] It includes a series of semi-structured in-depth interviews with representatives of local and state governments, international organisations, the local utility company, academics, local stakeholders and residents carried out as part of the PhD study. In addition, field studies and the assessment of grey literature, websites, historical and current official city and state documents, secondary literature and newspaper articles have played a crucial role.

Current debates on cities and infrastructure in the global South

It is one of the contributions of urban studies of technology to have conceptualised infrastructure systems as both shaping urban development and being shaped by urban conditions (e.g. Hughes 1987; Tarr and Dupuy 1988; Graham and Marvin 2001; Coutard *et al.* 2005; Coutard 2008; Monstadt 2009). These socio-technical systems are closely interrelated with the built environment, social arrangements, cultural contexts and the power relations in a city, and they function as material mediators between nature and the city. They "make up considerable portions of the material, economic and geopolitical fabric of contemporary cities" (Graham 2000: 115) permeating all urban functions and socio-spatial structures and representing long-term accumulations of finance, technology and organisational and geopolitical power in cities (Star 1999: 381; McFarlane and Rutherford 2008). The socio-technical constellations of urban infrastructures can accordingly be explained by the ways in which they are shaped by urban geographies, political cultures, economies, user practices and so on.

Debates on cities and networked infrastructure

Referring to historical studies on the dialectic relationship of cities and networked infrastructures (e.g. Hughes 1987), contemporary urban studies have primarily focused on urban contexts in the global North. These studies analyse the contemporary transition of networked infrastructures and their impact on urban development (Graham and Marvin 2001; Coutard *et al.* 2005; Coutard 2008; Graham 2000; Monstadt 2009). In these studies, attention has been focused primarily on the vital functions of infrastructure networks in the production and reproduction of "modern" cities in Europe and North America. This includes studies of the "rise of the networked city in Europe and America" (Tarr and Dupuy 1988), the "sanitary city" (Melosi 2008) and the "hydropolis" (Gandy and Frank 2006). These studies, particularly Dupuy's concept of "networked urbanism" (Dupuy 2008), emphasise the central role of networks in urban change. They highlight that infrastructure innovations and paradigms – –such as the "modern infrastructure ideal" of the universal, uniform and standardised supply of infrastructure services to all customers at all times and places (Graham and Marvin 2001) – have been fundamental in urban development. Moreover, they have conceptualised how urban infrastructures are shaped by physical and social geography, national and local cultures, politics, resource availability, economic practices, and organisational forms, among other factors, and thus they develop place-based "technological styles" (Hughes 1987). Recent debates on "circulation" and "appropriation" have addressed these technological styles (Hård and Misa 2008), highlighting the parallel processes of homogenisation and differentiation in the historical development of urban technological solutions across Europe. On the one hand, the "circulation" of artifacts, people and ideas by means of publications, fairs, visits, study tours, and other activities has acted as "an engine of homogenisation" for urban technology (ibid.: 10). On the other hand, these technologies were "appropriated" to suit the physical, institutional and cultural specifics of particular places (ibid.: 12), providing an explanation for the differences between hegemonic infrastructure ideals and technological standards and place-based socio-technical styles.

Although debates on socio-technical systems have considerably advanced the conceptualisation of networked infrastructures, they have paid little attention to urban infrastructures in the global South. These are often rooted in the urbanisation processes of the colonial era, achieved by the confinement of networked infrastructure provision to certain parts of the urban area (Gandy, 2006a). Although some scholars argue that the transfer and circulation of "Western" infrastructure ideals and concepts and a "technology driven colonization" (van Laak, 2004) have strongly driven urbanisation and shaped urbanity in the global South, their "appropriation" into place-specific socio-technical configurations and styles is still understudied. As studies of urban history show, the modern ideal of a unitary, orderly city, integrated by networked infrastructure, was thus often remodeled as a system of "spatial apartheid", symbolising the superiority of the colonial power-holders of colonised civilizations (Balbo, 1993). Until today that ideal has been realised only in certain urban areas, while others are bypassed by the networks. In those "non-grid" urban areas users, community groups and entrepreneurs have developed various practices supplementing, replacing or obstructing the provision of infrastructure services by utility companies and urban governments. The resulting hybridity of networked and non-networked technologies, of public, private and self-organised forms of infrastructure provision, of centrally planned networks and service provision beyond the state, have immense implications for urban development. While the analysis of infrastructure solutions in the global South has been a research field in engineering sciences for a long time, it is only recently that both empirical and theoretical social science studies have directed their attention to these distinct socio-spatial and socio-technical conditions of urban infrastructure (examples include McFarlane, 2008, 2010; Jaglin, 2008; Gandy, 2006a; Coutard, 2008).

Postcolonial studies have analyzed the colonial legacies of planning paradigms and models in many cities of the global South (e.g. Porter 2010; Watson 2009), while inherited infrastructure ideals from colonial times and their persistence in contemporary cities have hitherto not attracted much scholarly attention. However, the prevailing influence of the global North in shaping planning practice and investment strategies also occurs in the field of urban infrastructure: many urban governments in the global South apply persistent planning approaches – universal coverage of standardised infrastructure networks (Abbott, 2012), rational planning models and their master planning concept (cf. ibid.; Myers, 2011) often developed in the global North. As first studies demonstrate, under the place-specific conditions of often rapid urban growth, considerable urban inequalities and fragmentation, and severe infrastructural crises, approaches to engineering and planning require adaptation. A particular challenge for planning studies is to address urban sociotechnical change in the global South by analyzing how hegemonic infrastructure paradigms and planning models of a networked city are being adopted in the global South, how they co-evolve and hybridise with different forms of self-organisation and entrepreneurial activity in the co-production of infrastructure services. It is this coexistence, overlapping and hybridisation of networked and non-networked services, of public utilities, public-private partnerships, small entrepreneurs and civil self-organisation, of formal planning and informality, that has to be taken into account more systematically in both urban research and urban planning practice.

Debates on urban sanitation infrastructure in the Global South

In light of the fragmented infrastructures and the urban poor's low rates of access to adequate public services in many cities of the global South, there is a burgeoning – often alarmist – literature on urban sanitation, combining growing concern about the declining quality of life, public health and social equity with equal concern about urban environmental degradation. Much of the research focuses on *technological structures* and functional technical solutions to the lack of basic service provision. Increasingly it is being conceded that the model of a networked city, including a centralised topology and universal coverage of technical networks, not only does not exist in many cities of the global South, but is also neither feasible nor desirable. Various studies point to the high costs of building and maintaining these infrastructures, their deficiencies regarding sustainable resource use, their high degree of inflexibility to adapt to the rapidly changing urban conditions and the complexity of operating them professionally (e.g. Otterpohl, 2002). Many of these studies thus advocate a "paradigm shift" away from the networked city model. Instead of costly network extensions and centralised sewage plants, advocates of ecological sanitation (EcoSan) promote decentralised, low-tech, low cost and on-site solution considered to be more appropriate to solve the sanitary crisis in rapidly urbanising countries of the global South and to provide potentials in water reuse, wastewater recycling and the recovery of nutrients and energy (cf. Langergraber and Muellegger, 2005; Lamichhane, 2007; Schramm and Bieker, 2010). However, much of this research on appropriate technical solutions is disconnected from the analysis of the adapted organisational models of these systems, of their socio-spatial conditions, and of the corresponding requirements in urban and infrastructure planning.

With regard to the *industry and user structures*, there is an increasing awareness that a variety of actors, rather than municipal monopolists, provide sanitation services. While research has focused on the advantages and disadvantages of ownership privatisation and private sector participation (e.g. Prasad, 2006; Bakker, 2007), the debate on private versus public has largely neglected the actual trend towards the involvement of the private sector on all scales: from multinational utility companies to small local firms, local associations, individual entrepreneurs and different forms of user self-organisation (Budds and McGranahan, 2003).

While it has been recognised that the provision of infrastructure services in many cities of the global South cannot be characterised by their close relationship to the state to the same extent as in the global North (Coutard, 2008), surprisingly little attention has been paid to their urban *governance structures*. Rapid urbanisation under conditions of poverty and spatial fragmentation has given rise to peculiar land and infrastructure development patterns that often defy the application of master plans or comprehensive planning. Local governments in many cities of the global South are regularly considered both as having limited institutional, financial and human resources and as being under international pressure to commodify the provision of water and sanitation services, so that they can hardly regulate and provide for equitable infrastructure services (cf. UN-Habitat, 2005). Apart from these evaluations, knowledge as to the actual extent to which local governments contribute to the effective implementation of national policies, initiate their own policy innovations or creatively promote socio-technical innovations in the face of severe regulatory challenges is limited (an exception is Gopakumar, 2011).

It has been acknowledged by a considerable body of literature that the ideal of a networked city and corresponding investment plans and planning approaches are a key challenge and impediment to the implementation of sustainable infrastructure solutions (e.g. van Vliet *et al.*, 2010; Gopakumar, 2011). Moreover, various authors have critically analyzed co-production strategies, i.e. the cooperation of state and civil society to extend basic services (cf. for African contexts: Mitlin, 2008). Empirical cases have demonstrated that a variety of forms of self-organisation and entrepreneurial activity exist, ranging from the local, small-scale provision of water and sanitation services to organised user communities in poor neighbourhoods, to the buying-in of the professional provision of water and sanitation services by companies in wealthy neighbourhoods (cf. Fernandez-Malondado, 2008; Zérah, 2008; Jaglin, 2008). In many cities local utilities or public authorities are thus only partially involved in the provision or regulation of infrastructure services, and they do not have full control over the enforcement of legal standards. Urban infrastructures in the global South are thus not solely shaped by the networked services of incumbent public utilities, but also by dynamic and heterogeneous forms of civil self-organisation, small private entrepreneurs and public-private partnerships. These service providers operate "off the grid", supplementary to existing networks or through privatised small-scale networks. However, the resulting hybrid institutional and technological arrangements of networked and non-networked systems, their spatial variation within cities, their co-evolution with urbanisation patterns and their interplay with public planning and service provision have not yet been addressed systematically.

Between networked and off-the grid infrastructure: sanitation systems in Hanoi

The dynamic history of Hanoi, the capital of the Socialist Republic of Vietnam, located in the delta of the Red River, is characterised by a series of ruptures due to the succession of political regimes. The different political regimes are still apparent in today's urban fabric, with the urban quarters representing heterogeneous urban morphologies and socio-spatial structures. These influences – ranging from a millennium of Chinese rule, French colonial intervention, economic and spatial planning from the Soviet Union, and then to the current Western and Japanese financial and technical support situation – have also materialised in urban sanitation systems. *Doi moi*, the market and state reforms towards a "socialist-oriented market economy" introduced in 1986, are the latest cause of disruption in the development of a city shaken by wars, independence struggles and resistance movements (Logan, 2000). This most recent transformation of the economy has brought about changes in international relations, economic growth and rapid urbanisation. At the same time, it has energised attempts by the city government to realise the Western ideals of comprehensive planning, unitary infrastructure provision and centralised networks that have been pursued since French colonial times. In the following section we illustrate the dynamics of Hanoi's urban development and its co-evolution with sanitation systems. Although the introduction of these ideals in colonial times and their perpetuation over different political regimes have profoundly shaped urban infrastructure and the relations between water and land in the city of Hanoi, they could never be fully realised. Instead, limited network coverage, spatial differentiation and heterogeneous forms of service provision characterise today's urban infrastructure systems.

History of wastewater disposal – dynamics of urban space and stability of infrastructure paradigms

The paradigm of a centralised network with unitary service provision has been dominant in urban and infrastructure planning throughout every stage of Hanoi's history since colonial times. Before French colonisation, however, the feudal rulers of ancient Hanoi accepted the fragmentation of urban space into the royal enclosure and the people's city, with the former being exclusively equipped with networks for water supply and disposal, including even the separation of human manure and stormwater disposal (Vinh, 2004). In the royal enclosure, as well as in the people's city, human manure was collected as night soil to be used in agriculture, a practice that has been dismissed as unhygienic since colonial times but is still common today in large parts of the city and its periurban surroundings. Hanoi is located in the delta of the Red River, and the close relation of land and water traditionally characterise its urban space. In pre-colonial times, water bodies were even more present in the urban landscape than today: in the form of moats for protection, as a means of transport, for farming fish and cultivating rice, as a decoration for pagodas and sites of leisure, and as a place for disposing of household waste (Fayet, 1939). However, this philosophy of regarding water bodies as an integral part of the urban landscape and multifunctional infrastructure contradicted the infrastructure paradigm imported by the French. According to concepts of the sanitary or bacteriological city (Melosi, 2008; Gandy, 2006b), the fast removal of used water defined as waste was regarded a key prerequisite to improving the hygienic conditions of the city, requiring the introduction of new forms of technology and managerial expertise in urban government.

The modern infrastructure ideal imported by the French has shaped urban development and sanitation systems in specific ways but was achieved only temporarily (or even not at all), depending on the definition of the urban territory within which the citizens of Hanoi have access to universal networked infrastructure services. The definition of what belongs to the urban territory of Hanoi has varied depending upon different understandings of urbanisation processes, and it was shaped by historically close interrelations with periurban villages that were progressively included into the urban fabric. Decision-makers and citizens past and present have had different views as to whether these are to be regarded as part of the city and should therefore be provided with a network, or not. The colonial rulers imported the idea of a clearly bounded city to Hanoi, which included the people's city, known today as the "36-streets area", which was mostly inhabited by Vietnamese and some Chinese traders. However, they excluded neighbouring settlements – densely inhabited and mixed-use – that were consequently either defined as rural villages or demolished in order to provide space for the colonial city (cf. Pédelahore, 1986). The sanitation of the people's city, then called "*quartiers indigènes*" by the French, was a primary goal of colonial intervention. This was regarded as one central step toward the unification of the urban precinct. According to the colonial definition of its borders, the city encompassed quarters with a great variety of building structures, densities and uses, as well as inhabitants with multiethnic backgrounds, who were separated into "*indigène*" and "*européen*" groups by the colonialists (cf. Fayet, 1939). Obviously, Hanoi's diverse urban morphologies reflected the living conditions and cultures of its inhabitants in each quarter. These quarters, in turn, were influenced to differing degrees by the colonial policies on trade, housing, and other matters that were restrictive with regards to Vietnamese traditional practices. For example, large parts of the Vietnamese population were excluded from colonial housing markets due to a lack of financial means and differing preferences for housing styles and amenities (Interview Trần Đức Hạ , 2010). The profound socio-spatial fragmentation resulting from this divide could not be bridged simply by rolling out a unitary network for wastewater disposal. While the French actually installed a unitary network for wastewater disposal from all those quarters defined as urban, this uniform network did not function due to the diversity of the urban spaces in question. In the densely populated Vietnamese quarters the underground canals quickly became clogged, causing a backlash in the European quarters as well as in the suburban villages outside the perimeter of the colonial network (Vann, 2003).

The close link between the building of underground infrastructure networks and planned urbanisation in French colonial times slackened afterwards. During the war- and poverty-ridden decades after independence, the Soviet Union heavily supported the city's then state-led top-down urban planning. In remembrance of its assistance, the ambitious urban master plan drawn up in this period was later called the "*Leningrad Plan*" (Chi *et al.*, 2010). The modern infrastructure ideal fitted well into communist rhetoric and was therefore never openly negated. However, in this era, the planning ideals and urban realities of Hanoi drifted far apart (cf. Logan, 2000). This can be demonstrated, above all, by the contrast between the gigantic master plan, which far exceeded the urban perimeter of Hanoi, and the space actually being built by the state machinery. Living quarters were selectively erected, too sparsely spread throughout the designated urban extension to form a continuous urban fabric as suggested by the master plan. While social infrastructures were set up in each of the quarters, networked technical infrastructures were largely neglected. This was due to the strong restraints of financial and personnel resources that did not allow the implementation of universal infrastructure networks (Interview Lý Trực Dũng , 2011). After the local wastewater treatment plant, which was built in one of the first living quarters in Hanoi, quickly congested due to lack of maintenance, the living quarters built later on were not equipped with sanitation infrastructure other than decentralised septic tanks.

Hanoi's most prominent current project with regard to wastewater disposal, the "Hanoi sewerage and drainage environmental renovation project", focuses on the installation of "central large-scaled wastewater treatment plants" (SRV/HPC, 2005). It illustrates that the planning paradigm has basically remained stable throughout the various political regimes from colonial times until today despite constant urban change. It outlines a centralised network of pipes throughout the urban precinct, reaching into the periurban areas with treatment plants at strategic points. Especially since the flooding of Hanoi in November 2008, this plan, which was drafted with substantial support from the Japanese development agency JICA, has been to the subject of heated public controversy. Architects, planners and wastewater experts have criticised it, arguing that the installation of a unitary wastewater network cannot overcome Hanoi's spatial fragmentation (Trần , 2009). In conclusion, a historical perspective demonstrates how the development of Hanoi has been shaped since French colonial times by recurrent attempts to realise the modern infrastructure ideal.

Urban morphologies in Hanoi

Today, the dynamic and multifaceted urbanisation patterns of Hanoi contrast with the stability of the planning paradigm for wastewater infrastructures. Urbanisation processes shaping these patterns take place in the continuum between urban planning (following formal rules on planning procedures and outcomes) and incremental building activities based on negotiations between builders, neighbouring households and local administrations. Each process results in specific socio-technical arrangements in the provision of infrastructure services (see below). The reforms subsumed under *doi moi* have had considerable impact on those processes and the resulting built environment of Hanoi, especially those concerning ownership and trading of land (cf. Quang and Kammeier, 2002). In the first decade after *doi moi* Hanoi experienced a boom in informal urbanisation, with around 70% of the building activities taking place without any legal documents (cf. Logan, 2000: 250). Roughly since the turn of the century, urbanisation processes, led by coalitions of real estate investors and formal urban planning agencies, have become more and more dominant in shaping the urban form of Hanoi (cf. Han and Vu, 2008).

These processes have shaped the characteristic areas of Hanoi. The *historical centre*, comprising the ancient 36-streets area and the colonial town interlinked

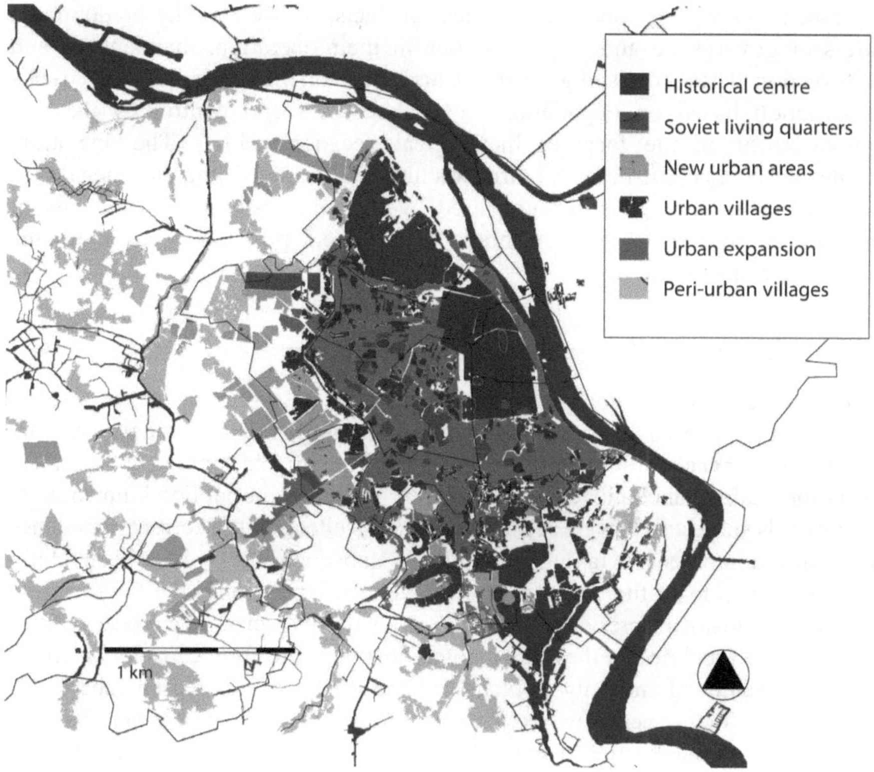

Figure 2.1 Characteristic areas of Hanoi
Source: Schramm, 2014.

by the wastewater network built up during the colonial era, has been substantially densified by the aggregate growth of buildings. The formerly selectively-built *socialist living quarters and urban village***s** in the extension have been interwoven into a dense urban fabric by incremental urbanisation. In contrast, the outskirts of the city are shaped by "a great diversity of intermixed landscapes, including walled residential estates, ad hoc densification of pre-existing villages, and the tight intermingling of small-scale industries with commercial, residential and agricultural activities." (Leaf, 2002: 29). *Periurban villages* that develop in incremental urbanisation processes, and residential estates – so-called *new urban areas* constructed within the formal process as detailed above – are not integrated spatially, but form independent mosaics.

Socio-spatial dimensions of wastewater systems in Hanoi

In this section, we elaborate on the sociotechnical characteristics of the wastewater systems in each of the characteristic areas. We present the socio-technical constellations of sanitation infrastructure in each area with regard to specifics in the urban morphology and the technical artifacts, as well as the predominant forms of governance and responsibilities in their operation, maintenance and construction. These forms of governance are placed on the spectrum between self-organisation based on negotiations by stakeholders and control by the city administration in the form of hierarchical decision-making. The operation, maintenance and construction of infrastructure facilities can either be delegated to professional private entrepreneurs hired by the administration, inhabitants or building companies, or else members of the local population carry out the respective tasks themselves.

Septic tanks as the predominant artifact in sanitation systems

In most urban areas, human wastes are pretreated in septic tanks. Such septic tanks can be regarded as the core sanitation technology in Hanoi today. They are installed under each building and are thus part of the private sphere – their design, operation, maintenance and renewal is barely regulated by the public administration, but rather is subject to private self-organisation. Reflecting the current prevalence of urbanisation processes taking place outside state control, they have been rebuilt and redesigned to fit the changing requirements arising from high building and population densities (Interview Việt Anh , 2009). However, in the process of this adaptation to local needs, they have largely lost their cleaning capacity, as filters have been removed and only a fraction of the wastewater is fed into the tanks. Furthermore, the sewer network features forms of governance other than the pretreatment of blackwater in septic tanks (Việt Anh 2007). As these networks connect several households and therefore extend into public space, their maintenance is traditionally defined as a public task, and is shaped by rather hierarchical forms of governance that differ in the characteristic areas, as explained below.

Historical centre

The built environment in the historical centre is mostly shaped by villas and, within an incremental building process, extremely densified row houses around five stories high. The emptying of the septic tanks is organised individually. In contrast to the septic tanks that are an integral part of the incremental rebuilding outside the formal planning framework, the wastewater disposal network is owned, constructed and maintained by a utility company, the Hanoi Sewerage and Drainage Company (HSDC), which in turn is owned by the city government (the Hanoi People's Committee). A 10% surcharge on the water tariffs partly finances the utility's investments (Việt Anh *et al.* 2005). As this is far too low to cover its expenditure, the national budget fundamentally subsidises the utility. However, due to the lack of public funds, the network has barely been renovated until recently.

Figure 2.2 Extremely densified historical centre
Source: photo by Sophie Schramm.

Urban villages

In the urban villages, septic tanks without filters are the prevailing means of pretreatment of household blackwater. The tanks are organised individually, as in the historical centre. The wastewater networks of the urban villages are in varying states depending on the overall stage of the urbanisation process in these incrementally built areas. The local drains and ditches follow the spatial structure of the urban villages, which is defined by a socio-spatial hierarchy of places differing with regard to the degree of privacy. The urban morphology structures this differentiation, with row houses built closely along streets that form a branched pattern, with smaller roads departing from bigger ones and eventually ending in very narrow cul-de-sacs.

Figure 2.3 Narrow alley with covered sewage network in the urban village Giap Bat
Source: photo by Sophie Schramm.

The status of the drains often correlates with the hierarchy within the road network, with the smallest streets having the least professionally built drains (Việt Anh *et al.* 2005). This hierarchy of spaces is also reflected in the provision of sanitation services. While the public utility maintains the main sewers in the urban villages, the maintenance in the secondary sewers is delegated from the city government to the level below, the districts, which are responsible for hiring companies to maintain those pipes (Interview Lê Văn Tàu, 2011). The spatial structure of the urban villages results in a further differentiation of service provision: the cul-de-sacs in those urban areas further from the centre are under the responsibility of the administrative level below the district, the ward or even neighbourhood groups, where citizens carry out the maintenance of the network themselves.

Soviet living quarters

As in the urban villages, the public utility maintains the main sewers around the living quarters, while the districts are responsible for subcontracting the maintenance of the secondary sewers to private companies. However, the local governance within these quarters differs from that in the villages, corresponding to the spatial form. The living quarters consist of apartment blocks, implying a rather sharp division between the public space outside the block and private space within the buildings. At the same time, the wastewater pipes within the buildings are more complex, as each pipe collects either the blackwater or the greywater from two neighbouring flats from each of the five stories and directs it either to the septic tank or directly into the drain.

Due to the aggregate expansion of the buildings, formerly open drains have now often been moved underground, hindering access. The maintenance of the pipes within the blocks as well as the septic tanks is not provided by the local governments but is self-organised by neighbourhood groups. Due to the technological complexity of the operation of these networks, these groups are hardly able to fulfill the task and are therefore frustrated by the current neglect of these quarters by the government (Interview TDP C1, 2011).

Periurban villages

Periurban villages feature spatial morphologies similar to those of the urban villages, but they are less dense. Similar to rural areas, they are characterised by the separation between the agricultural use of blackwater and the treatment of other household wastewaters in septic tanks. The households themselves bring the blackwater to the fields, while the emptying of pit latrines or septic tanks is delegated to private or public companies.

As the periurban villages are located outside the urban districts of Hanoi, the local authorities below the district level control the maintenance of the open wastewater drains, which is carried out by inhabitants themselves beyond the responsibility of Hanoi's public utility (Việt Anh *et al.* 2005). The local community

Figure 2.4 Retroactively extended flats in the Soviet living quarter Nghia Tan
Source: photo by Sophie Schramm.

has a direct interest in maintaining the sewers, as they serve as irrigation channels in the agricultural areas. The main challenge for the periurban villages is the ongoing transition from rural to urban, accompanied by a change in income generation from mostly agriculture to the industrial or service sectors.

New urban areas

The new urban areas are mostly equipped with modern wastewater networks, disposing of the wastewater into the surrounding areas. With some exceptions, treatment is technically still often decentralised, with advanced forms of septic tanks serving single buildings or groups of buildings. However, the maintenance of the tanks is more centrally governed, as it is part of the management contract of the whole area. Therefore, it is more strictly regulated by the city government and is largely outside the scope of the local population (Inteview Hoàng Anh Tuấn , 2008). Local management boards carry out the emptying themselves or they delegate the task to private enterprises.

Furthermore, there is a tendency towards the installation of centralised plants typical of urban contexts in the global North. Here, a clear relation between the process of urbanisation and the socio-technical system of wastewater disposal and treatment is visible, as the process of urbanisation is led by a coalition of the formal planning administration at the city level and (mostly international) investors, who initiate the implementation of a deal known as "land for plant". In this type of deal, investors get the land use rights for the new urban area free of charge if they build a wastewater treatment plant in return (Interview Việt Anh , 2009). Considering the exploding land prices in Hanoi, which in the last few years have been comparable to those in London or Paris, project investors benefit considerably from this kind of deal. At the same time, the city government benefits from investments in centralised wastewater systems, which are improved and made comparable to Western standards. Although the city government promotes the HSDC in taking responsibility for the overall planning and operation of

Figure 2.5 The periurban village Lai Xa in the process of urbanisation
Source: photo by Sophie Schramm.

sewage facilities, domestic investment companies compete for the exclusive provision of services within the new urban areas that they have built. Most prominent in this regard is a state-owned housing development company, which also provides solutions for technical infrastructures. Due to contractual disputes with the city government concerning the costs of connecting their project sites to the city water supply and sewerage network, they manage all water-related technical infrastructures within their new urban areas themselves, including wells for tap water.

In summary, it is evident that Hanoi's wastewater management is characterised by a diversity of place-specific socio-technical constellations (see Table 2.1), which – apart from the periurban areas – combine septic tanks and centralised networks in different forms. However, there are various similarities, especially in the management of septic tanks – apart from those septic tanks installed in new urban areas – which are shaped by the self-organisation of the emptying process. However, this common practice impedes the cleaning efficiency of the tanks, as local users focus on the disposal of the wastewater and are less interested in broader environmental implications. Therefore, recent initiatives towards more effective public regulation of the septic tank emptying process in the new urban areas could be exemplary for the other urban areas. Generally, the new urban areas represent a disjuncture from the traditional wastewater system in Hanoi, as they display a greater variation in governance and service provision related to the more diverse technical systems. In contrast to the intense competition of private and public service providers in new urban areas, professional providers are practically absent in the periurban areas.

Figure 2.6 Linh Dam, the first new urban area in Hanoi
Source: photo by Sophie Schramm.

Table 2.1 Socio-technical characteristics of sanitation infrastructures in different areas of Hanoi

	Technostructure	*Service providers*	*Forms of governance*
Historical centre	• Decentralised septic tanks • Underground network	• Septic tanks: small entrepreneurs • Network: public utility (HSDC)	• Septic tanks: negotiation, self-organisation • Network: hierarchical decision-making
Urban village	• Decentralised septic tanks • Underground network • Open drains	• Septic tanks: small entrepreneurs • Network: public utility • Drains: district or ward administration, household groups	• Septic tanks: negotiation; self-organisation • Network: hierarchical decision-making
Soviet living quarter	• Decentralised septic tanks • Underground network • Open drains • Local underground pipes	• Septic tanks: small entrepreneurs • Underground network: public utility • Open drains: district or ward administration • Pipes: household groups	• Septic tanks and local pipes: negotiation, self-organisation • Network: hierarchical decision-making
Periurban village	• Decentralised pit latrines • Open drains for irrigation	• Local use as fertilizer in agriculture: self-organisation • Irrigation channels: ward administration, household groups	• Local use as fertilizer in agriculture: self-organisation • Irrigation channels: hierarchical decision-making
New urban area	• Advanced de- or semi-centralised sewage plants • Underground network	• Advanced decentralised plants: building or investment company and subcontractors • Semi-centralised plants: underground network: building or investment company and subcontractors, public utility	• Hierarchical decision-making • Public-private partnerships

Source: authors.

In contrast to official announcements and political programmes, one of the key problems is that the HSDC has been unable to fulfill its core responsibility of network maintenance and extension. Due to the rapid and largely unplanned growth of Hanoi, coupled with financial restrictions, the responsibility of the public utility company has been limited in the urban villages and the Soviet living quarters to the main sewers and canals, and is practically absent in the periurban villages. Despite political ambitions and statements, the modern infrastructure ideal – a specific constellation of technical artifacts and social arrangements, including a universal network maintained by a public utility financed by tariffs and taxes – has, at best, only been implemented in one specific part of Hanoi, namely the historical centre.

Planning centralised urban sanitation system in the light of diverse urban structures

As mentioned above, current wastewater plans envisage an expansive and costly extension of the wastewater network, with a sewerage system separating stormwater and household wastewaters and, as the title of a recent feasibility study suggests, "central large-scale wastewater treatment plants" that will render septic tanks dispensable (SRV/HPC, 2009). Hanoi's public utility, the HSDC, will maintain the central network. In the light of the persisting diversity of urban morphologies and of wastewater systems in Hanoi, this attempt to realise the modern infrastructure ideal seems questionable. First of all, problems concerning the feasibility of doing so have become apparent regarding the funding for these huge investments – these costs are calculated at the comparatively high amount of roughly one billon USD, which is to be loaned from a Japanese development bank (SRV/HPC, 2009). If the central state of Vietnam enters into this contract, it will also enter into a long-lasting financial dependency on Japan, which will bind resources that consequently cannot be used for other investments.

Even more problematic than this financial burden are the long-term repair and maintenance costs of the projected network with central plants. As mentioned above, even those plants existing today are prone to malfunction due to the lack of qualified staff for the operation and maintenance of these facilities. Due to the restricted financial capacities of the public utility company, the maintenance of the network in large parts of the city has been delegated to local authorities or even inhabitants. These problems have implications for the specific areas, as the funds for locally required repair and maintenance are lacking – especially in the Soviet living quarters and the historical centre, where an underground network has been built. For the mostly agricultural periurban villages, the installation of an underground network currently displeases farmers, as the water in open drains is used for irrigation. Therefore, network construction in these villages is better adapted to the current changes in income generation from the agricultural to the industrial and service sectors in the course of the urban transition. Since agriculture, as opposed to other forms of income generation, depends on open drains for irrigation, this transition of income generation is related to the spatial configurations of wastewater and sanitation systems.

This demonstrates that even under current circumstances the projected uniform expansion of the network is questionable, and the long-term dynamics of this dependence on a centralised uniform network are another huge problem. As the historical perspective has shown, the repeated attempts to install a central network during more prosperous times were followed by periods of financial constraints and poverty, leading in turn to the challenges of maintaining costly infrastructures. In particular, the repeated efforts to rebuild the network in the former French colonial quarters and the current decay of the network in the Soviet living quarters demonstrate that once a central underground network is installed, it needs to be maintained professionally in order to continue to function. Once constructed, its technological complexity can no longer be flexibly reduced without major investments or the loss of its functionality. As opposed to the networks centrally planned and constructed in the historical centre and the Soviet living quarters, the locally built and incrementally expanded networks in the urban villages display a smoother process of transition from self-organised construction and maintenance to the public provision of services, with various stages in between. While the support and professionalisation of these arrangements could be the focus of urban and infrastructure planning in Hanoi, they are completely neglected in current plans, which simply project the centralised network far into the periurban space. Considering the enormous investments needed to realise these plans for a networked city, it seems more reasonable to invest in the improvement of the existing decentralised systems based on septic tanks, and in the coordination of the diverse, place-specific stakeholders involved in the construction, operation and maintenance of sanitation systems.

Conclusion

In this chapter we have discussed current debates on urban infrastructure systems that reveal the need for reflecting on the specific spatial conditions and socio-technical arrangements in the provision of urban services in the global South. Hitherto, most of the studies point to the diverse and often heterogeneous socio-technical arrangements that shape splintered urban spaces in the global South. Similar to this case study on Hanoi, many other studies indicate that the perpetual pursuit of a networked city and the modern infrastructure ideal of the uniform and ubiquitous provision of infrastructure services has not only never been realised, but paradoxically their transfer has even resulted in fragmented urban landscapes with networked spaces separated from "non-grid" areas. However, this case study on sanitation systems in Hanoi suggests that these findings need to be specified. As the debates summarised above suggest, the modern infrastructure ideal introduced in colonial times has fundamentally shaped Hanoi's sanitation systems. Following independence, this ideal remained the model for urban and infrastructure planning in Hanoi and was promoted by all the foreign engineers and planners who have advised the city since: the planners from the USSR as well as the Japanese development agency that is at the forefront of wastewater planning in Hanoi today. This stability of the paradigm, which is put forward by a range of planners from varying political regimes and national backgrounds, exemplifies that the findings of international circulation and local appropriation not only apply to technologies, as established by Hård and Misa (2008), but, beyond that, also to circulating planning models, infrastructure ideals, hegemonic ideals of modernity and ideals of "cityness". However, the interesting issue that needs further empirical and theoretical exploration is how this adaptation takes place, and how these models and ideals are localised into webs of planning ideals, webs of institutions and material infrastructures.

The case study of Hanoi demonstrates that while a differentiation of the urban infrastructure landscape is apparent, this differentiation is more complex than current debates suggest. It can neither be reduced to an analytical antagonism between networked urban spaces and "non-grid" areas co-existing largely independently within the city, nor can we clearly identify a "battle of the systems" (Hughes, 1987) between competing technological frameworks in which "one side" wins a victory. Instead, Hanoi's sanitation systems display a more complex topology, characterised by hybridised forms of service provision. Depending on the place-specific conditions, hegemonic technological models of centralised underground networks and sewage plants co-exist and hybridise with open drains and non-networked technologies, septic tanks and pit latrines. In other places, these decentralised solutions occur in place of existing, malfunctioning or absent centralised networks. The same is true of the complex forms of wastewater management within different urban areas in which the public utility, household groups, private service providers and civil self-organisations co-produce critical infrastructure services in place-specific ways, and/or compete with one another. In other words, the case of Hanoi demonstrates how so-called modern and alternative socio-technical frameworks, technological solutions and management approaches persist together in urban space, and the way that they have been creatively amalgamated and hybridised or rejected in place-specific ways to form a hybrid system of infrastructure provision. As a consequence, the topology of the wastewater systems in Hanoi is neither consistent with the modern infrastructure ideal of uniform and unitary service provision, nor can it be characterised as fragmented in the sense that two or more systems (centralised networks with public service provision, and decentralised disposal through self-organisation) define clear spatial or socio-technical boundaries. Together, these socio-technical constellations within each area add to the diversity of the city, and have tremendous implications for urban development, urban functioning and everyday life in cities. Alternatively, arguing alongside Anderson's (2002) approach to postcolonial technoscience, the challenge for both urban planning and research is to more systematically recognise and address the "hybridities, borderlands and in-between conditions" (ibid., 2002: 643) of urban infrastructure landscapes.

Notes

1 The authors would like to thank the German Research Foundation (DFG) for funding under the "Topology of Technology" graduate programme (GK 1343) and the

Canadian Social Science and Humanities Research Council for funding under the "Global Suburbanisms: Governance, land and infrastructure" major collaborative research initiative. Much of our thinking on the networked city in this essay has been inspired by the workshop "From Networked to Post-Networked Urbanism" in Autun, France in Summer 2012. We thank all participants and especially the organizers, Olivier Coutard and Jonathan Rutherford, for stimulating discussions.

References

Abbott, J. (2012) *Green Infrastructure for Sustainable Urban Development in Africa.* London: Routledge.

Anderson, W. (2002) Introduction: postcolonial technoscience. *Social Studies of Science* **32**(5–6): 643–658.

Bakker, K. (2007) Trickle down? Private sector participation and the pro-poor water supply debate in Jakarta, Indonesia. *Geoforum* **38**(5): 855–868.

Balbo, M. (1993) Urban planning and the fragmented city of developing countries. *Third World Planning Review* **15**(3): 23–35.

Budds, J. and G. McGranahan. (2003) Are the debates on water privatization missing the point? Experiences from Africa, Asia and Latin America. *Environment and Urbanization* **15**(1): 87–114.

Chi, T. T., N. Luan, N. Z. Luyen, N. H. Thai and N. Q. Thong (2010) *Half a Century of Vietnamese Architecture*. Hanoi: Time Publishing House.

Coutard, O. (2008) "Placing Splintering Urbanism: Introduction." *Geoforum* **39**(6): 1815–1820.

Coutard, O., R. Hanley and R. Zimmerman (2005) Networks systems revisited: the confounding nature of universal systems. In O. Coutard, R. Hanley and R. Zimmerman, Eds. *Sustaining Urban Networks: The Social Diffusion of Large Technical Systems.* London and New York, Routledge.

Coutard, O. and J. Rutherford (2011) Post-networked cities: recombining infrastructural, ecological and urban transitions. In H. Bulkeley, V. Castan Broto, M. Hodson and S. Marvin (Eds.), *Cities and Low Carbon Transitions*. London: Routledge,.

Dupuy, G. (2008) *Urban Networks: Network Urbanism*. Amsterdam, Techne Press.

Fayet, L. (1939) *Avant Projet sur les égouts de Hanoi.* Hanoi: Imprimerie d'Extrême-Orient.

Férnandez-Maldonado, A. M. (2008) Expanding networks for the urban poor: water and telecommunications services in Lima, Peru. *Geoforum* **39**(6): 1–13.

Gandy, M. (2006a) Planning, anti-planning and the infrastructure crisis facing metropolitan Lagos. *Urban Studies* **43**(2): 371–396.

Gandy, M. (2006b) The Bacteriological City and its Discontents. *Historical Medical Geography* **34**(1): 14–25.

Gandy, M. and S. Frank, Eds. (2006) *Hydropolis*. Frankfurt am Mainz, Campus.

Gopakumar, G. (2011) *Transforming Urban Water Supplies in India: The Role of Reform and Partnerships in Globalization*. London: Routledge.

Graham, S. (2000) "Introduction: cities and infrastructure networks." *International Journal of Urban and Regional Research* **24**(1): 114–119.

Graham, S. and S. Marvin (2001) *Splintering Urbanism. Networked Infrastructures, Technological Mobilities and the Urban Condition*. London, Routledge.

Han, S. S. and K. T. Vu (2008) Land acquisition in transitional Hanoi, Vietnam. *Urban Studies*, **45**(5–6) 1097–1117.

Hård, M. and T. J. Misa, Eds. (2008) *Urban Machinery Inside Modern European Cities*. Boston, MIT Press.

Hughes, T. P. (1987) The Evolution of Large Technical Systems. In W. E. Bijker, T. P. Hughes and T. J. Pinch (Eds.), *The Social Construction of Large Technological Systems*. Cambridge, MIT Press.

Jaglin, S. (2008) Differentiating networked services in Cape Town: echoes of splintering urbanism? *Geoforum* **39**(6): 1897–1906.

Lamichhane, K. M. (2007) On-site sanitation: a viable alternative to modern wastewater treatment plants. *Water Science and Technology* **55**(1–2): 433–440.

Langergraber, G. and E. Muellegger (2005) Ecological sanitation – a way to solve global sanitation problems? *Environment International* **31**(3): 433–444.

Leaf, M. (2002) A tale of two villages. *Cities* **19**(1): 23–31.

Logan, W. S. (2000) *Hanoi – Bibliography of a City*. Seattle: University of Washington Press.

McFarlane, C. (2008) Governing the contaminated city: infrastructure and sanitation in colonial and postcolonial Bombay. *International Journal of Urban and Regional Research* **32**(2): 415–435.

McFarlane, C. (2009) Everyday interruptions: urban life in the global South. In S. Graham, (Ed.), *Disrupted Cities: When Infrastructure Fails*. New York: Routledge.

McFarlane, C. (2010) Infrastructure, Interruption, and Inequality: Urban Life in the Global South. In Graham, S. (Ed.), *Disrupted Cities: When Infrastructure Fails*. London: Routledge.

McFarlane, C. and J. Rutherford (2008) Political Infrastructures: Governing and Experiencing the Fabric of the City. *International Journal of Urban and Regional Research* **32**(2): 363–374.

Melosi, M. (2008) *The Sanitary City: Urban Infrastructure in America from Colonial Times to the Present*. Baltimore, John Hopkins University Press.

Mitlin, D. (2008) With and beyond the state – co-production as a route to political influence, power and trans-formation for grassroots organizations. *Environment and Urbanization* **20**(2): 339–360.

Monstadt, J. (2009) "Conceptualizing the political ecology of urban infrastructures: insights from technology and urban studies." *Environment and Planning A* **41**(8): 1924–1942.

Monstadt, J. and S. Schramm (2013) Beyond the Networked City? Suburban Constellations in Water and Sanitation. In R. Keil (Ed.): *Suburban Constellations: Governance, Land and Infrastructure in the 21st Century*. Berlin: Jovis, 85–94.

Myers, G. (2011) *African Cities: Alternative Visions of Urban Theory and Practice*. London: Zed Books.

Odendaal, N. (2012) Reality check: planning education in the African urban century, *Cities* **29**: 174–182.

Otterpohl, R. (2002) Options for alternative types of sewerage and treatment systems directed to improvement of the overall performance. *Water Science & Technology* **45**(3): 149–158.

Pédelahore, C. (1986) Constituent elements of Hanoi City. *Vietnamese Studies* **12**: 104–159.

Porter, L. (2010) *Unlearning the Colonial Cultures of Planning*. Farnham and Burlington: Ashgate.

Prasad, N. (2006) Privatisation results: private sector participation in water services after 15 years. *Development Policy Review* **24**(6): 669–692.

Quang, N. and H. D. Kammeier (2002) Changes in the political economy of Vietnam and their impacts on the built environment of Hanoi. *Cities* **19**(6): 373–388.

Schramm, S. (2014) *Stadt im Fluss: Die Abwasserentsorgung Hanois im Lichte sozialer und räumlicher Transformationen*. Stuttgart: Franz Steiner Verlag.

Schramm, S. and S. Bieker (2010) Urban semicentralised supply and disposal: innovations and challenges for Hanoi, Vietnam. *International Journal of Sustainable Development* **13**(1): 97–110.

Silva, C. N. (2012) Urban planning in sub-Saharan Africa: a new role in the urban transition. *Cities* **29**: 155–157.

SRV/HPC (Socialist Republic of Vietnam/People's Committees of Hanoi) (2005) *Feasibility Study Report and Program Implementation for Hanoi Sewerage and Drainage Environmental Renovation Project – Phase II Volume I Main Report*. Hanoi: NIPPON KOEI; Socialist Republic of Vietnam; People's Committees of Hanoi.

SRV/HPC (Socialist Republic of Vietnam/People's Committees of Hanoi) (2009) *Feasibility Study for the Construction Project of Central Large-scaled Wastewater Treatment Plants for Hanoi Environmental Improvement – Final Report Summary*. Hanoi: NIPPON KOEI Co., Ltd. VIWASE.

Star, S. L. (1999) "The ethnography of infrastructure." *American Behavioral Scientist* **43**(3): 377–391.

Trần , H. A. 2009. Những nhà máy xử lý nước thải tỷ đô có làm sông hồ Hà Nội sạch trong? [Have the wastewater treatment plants made billions of dollars by cleaning rivers and lakes in Hanoi? Article from 24.01.2009 [WWW document]. URL http://mag.ashui.com/index.php/chuyenmuc/moitruong/70-moitruong/670-nhung-nha-may-xu-ly-nuoc-thai-ty-do-co-lam-song-ho-ha-noi-sach-trong.html (Accessed 12 April 2011).

Tarr, J. A. and G. Dupuy, Eds. (1988) *Technology and the Rise of the Networked City in Europe and America*. Philadephia, Temple University Press.

UN-Habitat (2005) *Water and Sanitation in the World's Cities*. United Nations Human Settlements Programme, Nairobi.

Van Laak, D. (2004*) Imperiale Infrastruktur: deutsche Planungen für eine Erschließung Afrikas 1880 bis 1960*. Paderborn, München [u.a.], Schöningh.

van Vliet, B., G. Spaargaren and P. Oosterveer (Eds.) (2010) *Social Perspectives on the Sanitation Challenge*. Springer, Dordrecht.

Vann, M. G. (2003) Of rats, rice, and race: the great Hanoi rat massacre, an episode in French colonial history. *French Colonial History* **4**: 191–203.

Việt Anh N., A. Morel, D. H. Anh, T. T. H. Hoa and P. T. Nga (2005) *Overview of Water and Sanitation Sector in Vietnam*. Hanoi: CEETIA, SANDEC.

Việt Anh , N. (2007) *Bể tự hoại và vá bể tự hoại cải tiến* [*Septic Tanks and Baffled Septic Tanks with Anaerobic Filters (BASTAF*)]. Hanoi: Nhà xuất bản xây dựng [Ministry of Construction Publishing House].

Vinh, P. N. (2004) *Hanoi Past and Present.* Hanoi: The Gioi Publishers.

von Sperling, M. and C. A. de Lemos Chernicharo (2005) *Biological Wastewater Treatment in Warm Climate Regions*. London: IWA Publishing.

Watson, V. (2009) Seeing from the south: refocusing urban planning on the globe's central urban issues. *Urban Studies* **46**(11): 2259–2257.

Wright, G. (1991) *The Politics of Design in French Colonial Urbanism*. Chicago: University of Chicago Press.

Zérah, M.-H. (2008) Splintering urbanism in Mumbai: contrasting trends in a multilayered society. *Geoforum* **39**(6): 1815–1820.

Interviews

Hoàng Anh Tuấn , Hanoi Authority for Urban Planning and Architecture, Vice-Director. Interview on 4 September, 2008.

Lê Văn Tàu, Development Center for Maintenance and Administration of Urban Infrastructure of Cầu Giấy, Cầu Giấy , Hanoi. Interview on 17 November, 2011.

Lý Trực Dũng , Architect and Journalist Hanoi. Interview on 24 October, 2011.

Nam Ho Park, Project Director PPJ Masterplanung Hanoi. Interview on 15 April, 2009.

TDP-C1, Leader To Dan Pho (neighborhood group) No. C1 Nghĩa Tân, Cầu Giấy, Hanoi. Interview on 29 October, 2011.

Trần Đức Hạ, Professor Institute for Environmental Science and Engineering (IESE), Hanoi University of Civil Engineering (HUCE). Interview on 12 June, 2010.

Trần Huy Ánh , Urban Planner Hanoi Data. Interviews on 9 April, 2009; 31 May, 2010; 20 October, 2011.

Việt Anh Nguyễn , Professor University of Civil Engineering, Institute of Environmental Science and Engineering (IESE). Interview on 12 April, 2009.

3 Beyond the networked city, the hyper-networked city?

Decline and renaissance of the Parisian non-potable water system (1820–2020)

Sabine Barles, Olivier Coutard and André Guillerme

Introduction

In the last ten years or so the organisation of the Paris water supply, a legacy of the nineteenth century, has once again moved to centre stage. This fact is notably a result of two initiatives taken by the majority local government that took office following the 2001 municipal elections, headed by the socialist Bertrand Delanoë. First, the municipal government decided to take back municipal control of the water service, which in 1984 had been contracted out by the previous team under Jacques Chirac[1] to two private companies (one for the left bank, the other for the right bank). Second, it wanted to reverse the policy implicitly pursued by its predecessors, which had led to the progressive decline of the municipality's "second water network" – in fact, historically the first – which supplies non-potable water to the whole municipal area.

Our concern in this article is the latter policy and the future of the non-potable water system as it is currently evolving. We show that the choices regarding the non-potable water system are primarily associated, by local government, with regional and local environmental policies, in response to forecasts of growing hydrological stress in the Seine basin, rising quantitative and qualitative pressures on water resources, a deterioration in the urban heat island phenomenon, and a crisis of regional biodiversity. We argue that the municipal policy introduced by the Delanoë team signals a new political economy of infrastructures. In it, the environmental priorities associated with water resources, and with the quality (or more specifically the different levels of quality) and the uses of water, have taken on new importance, which legitimises the maintenance of a dual system justified by the benefits of managing water differentially in terms of its uses, despite the *de facto* competition between the two water systems in a context of infrastructural overcapacity.[2]

Following this analysis, we conclude by considering the possible implications of this new political economy of infrastructures for the "qualitative management" of water resources and aquatic environments, for the control of the sources and uses of non-potable water, and for the organisation and functioning of the urban areas concerned.

An infrastructure with diverse functions: the non-potable water system 1820–2020

In this first section we put recent discussions about the utility and future of the non-potable water system into their historical context, by returning to the origins of that infrastructure which, before becoming the "second municipal water system", was for a long time the first. This historical perspective allows us to cast light on the successive functions performed by the system over its history, and the reasons for its neglect and decline in recent decades, as well as the implicit policy that led to this decline. We then describe the process whereby, from 2009, the municipal team finally came to the decision to revive the non-potable water network through a policy of investment, the pooling of diverse raw water sources,[3] the "readoption" of historical uses of the non-potable water network and the development of new uses.

From symbol of urban modernity to declining infrastructure: functions and fortunes of the non-potable water system (1802–2009)

1802–1860: a capital system

In this period of the city's history, the alluvial soils on which Paris stands were naturally permeated with run-off and infiltration water, whose taste smacked of the very wealth of the city: imported sea salt used in food preparation was present in large proportions in the urine and residual water that seeped into the water table that fed the more than 32,000 wells within the Thiers Wall. The few spring water fountains provided only a trickle of crystalline water, much sought after for drinking and food, as was the water pumped out of the Seine and distributed by the water carriers.

At the end of the 18th century, the main characteristic of urban water was the plurality of its forms – rain, run-off, reservoir, barrel, pipe – and of its uses – crafts, industry, energy, food, hygiene, recreation, transport. This was true in most European cities. Several river diversion projects – on the Eure, Yvette, Hyères, Ourcq – sought to bring raw, but clean, water to the Queen of Capitals, first used to clean the streets, to supply the proletariat and the bourgeois, and to beautify the city. However, the diversity of the uses of water diminished with industrialisation: the new factories making paper, leather, starch, painted fabrics and so on fostered faster and less unpredictable dry manufacturing processes, using chemical products, the best example being dry cleaning (with chlorine). The protective ditches were filled in to make way for shops full of new products.

First Consul Bonaparte made his decision:

> The goal I would like to achieve above all would be to make water flow in as many fountains as possible, day and night. For four summer months, this is essential in Paris. We should not be quibbling about 100,000 francs when it is a question of the attractiveness and health of such an immense city. The goal I wish to attain is: 1° that the 56 monumental fountains of Paris should

> flow day and night…; that water should no longer be for sale and that everyone should be able to take as much as they like; 2° that the other fountains in Paris should be brought as quickly as possible into a condition to furnish water."[4]

In 1817 the waters of the Ourcq were channelled to the lake at La Villette, where they were divided into three pipelines and a canal feeding into the sluiced Arsenal Basin. Each of the three master pipes was subdivided into secondary pipes to feed the monumental fountains, and into tertiary pipes for subscribers and the pavement outlets which the fountain keepers periodically opened with a wrench to fill the receptacles of the poor. The Ourcq system was a political and social success, although it cost three times more than expected.

However, it very soon became overwhelmed: demand was too high, there were daily cutoffs, the pipes broke from excessive pressure. In 1826, five years after the first home installations, the Ourcq system was obsolete, suffering from lack of pressure, flow and energy losses and leaks. Nonetheless, wells continued to be sunk, mainly for the tradesmen in the industrial capital.

1860–1984: the two services

In the second half of the nineteenth century, with the need to supply the new Parisian districts annexed in 1860 domestic water provision became one of the major service priorities. Given the failings of the Ourcq canal system, the Paris administration considered several solutions – in particular pumping water from the Seine or diverting distant springs – and quickly chose the second option, with the aim of supplying water that was "pure, limpid and cool" (Belgrand, 1854, 1, 28). The outcome was a service divided between good quality spring water, supplied to the buildings' upper floors, and river water (from the Ourcq, the Seine and the Marne) which remained at ground level and was used for cleaning and sprinkling, as well as supplying the poor of Paris and its suburbs. This division led to a not wholly unambiguous distinction between the private and the public service (Table 1), the former dedicated to "drinking water" and the latter to "washing water, water for communal use, water as a raw material, *industrial water*" (Bechmann, 1899, 1, 36). The City of Paris kept all the systems under its control, though commercialisation was assigned to a private operator, the Compagnie Générale des Eaux (CGE).

These principles underwent little change, except – from the late nineteenth century – for the use of treated river water to meet the ever-growing demands of the so-called private service, which in the next century came to account for up to 50% of the drinking water distributed. In addition, in the second half of the twentieth century the non-potable water system became increasingly specialised: it moved gradually out of the private domain, where it was used in the supply of toilet cisterns, to be replaced by the potable water network, a solution that seemed hygienically more dependable. At the same time, industrial uses declined. The non-potable water system gradually became restricted to outdoor functions and, increasingly, to the public domain, generating almost no revenues.

Table 3.1 Public service vs. private service, Paris, 1860–1980. (Bechmann 1899; Martin, 1902; Canneva, 2009; APUR, 2010).

	Public service	*Private service*
Water source	Rivers and artesian wells	Springs and, from 1896, rivers (excluding the Ourcq)
Treatment	None or clarification	Springs: none or chlorination River: Filtration, then filtration and chlorination
Pressure	Low, not guaranteed	High, guaranteed
Price	Low	High
Use	Interior (toilet flushes) and exterior of buildings, then (from the 1960s) only exterior (private domain and, primarily, public domain)	Essentially inside buildings, then (from the 1960s) interior and exterior in the private domain
Operation	Public, private for commercialisation only	Public, private for commercialisation only

1984–2009: the decline

The Parisian model of water management underwent a profound transformation in 1984, when the City contracted out the service (no longer the commercialisation alone) to two private companies, Compagnie des Eaux de Paris (CGE, later Vivendi, later Veolia) for the right bank of the Seine, and Eau et Force Parisienne des Eaux (Société Lyonnaise des Eaux, then Suez) for the left bank. The Société Anonyme de Gestion des Eaux de Paris (SAGEP, renamed Eau de Paris (EDP) in 2005), a semipublic company[5] set up in 1987, took over the production of water. These companies were responsible for the two water systems under a twenty-five-year contract.

From this point on, the different water actors in Paris, whether public or private, made it their main objective to reduce the consumption of non-potable water, a substance that brought no profit. One of its major uses was the flushing of Paris's unitary sewage network, 200,000m^3 per day, representing 65% of consumption in 1986. 6000 tanks, located at the head of the secondary pipes, filled continuously, and once full (i.e. six times a day), released their 5m^3 of water. More than half were eliminated under a modernisation plan introduced between 1999 and 2004, while the remaining tanks were fitted with electronic timers that limited the number of releases to one per day, reducing the daily consumption of flush water to 13,500m^3. At the same time, in non-parking zones the street cleaning service replaced the drain washout system – under which an operator would open the pavement outlet and manually sweep out the drainage channels – with road sweeping machines that used much less water. One of the stated objectives of these changes was to limit the volumes of clear water in the sewers, since it diluted the wastewater, causing malfunctions in the water treatment plants. In addition, the city parks department was looking to cut consumption with more economical watering systems. Some gardens and parks opted for potable water, seen as the only method compatible with automatic and trickle watering methods. Coinciding with the quest to reduce non-potable water consumption, the maintenance of the "second network" was almost completely abandoned, leading to local reductions in service quality. Because of the resulting pressure irregularities, some users – in particular the fire services and the Jardin du Luxembourg – switched to potable water. Finally, as of 1985, no further connection requests were accepted from private customers.

The outcome was swift and spectacular: production and consumption of non-potable water declined sharply (Figure 3.1). As a result, the non-potable water system came to be seen as an old, obsolete and useless infrastructure.

Chronicle of a proclaimed renaissance (2009-)

A new political and institutional context

The election of Bertrand Delanoë as Mayor of Paris in 2001 (re-elected in 2008) at the head of a team of Socialists and Greens, the growing criticism of the proliferation in the contracting out of public service functions in France since the 1980s, and in particular the publication of a 2003 report by the Court of Accounts exposing the Paris water services on several counts, created new conditions just at the time when the licence contracts signed in 1984 were coming to an end. The result was that on 1 November 2008 the Municipal Council voted to bring the water service back under municipal control.

The question of the future of the non-potable water network had already been brought to the table by right-wing politicians in 2005 – questioning its utility in a context of a general decline in consumption – and then by Green politicians in 2008, leading to the Council's decision to maintain the network. Four technical reports produced between 2008 and 2009 reached very unfavourable conclusions (see Box 1, p. 57), ultimately concurring with the view of the technical water and sewerage department that the network should be abandoned.

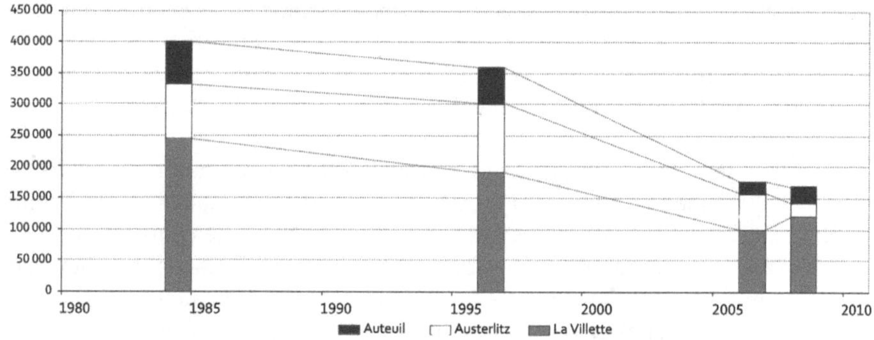

Figure 3.1 Non-potable water production, 1984–2009, m³ per day. Auteuil, Austerlitz and La Villette are three pumping stations. (APUR, 2010, 1, 43).

Source: Eau de Paris.

The 2009 consensus conference

Against this background, the approach taken by the Deputy Mayor responsible for water, Anne Le Strat, was to take the discussion beyond the technical and municipal spheres by setting up a consensus conference in autumn 2009. Most of the jurors, a majority of whom were initially in favour of abandoning the network, had received their training from municipal departments. It emerged quite quickly that the case presented to the jurors was one-sided, legitimised by recent biased studies. The sheer weight of emphasis, together with very different impressions drawn from non-Parisian experiences (in London, Barcelona and Madrid), led the jury to request additional information on the future priorities of water management, especially with regard to climate change.[6] It ultimately emerged that, seen in a broader light, the non-potable water system was worth retaining.[7]

In the end, to general surprise, in late 2009 the jury ruled by twelve votes to two in favour of retaining the network: "in preparation for the future (climate change, urban warming, social risks, …) and for the purpose of better resource management, the non-potable water system constitutes an opportunity to diversify water supplies and water qualities for different uses. The system should be maintained and the service modernised. The old uses should be revived and new, innovative uses developed." The Deputy Mayor responsible for water would draw on these conclusions in 2012 to win a vote along the same lines in the Paris Council.

The APUR studies of 2010 and 2013

The need for further analyses prompted the City of Paris to assign the conduct of subsequent studies of the non-potable water network to an unusual player in this sector, the Atelier Parisien d'Urbanisme, generally more concerned with urban planning than urban engineering.[8] It conducted an initial series of studies in 2010–2011, which began by analysing – not to say deconstructing – all the existing studies, reports and other documents on the non-potable water system, while hearing experts from very different disciplines (hydrology, climatology, architecture, technical history, rainwater management, water treatment etc.), as well as politicians and technicians from cities involved in the quest for resources other than potable water. This first exploratory analysis confirmed the conclusions of the consensus conference regarding both the one sidedness of the original case and the potential of the system.

Following the Municipal Council's approval of the retention of the network on 19 March 2012 (see below), in 2013 APUR began a further study, a combination of additional analyses and experiments. Broadly speaking, the non-potable water system was now seen as the basis for the exploitation of resources other than potable water, alternatives that were to be identified and employed against the likelihood of water shortages arising from climate change. In relation to this, the objective was to promote the redevelopment of old uses of the non-potable water system and to introduce new uses, as part of a drive to exploit urban amenities and combat the urban heat island phenomenon.

From abandonment to revival: chronology.

1986	Report of the Merlin cabinet on reducing the consumption of non-potable water: potential reductions of 33% identified
1997	Hydratech report examining different scenarios for the non-potable water system: in economic terms, complete abandonment is the best solution
2001	New municipal coalition (socialist-green) headed by the socialist Bertrand Delanoë, re-elected in 2008
2003	Court of Accounts report on the management of the public water and sewerage services, identifying failures in the Parisian services, in particular with regard to maintenance and financing (Cour des Comptes, 2003)
2005 (5 Jan)	Paris Municipal Council asks for a study to be conducted in order to assess the relevance of maintaining the non-potable water system (instigated by the UMP group)
2008	SAFEGE report examining different scenarios for the non-potable water system: in economic terms, complete abandonment is the best solution
2008 (5 Dec)	The Paris Municipal Council asks for the non-potable water system to be maintained, consolidated and developed (instigated by the Green group)
2009	Report by Prolog Ingénierie assessing the SAFEGE report: SAFEGE's conclusions described as very pessimistic, but a second report by the same office on new uses of potable water concludes that they would be difficult to develop and the consumption likely to be generated was very small Report by PriceWaterhouseCoopers comparing the environmental impact (by life-cycle analysis) of maintaining or abandoning the system: abandonment slightly preferable
2009 (3–4 Dec)	Consensus conference on the future of the non-potable water system: 12 of the 14 jury members vote for its maintenance and development
2010 (1 Jan)	End of the public licence contract for the City of Paris public water service, which is brought back under municipal control
2010 (5 Apr)	Press alert: André Guillerme, "Paris perd ses eaux" (Paris is losing its water) published in *Le Monde*

2010-11	First study on the non-potable water system commissioned by the City of Paris from Atelier Parisien d'Urbanisme (APUR): the assessment and forward analysis conclude in favour of maintaining and even extending the network, on the basis of additional studies
2012 (19–20 Mar)	Ruling of the Municipal Council in favour of maintaining and improving the non-potable water system
2013	First study on the non-potable water system commissioned from APUR: additional experiments and analyses confirm the conclusions of the first report.
2014 (July)	Unanimous support expressed by the Paris Council for the strategy of (re)developing the non-potable water system

The strategy of "maintaining" and "optimising" the non-potable water system

In its session of 19 and 20 March 2012, the Paris Municipal Council decided:

- to approve "the maintenance of the non-potable water system and its optimisation";
- to ask Eau de Paris "to make the most urgent investments and to begin a programme to maintain the existing non-potable water infrastructure";
- to include the "expenditure needed to release landholdings not essential to maintaining the production and distribution of non-potable water (...) in the land-use assessment for each operation";[9]
- to ask Eau de Paris "to continue studies, in concert with the municipal services and APUR, on the three major priorities of the medium or long-term use [and development] of the non-potable water system: resource diversification and the bioclimatic and metropolitan dimensions."[10]

This final provision sets out the municipal priorities for the development of new uses of non-potable water. However, the City's reflections also concern the development ("revival") of the traditional uses of non-potable water, which had gradually been or were in the process of being abandoned (cf. APUR, 2013a).

The municipal strategy thus rests on two additional elements: diversifying the sources of supply to the system; and developing uses ("reviving" old uses and eliciting new ones) within the territory of Paris and at the scale of the Paris conurbation. The implementation of this strategy began in 2013, notably with experiments designed to test the relevance and viability of particular measures before their possible full-scale implementation.

In addition, in July 2014, the Paris Council unanimously adopted a proposal by the Paris Ecologist Group that this strategy should be further explored, notably recommending that "the Municipality should make it a priority to develop the use of non-potable water wherever possible in order to save resources and manage quality", that "the Mayor of Paris should work with the State authorities towards more extensive use of raw water and towards the alignment of regulations for the use of non-potable water in the Parisian system onto regulations pertaining to the use of rainwater", that "the City should draw up a master plan for non-potable water for the purpose of developing the use of non-potable water by the community and professionals", that "additional studies should be undertaken in order to identify avenues for increasing the uses of non-potable water under the Grand Paris programme", and that "pending the conclusions of the master plan for non-potable water, all components of the non-potable water system should be kept in operation, including all the existing reservoirs".

The network as a shaker for the raw water "cocktail"[11]

A first plank of this strategy is to seek to turn the non-potable water infrastructure into a truly networked system fed from different sources, through which water of different kinds, sometimes described as "hidden resources" – "treated wastewater, exhaust water, heating and air conditioning water, rainwater, greywater…" (Ville de Paris, 2012) – would pass in a single unified flow. Indeed, a detailed inventory was made of potential sources across the Paris conurbation (APUR, 2013a, p. 30), along with an estimate of the potential volumes of each type of resource (see Figure 3.2). Exhaust water and, above all, re-treated wastewater emerged as the largest resources, with the potential to very substantially increase the uses of the non-potable water system, without placing further significant direct pressure on water resources.

This inventory reveals the multiplicity of the sources and the diversity of their characteristics (water quality, available flow, continuous or intermittent source...). The aim of the strategy is to "pool and optimise existing resources in the territory in order to tackle the possibility of a shortage of available resources" (APUR, 2013a, p. 31), and in particular to protect potentially potable water and limit the emission of pollutants that might affect it (*op. cit.*, p. 35).

A range of experiments were undertaken or planned in 2013. With regard to the reinjection of exhaust water, for example, tests were planned for two buildings (Meyerbeer car park in the 9th arrondissement, coinciding with the renewal of the operating licence, and the Palais de Justice in the Clichy-Batignolles urban development zone, during construction work in 2015), where the exploitation and quality of the water collected will be monitored.[12] In the longer term, a plan for the reuse of large volumes of retreated wastewater is envisaged, entailing the installation of a pipeline between the Valenton treatment plant and the Villejuif non-potable water reservoir, taking advantage of the existence of a nearby SIAAP emissary[13] (APUR 2013a, pp. 42–43). These projects were originally met with reservations from sanitary departments which, having expressed strong doubts, nevertheless seemed to accept the principle of the experiments.

Nonetheless, there does not seem to be unanimity regarding the "all-network" option among the municipal services. APUR is therefore considering the

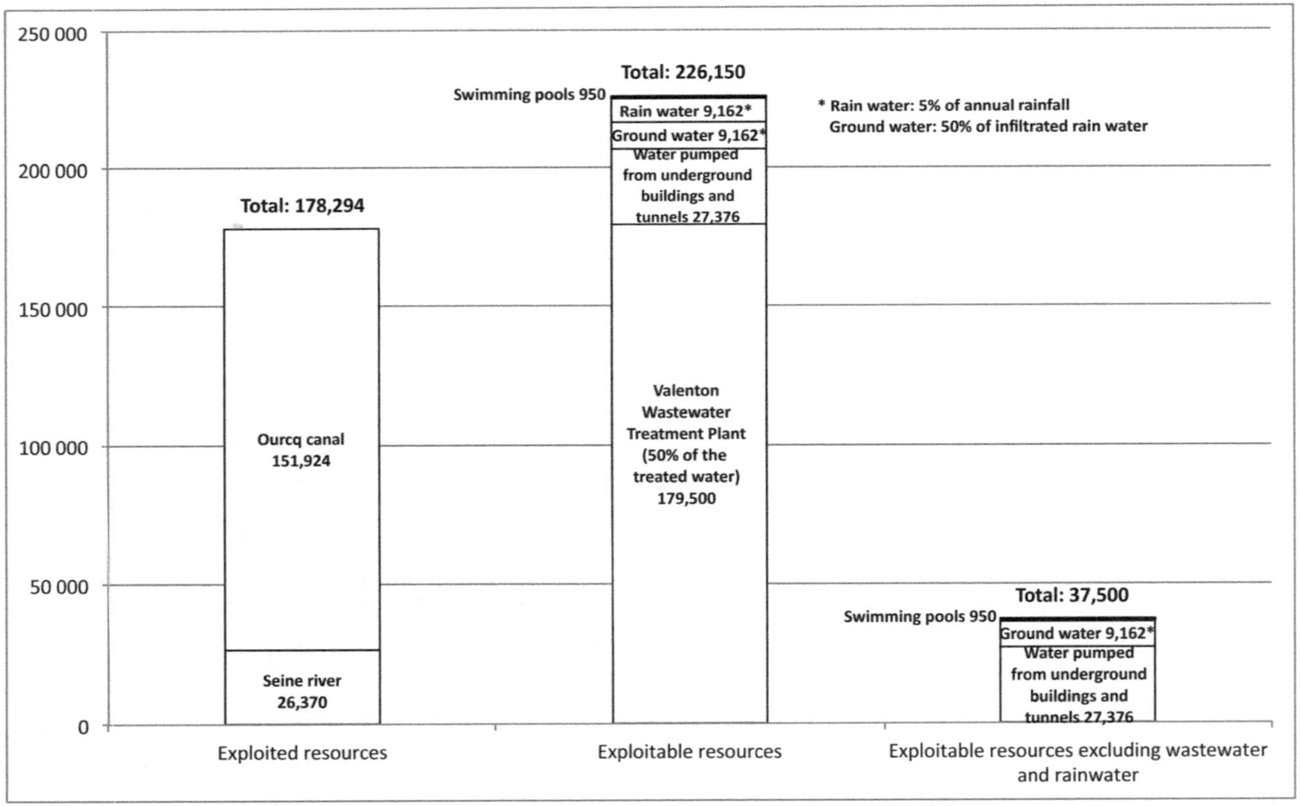

Figure 3.2 Nature and volume of exploited and exploitable resources (Source: adapted form Apur, 2013a, p. 27. Source: DPE, RATP, SIAAP)

development of autonomous connections (i.e. connections that escape control by the City of Paris non-potable water department) for users located near the Saint-Denis Canal and the Ourcq Canal,[14] or accessible galleries on the SIAAP network (in which non-potable water pipes could be installed).[15]. There are sometimes even disagreements between departments over alternatives between "all-network" and other configurations. APUR, for example, sees the network as suitable for sources that represent large and continuously available volumes, whereas "rainwater and water from swimming pools and the water table could provide indirect additions to the non-potable water system, managed by independent technical mechanisms such as drain inlets directing rainwater into the flush tanks..." (APUR, 2013a, p. 31; see also Figure 3.3). For EDP, on the other hand, "the ideal non-potable water system would recover all non-potable water (exhaust water, rainwater…) and thereby reduce drawings on the resource".[16]

Developing the uses of non-potable water

The second plank of the municipal strategy is to develop the uses of non-potable water. The aim is firstly to redevelop, or "revive", traditional uses of the network, and secondly to develop new uses and to connect new users both within the Paris municipal area and across the rest of the conurbation. Initially, the purpose of this choice was to give an economic justification for maintaining the system by increasing the revenues it generates, hence the idea of using it to support Paris' existing cold network (operated by a firm called Climespace). Subsequently, it emerged not only that the decline in certain uses was linked with malfunctions in the service that could be overcome[17] or with choices that were no longer justified,[18] but also that the purpose of the non-potable water service could change in the light of factors such as climate change or the development of certain water-related urban amenities.

The traditional uses to be redeveloped are therefore primarily of two kinds: the cleaning of communal areas with running water, both in public spaces (including the sewers) and private spaces (condominiums, social housing...), watering of parks and gardens, and supplying fountains, watercourses, lakes and cascades. Added to these are more specific uses.

The new uses notably include: vehicle washing (outside Paris); laundry services (across the centre of the conurbation); and different bioclimatic services, such as the creation of water mirrors, water-spray-based street cooling, extending the water network into the Boulogne and Vincennes woods, greening the urban space, and even opening Bassin de la Villette to bathers. The development of these uses depends, among other things, on the targeting of users with particular potential (condominiums, social landlords, hospitals). It also fits partly within the perspective of adapting the French capital to climate change (in particular through work to prevent the development of an urban heat island).

Consistent with the above considerations, the development of the non-potable water system is, in addition, highlighted as one of the City of Paris's 30 metropolitan projects defined in 2012,[19] i.e. as a priority instrument of Parisian policy for intermunicipal cooperation.

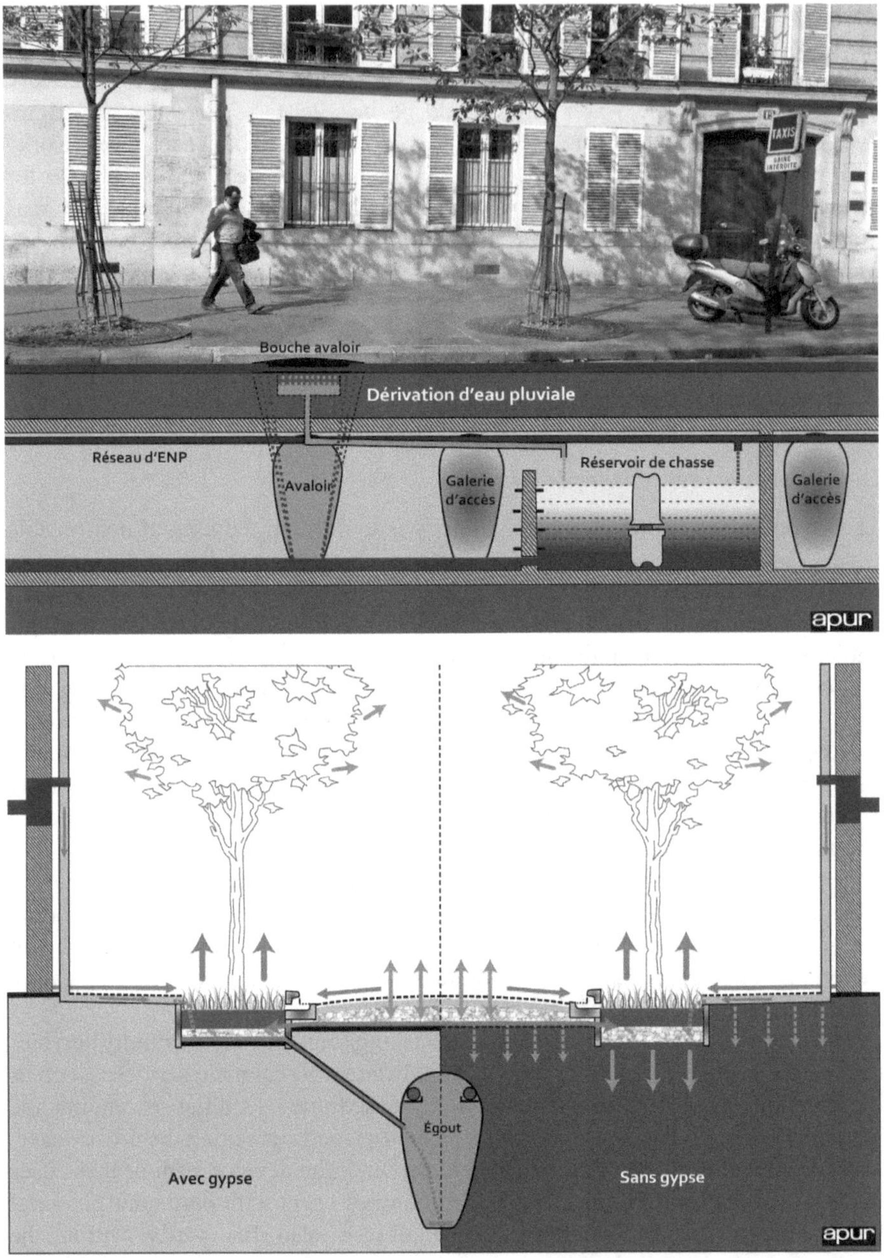

Figure 3.3 Drain inlets feeding into flush tanks on the sewage network

Key:

(top) Rainwater collected in inlet (avaloir) is derived to the sewer system's flush tanks, also supplied by the non potable water network (réseau d'ENP). Sewer system adits (galeries d'accès) are represented on both sides of the tank. A possible water cycle in the city.

(bottom) Sewers; subsoil with/without gypsum.

Source: Apur, 2013a, p. 80; Apur, 2013c, p. 97.

A hybrid strategy

In this section, we discuss the principles put forward and the "levers" employed or envisaged for the resuscitation of the non-potable water system. We show that the strategy developed is a hybrid one. Indeed, the decision to maintain the non-potable water system reflects the strong influence of a networking principle, although the legacy dimension of the infrastructure (Guillerme, 2010) also comes into it. By contrast, this decision goes hand-in-hand with the deliberate promotion of technical systems autonomous from the non-potable water system and with a profound redefinition of the principles of water management in Paris.

The interconnection function at the heart of networking processes

Let us begin by specifying what we mean by the "networking, or interconnection, principle", which historically plays an important role in the development of networked systems (see Box 2, p. 64). In his seminal study of electrification in Western societies, Thomas P. Hughes (1983) identifies the techno-economic origins of the development of electricity grids, thereby highlighting the triple function of interconnection, to use an anachronistic term: interconnection of different sources of electricity production; interconnection of different uses; and interconnection between the former and the latter.

The property that Samuel Insull – director of the Edison power company in Chicago – understood and began to exploit systematically in the final years of the nineteenth century was the role of the electricity interconnection grid as "spatial storing device" for a product, electricity, which cannot be stored (over time, therefore) on economically viable terms.

Interconnection is a way to:

- exploit different sources of electricity production by combining, in the most economically advantageous way possible, base-load supply sources (costly in investment, but economical to operate, such as a hydroelectric dam) and peak-load supply sources (requiring relatively low investment, but costly to operate, such as a coal-fired power station);
- take advantage of the complementarities between the load curves (which describe the trends in consumption over time) of different electricity consumers and the fact that aggregated consumption varies less sharply than individual consumption;
- maintain security of supply through interlinked networks.

So interconnection is a means of taking maximum advantage of the scale, size and network effects of large networked technical systems.

This interconnection function is at work in all networked services. As regards the Paris water supply, its historical origin can be precisely identified around 1830, in the design of the Canal de l'Ourcq scheme. It is at the heart of what Jean-Marc Offner (1993, 1999) has suggested should be seen as the two large and often concomitant forms of network development: *extensive* development,

The network form (cf. Coutard, 2010)

A networked system can be defined as an ensemble of interconnected equipment, planned and managed in a centralised manner (sometimes locally, sometimes at wider scale) offering a more or less homogeneous service over a territory which it thereby helps to unify.

The keys to the success of networks:

- Networks as a response to the problems caused by the spatial concentration of populations and activities and by urban interdependencies (de Swaan, Tarr)
- The techno-economic properties of networks (Hughes, Curien) and their emergence as "growth technologies"
- The mutual reinforcement between network organisation and territoriality in the forms of political power (Raffestin, Mayntz)
- The mutual reinforcement between the spatially distributed form of networks and the networked forms of territoriality ("generalised urban", "generalised metropolitanisation") (Dupuy, Offner, Pumain)

The ideology of the network:

- The network is the most effective way to provide collective urban services
- Network performance grows with size (spatial coverage, number and diversity of users connected)
- The solution to the problems created by networks lies in the networks themselves, in their greater extension, their more centralised management, their increased technical sophistication

A few questions on the links between networks and "urban ecological transition":

Delinking/hyperlinking

- The rise of "small" networks
- "Smartification"

Sobriety and circularity vs. autonomy

- The example of nonlocal renewable energy
- Long-range circularities
- The role of networks in virtualisation (of the economy) and decoupling (between development and natural resource consumption)

Decentralisation, symbioses and infrastructures: the new economy/political ecology of infrastructures "through the topological connection of similar networks" and, above all, *intensive* development, "through the grouping of flows in a process of homogenisation" (Offner, 1993, p. 15).

The decision to reinvest in the non-potable water system: a networking rationale

Intensive and extensive development towards a logic of interconnection

As it happens, the function of interconnection is central to the strategy adopted for the maintenance and redevelopment of the non-potable water system.

With regard to the sources of supply to the non-potable water system, the principle of interconnection directly feeds into the notion of a "cocktail of water(s)", i.e. of pooling sources of raw water with different characteristics: abundant or limited, concentrated or scattered, constant or intermittent, requiring or not requiring pretreatment before injection into the system, and so on.

With regard to uses, the interconnection of a large number of users should foster the (re)development of network use in two ways: by multiplying and diversifying uses within the traditional territory of the non-potable water system, i.e. the Paris municipal area (intensive development); and, to a certain degree, by extending the system's coverage to other areas in the metropolitan conurbation (extensive development).

Highly centralised planning and operation

The City of Paris, by basing its strategy on the (re)development of the system, has chosen to develop new resources and new uses of non-potable water within a centralised planning and operational framework, rather than through a multiplicity of autonomous systems: individual wells, autonomous rain or exhaust water recovery systems, etc.

In fact, priority goes to resources and uses that fit within an integrated sociotechnical system (the network), under the control of the municipality. The latter can therefore decide which sources of raw water to use preferentially and which uses of non-potable water to promote or emphasise.[20] The centralising dimension is further reinforced by ideas to extend the network outside Paris, using infrastructures owned by the city outside its walls: Ourcq and Saint-Denis canals supplied with raw water, reservoirs built in the suburbs etc., giving Paris greater extraterritorial influence.[21]

In addition, this model establishes solidarity between the three big water infrastructures (potable water, sewage, non-potable water). Thus, for example, if the municipality signs an agreement with a car park operator to recover its exhaust water and reinject it into the non-potable water system, the financial cost to the City (in the loss of the water treatment fee) is directly offset by the raw water resource that this exhaust water provides.

Management of water resources and uses: a post-network rationale?

As we have just seen, the networking principle is crucial in the strategy developed by the City of Paris to manage its water resources and aquatic environments effectively and to increase the uses of non-potable water. At the same time, however, this strategy is based on decisions about technical architecture and flow management that differ from those that have historically characterised water management in Paris.

A multiplication of infrastructures for differential flow management

One major departure from the traditional network model is that the plan to redevelop the non-potable water system relies on a principle of long-term coexistence between two networks that serve a single territory, with the objective of differential flow management (between non-potable water and treated potable water). Here, we are right at the core of the emergence of a new political economy of flows, where resource and environment management is disrupting the traditional monopolistic economy of infrastructure networks which, for the last twenty years, underpinned the implicit policy for the extinction of the non-potable water system. This principle also challenges the hygienist foundations of the systems, under which potable water gradually became the distribution norm regardless of purpose, and the maintenance of a non-potable water system was perceived as an anomaly or historical accident (resulting from the lack of foresight of the designers of the first network).

However, this decision in favour of infrastructural differentiation is not without paradox: the non-potable water system is advanced as an alternative to potable water on the basis of the observation that not all uses require such quality. However, the consequence is that the very varied needs of "non-potable" users are amalgamated in terms of quality (and pressure), and for this reason the system needs to provide a service of sufficient quality for its most demanding customers, while remaining competitive with the potable water system.

In addition, the proposed strategy includes the possibility of releasing extraction rights on the Canal de l'Ourcq below the Sevran lock, up to a potential volume of 200,000m^3 per day: "the very operation of Canal de l'Ourcq requires the extraction of around 200,000m^3 per day in order to maintain its navigation [function]. [If the canal ceased to feed the non-potable water network] from Sevran lock, those 200,000m^3 could be shared with the communities along the canal. This measure would be in keeping with an approach recently introduced" (APUR, 2013a, p. 13, 15).

A systematic quest to exploit local resources and a desire to (re)localise and (re) close the urban water cycle

The network model as it developed historically is based on a *linear* configuration of the urban metabolism: natural resources are extracted at a certain distance and channelled to the cities; those resources are exploited within the urban space;

waste is evacuated outside that space. The linearisation of Paris' metabolism has been clear since the interwar period and the almost total withdrawal from the recovery and recycling of urban byproducts (Barles, 2005). Water management in Paris reflects this process, notably in the absence of a tradition of recovering or reusing rain-, waste or exhaust water or the heat contained in wastewater, to cite just a few examples.

By contrast, the proposed approach to the redevelopment of the non-potable water system is based primarily on the exploitation (or recovery) and pooling of *local resources*, on *loop (re)closing* of water flows and on *cascade* uses of water: rainwater, exhaust water, grey water. However, the issue here is less to do with sobriety (using less water) – in many cases the first priority – than with preserving resources and protecting the environment, by developing the uses that have the least impact on the surrounding milieu (notably the Seine basin).

Environmentalisation of infrastructural development, hyperlinking and the post-network city

This chapter describes an infrastructure policy that seeks an "optimised" use of water resources, in both quantity and quality, reflected in the decision to maintain two coexisting centralised infrastructures, despite the *de facto* competition between them against a background of steady decline in the consumption of potable water and the resulting overcapacity of water supply infrastructures.

We thus argue that the principle at work is hybrid in nature, a combination of a type of hyperlinking with the promotion of non-networked systems that, together, challenge the historical rules of water management in Paris. On the one hand, it is true that the redevelopment of the non-potable water system obeys the generic principles of intensive and extensive network development, and this approach is clearly distinct from one that entails a proliferation of local, autonomous systems. On the other hand, however, it also operates through methods and a configuration that differ from those of the traditional network model, in particular in the following respects:

- first, only partial intensive development, within deliberately planned limits, of two distinct infrastructures supplying a single territory; there is a sustained division of potable and non-potable water into separate systems of management, thereby reversing a long period (at least twenty years) in which priority was given exclusively to the potable water system; the plan therefore does not promote a single, exclusive, monopolistic infrastructure;
- next, a network (the non-potable water system) that obeys a logic of loop (re) closing and (re)localisation rather than a linear metabolism, and that supports long-range solidarities; this choice represents a significant change from previous practice (a growing dependency on remote sources of potable water; almost zero use of retreated wastewater, exhaust water, rainwater…);

- finally, a challenge to the ideology of the network:
    - the network is not (any more) considered a panacea; true, it remains the principal sociotechnical form, but it now accommodates autonomous systems (micro-networks, isolated connections) or end-of-pipe systems (filters for high-pressure jets etc. in order to adapt water quality to uses);
    - the conditions and limitations of network effectiveness are explicitly recognised;
    - effectiveness is now not only techno-economic, but also environmental.

The choice of this "hyperlinked" configuration no doubt reflects the specificities of Paris as a territory, in particular its very high density of population and activities and the legacy of a dual water supply system. It nevertheless reveals, in a very specific form, a new political economy of infrastructures, which challenges the historical primacy of pipes over flows. This "transition" seems to be taking place, in different ways, across a very wide range of urban contexts and infrastructure sectors. In closing, we therefore wish summarise the main aspects of this emerging political economy of infrastructures.

In environmental terms, the Parisian strategy for non-potable water management does not seek maximum *sobriety*, i.e. to draw as little as possible on the resource: it would be possible to develop a much more sober sociotechnical system. The strategy adopted seeks not only to replace drinking water with raw water wherever possible, but also to develop new uses for the non-potable water system. More than sobriety, this strategy hence strives rather to promote "qualitative management" of water resources and aquatic environments, by fostering extractions that do the least damage, in order to meet demand and, indeed, to promote an increase in demand. Water management is freed from a strict network logic, and now privileges the logic of the ecosystem of Paris and the Seine basin, since it includes priorities that were previously foreign to it: adaptation to climate change, and the incorporation of the water infrastructures into the green and blue network and into biodiversity policy.

In terms of territorial interdependence, the strategic choice is to prioritise the use of very local resources rather than increase the dependence on the potable water system's more distant sources, in circumstances where the city's use of those distant sources at a time of hydrological stress is, or could become, increasingly contested.

In sociopolitical terms, the strategy adopted undoubtedly reinforces control by the City of Paris and, more broadly, by local authorities, over all raw water resources and their uses; this differs from a more private sector or commercial (and more decentralised) model which would be conceivable (e.g. through the purchase of exhaust water). In this sense, the non-potable water system appears to be an important instrument of the policy of promoting the public – rather than private and commercial – management and use of raw water sources.

While this strategy is beginning to be implemented, several questions nevertheless remain:

- Who, in the future, will guarantee the low water flows of the Seine and the Marne rivers in summer, when uses are (or likely to be) the greatest? The dams/reservoirs on the potable water system, or the non-potable water system?
- What happens to the notion of urban "water transparency", promoted to reconcile the small and large water cycles? The proposed strategy in a sense goes the opposite way, with maximum collection, diversion and recycling of all water sources.
- What can be done to manage the fall in consumption of potable water, which will undoubtedly be exacerbated by the development of the non-potable water system?

Notes

1 Jacques Chirac, Mayor of Paris (1977–1995), President of France (1995–2007), Chairman (1976–1994) of the RPR, one of France's two main right-wing French parliamentary parties in the final quarter of the twentieth century.

2 As evidenced by the closure of the Ivry potable water production plant in 2010, some fifteen years after it was totally renovated (1988–1994).

3 In this text, the expressions non-potable water and raw water have similar meanings, but are not synonyms. By non-potable water, we mean water that has undergone rudimentary treatment (in particular filtration) and which flows through the non-potable system. By raw water, we mean all of the actual or potential sources of supply to the non-potable water system. The term distinguishes these sources from the smaller range of sources considered as suitable for the supply of the potable water system.

4 Letter from Napoleon to Cretet, Minister of the Interior, on 10 April 1806 (Fondation Napoléon, 2009).

5 With the following structure of ownership: 70% Paris municipality, 14% CGE, 14% Lyonnaise des Eaux, 2% Caisse des Dépôts et Consignations.

6 The authors of this chapter were involved in different ways in the research conducted by the City of Paris prior to the conclusions relating to the decision to reinvest in the non-potable water network. They argued in favour of the decision finally taken. Sabine Barles and Olivier Coutard were members of the 2009 consensus conference jury, and Sabine Barles and André Guillerme were part of the team of experts put together by the Atelier Parisien d'Urbanisme in 2010, and then again in 2013.

7 The other option, which attracted two votes, "considered that the quantities of water currently consumed no longer justify the existence of two different systems, both for economic reasons and because the existence of these two systems leads to greater resource wastage due to cumulative leaks" (Ville de Paris, 2009, p. 90).

8 Indeed, this point is itself symptomatic of a new approach to the question.

9 This provision needs some clarification. The City departments proposed that certain pieces of land connected to the non-potable water system should be released and "financially exploited", in particular one production plant and certain reservoirs (Auteuil plant, Grenelle reservoir). However, members of both the municipal majority and the opposition objected to this proposal. The provision adopted represents a compromise between the "financial" position of the municipal departments and the "ecological-patrimonial" position of its detractors. It followed discussion of a proposed amendment put forward by the Paris Ecologists Group, argued as follows: "The DPE 2 ruling proposes the closure of a production plant and non-potable water reservoirs, in order to gain financially from these assets. However, thinking about the potential uses of non-potable water is in its early stages, and climate changes are

undoubtedly going to lead to greater use of water in the city. EELVA members would like the City of Paris to consider this choice and its financial impacts with care."

10 Ruling 2012 DPE 2.

11 APUR, 2013a.

12 Presentation by Eau de Paris at the meeting of the Non-Potable Water steering committee, on 20 November 2013.

13 SIAAP is the interdepartmental water treatment federation for the Paris conurbation, responsible for the primary sewage and wastewater treatment network.

14 The Ourcq Canal, which extends 80km out of Paris, and the 6.6km Saint-Denis Canal, which feeds into the Ourcq in a loop of the Seine outside Paris, are owned by the City of Paris.

15 Presentation by Eau de Paris at the meeting of the Non-Potable Water steering committee, on 20 November 2013.

16 Presentation by Eau de Paris at the meeting of the Non-Potable Water steering committee, on 20 November 2013.

17 In particular for washing out the sewers, which the APUR study showed was not working properly, because the electronic timers installed on the flush tanks, damaged by the hostile conditions in the sewers and by rats, were simply not functioning.

18 The attempt to prevent clear water entering the sewers reduced the amount of matter drawn into them, with the result that they silted up and released odours into the streets. "[Project] 19. To employ technical studies and exchanges of experience at metropolitan level to develop further ideas about the management and uses of non-potable water."

19 "In the light of the bioclimatic issues and the existence of new uses for the non-potable water system in Paris, it would seem necessary to modernise, taking into account the experiences and needs of the city's different municipalities." http://www.paris.fr/politiques/gouvernance/une-vision-de-la-metropole/30-projets-pour-2014/rub_10094_stand_111478_port_25320 (accessed 5 February 2015).

20 By contrast with other local authorities which have made other choices. For example, some municipalities subsidise the installation of rainwater recovery and retention tanks; and the Hauts-de-Seine General Council contributes up to 50% of the investment required for the acquisition of a rainwater recovery tank. http://applis.hauts-de-seine.net/v3fichiers/brochure_subvention.pdf (accessed 27 February 2015).

21 On the question of Parisian extraterritoriality, see Barles (forthcoming).

References

APUR (2010) *Étude sur le devenir du réseau d'eau non potable*. Rapport pour la Ville de Paris, 2 vol.

APUR (2013a) *Du réseau d'eau non potable à l'optimisation de la ressource en eau. Partie 1: Cocktail d'eau et valorisation de la ressource*. Atelier Parisien d'Urbanisme, July 2013.

APUR (2013b) *Du réseau d'eau non potable à l'optimisation de la ressource en eau. Partie 2: Restitution des ateliers des 11 et 12 février 2013*. Atelier Parisien d'Urbanisme, April 2013.

APUR (2013c) *Du réseau d'eau non potable à l'optimisation de la ressource en eau. Partie 3: Reconquête des usages de l'ENP*. Atelier Parisien d'Urbanisme, December 2013.

Barles, S. (2005) *L'invention des déchets urbains, France, 1790–1970*. Seyssel: Champ Vallon.

Barles, S. (forthcoming) The Seine as a Parisian River: its Imprint, its Ascendency and its Dependencies in the late 19th and 20th centuries. In M. Knoll, U. Luebken and D. Schott

(Eds.), *Rivers Lost – Rivers Regained? How cities use, abuse and adore their rivers*. Pittsburgh: The University of Pittsburgh Press.

Bechmann, G. (1899) *Salubrité urbaine. Distribution d'eau et assainissement*, 2nd ed. Paris: Baudry et Cie, 2 vol.

Belgrand, E. (1854) *Recherches statistiques sur les sources du bassin de la Seine qu'il est possible de conduire à Paris (...)*. Paris: Vinchon.

Canneva, G. (2009) *Réseaux d'eau non potable: Enjeux et perspectives. Document pédagogique préparatoire pour la conférence de consensus*. Montpellier: AgroParisTech-ENGREF.

Cour des Comptes (2003) *Rapport sur la gestion des services publics d'eau et d'assainissement*. Rapport au président de la République suivi des réponses des administrations et organismes intéressés. *http://www.cace.fr/jurisprudence/ccomptes/cour/cdc2003.html* (accessed 10 July 2012).

Coutard, O. (2010) Services urbains: la fin des grands réseaux? In O. Coutard and J.-P. Lévy (Eds.), *Ecologies urbaines, Paris, Anthropos-Economica* (Coll. Villes), 102–129.

Fondation Napoléon (2009) *Napoléon Bonaparte. Correspondance générale, t VI, 1806: vers le Grand Empire*. Paris: Fayard.

Guillerme, A. (2010) "Paris perd ses eaux", *Le Monde*, 5 April 2010. *http://www.lemonde.fr/idees/article/2010/04/05/paris-perd-ses-eaux-par-andre-guillerme_1329118_3232.html* (accessed 8 August 2014).

Hughes, T. P. (1983) *Networks of Power: Electrification in Western Society, 1880–1930*. Baltimore: Johns Hopkins University Press.

Martin, L. (dir.) (1902) *Encyclopédie municipale de la ville de Paris*. Paris: Neger Reeb.

Offner, J.-M. (1993) "Le développement des réseaux techniques: un modèle générique." *Flux*, **13–14**: 11–18.

Offner, J.-M. (1999) "Are there such things as small networks?" In O. Coutard (Ed.), *The Governance of Large Technical Systems*. London: Routledge, 217–238.

Ville de Paris (2009) *Quel avenir pour le réseau d'eau non potable de Paris? Actes de la conférence de consensus des 3 et 4 décembre 2009*. Mairie de Paris, Direction de la Propreté et de l'Eau.

Ville de Paris (2012) *Projet de délibération 2012 DPE 02: Maintien du réseau d'eau non potable et perspectives de développement de ses usages et de son extension géographique. Exposé des motifs*. Ville de Paris, Direction de la Propreté et de l'Eau. 19 March 2012.

4 District heating comes to Ecotown

Zero carbon housing and the rescaling of UK energy provision

Simon Guy and Andrew Karvonen

> At first glance, heat plants and pipelines *per se* may seem of little interest from a social science perspective. What makes them truly intriguing is what they embody: the tensions and tactics behind their emergence, the complexity of the social organization that supports them, and their long-term implications for the actors they link and the communities they serve.
>
> Summerton 1992: 62

It is now more than two decades since Jane Summerton published her landmark book *District Heating Comes to Town* about the successful development of a district heating system in an 'ordinary' town in Sweden. It was published when sociotechnical studies of urban technical systems were still relatively rare, and arrived during a fertile period of intellectual development that was striving to connect studies of large technical systems with urban studies, inspired by new thinking in STS studies (Guy and Karvonen 2011).[1] Her subtitle, 'the social shaping of an energy system,' continues to be a fresh (and to some, a provocative) perspective suggesting that the development of urban infrastructure systems depends as much on the 'cooperation and coordination of many diverse actors' as on the technological and financial means of production. This is a direct challenge to the prevailing epistemological order in energy studies and encourages a relational perspective on the co-production of energy provision and consumption.

While such sociotechnical studies still arguably represent a minority perspective, they are now very much part of the theoretical and policy debate with disciplines across the social sciences developing a keen interest in energy systems (Guy 2006) and an increasing number of recent energy research projects funded by British research councils claiming to be framed by a sociotechnical analytical frame, including a landmark Government Foresight study of sustainable energy management and the built environment, *Powering Our Lives* (Government Office for Science 2008).[2] More recently, Bridge and colleagues have sought to identify connections between energy studies and human geography, foregrounding questions about *spatial difference* and the 'co-existence of multiple transition pathways and possibilities' (Bridge *et al.* 2013: 339). They argue that the research goal for energy studies should not simply be 'mapping the consequences of policy' or 'understanding the implications of different policy options for particular places', but should rather be to 'understand how energy transitions are *spatially-constituted*' (p. 339, emphasis in original).

The inspiration for this chapter is not simply the sociotechnical analytical approach, but also the subject matter of British district heating systems. Here there has been rather less development, perhaps due to heat networks only accounting for approximately 2% of domestic, public sector and industrial heat demand in the United Kingdom (DECC/DCLG 2009), so they have not been seen as a productive focus of research concern. However, that has started to shift with new projects and papers emerging that are exploring the potential contribution of heat networks to meeting national low carbon targets while also strengthening local and regional influence over environmental governance (Hawkey 2009, Russell 2010, Hawkey 2012, Bolton and Foxon 2013, Hawkey *et al.* 2013, Webb 2013, Bolton and Foxon 2014, and Lund *et al.* 2014). For example, Webb has been exploring how locally-owned, small-scale heat networks in Aberdeen might become 'scalable models for low carbon, affordable, and locally-accountable urban energy' (2013: 24), while Bolton and Foxon (2014) have been examining the interaction between district heating systems and the regulatory regime that underpins the liberalisation of energy markets. In short, the re-emergence of district heat networks as a response to national low carbon targets involves a multitude of tensions and tactics in the reconfiguration of infrastructure networks.

We contribute to this debate by exploring how medium- to large-scale zero carbon housing developments are employing district heating and combined heat and power networks to provide multiple households with renewable energy. Our motivation to focus on district heating is linked to the policy shift around carbon neutrality, with its menu of visions and targets that is now exercising the minds and calculations of planners, designers and developers in the built environment. Hodson and Marvin (2013) have recently asked whether a transition to a low-carbon Britain creates new 'transformational opportunities', or simply means 'business as usual'. For the purposes of this chapter, we similarly ask if the low-carbon turn presents an opportunity for heat networks to establish themselves in Britain as they have elsewhere in Europe.

To address this question we focus on new housing as a sector, following the emergence of a new generation of urban development schemes that are seeking to achieve carbon neutral status to meet new housing design and construction standards. The journey from policy objective through demonstration house to mainstream commercial development is a process we have been following through an ESRC/DEFRA/Scottish Government funded project that is part of a larger programme of studies around theories of social practice.[3] Our project followed both the formation of a policy discourse around carbon neutral housing and attempts to realise this policy in mainstream commercial development. Not surprisingly for a study influenced by STS, the stories we followed involved deep entanglements between human and non-human actors, strategies of translation, tactics of enrolment, and attempts to configure users, all of which are highly contested. Elsewhere we have highlighted the interpretive response of commercial designers and developers to the policy prescription of carbon neutrality (Fischer and Guy 2009) and the emergence of community-based partnerships for domestic retrofit (Karvonen 2013).

In this chapter we examine the role of energy systems in the zero carbon housing debate, looking at how a new impetus around district heating is being realised as well as contested on the ground in debates and decisions about design and planning. In doing so, we draw inspiration both from Summerton's attempt to understand the contingencies of successful innovation in energy systems and the more recent sociotechnical thinking about the dynamics and tactics of rescaling energy systems as they move through moments of transition.[4]

The chapter is in four parts. First we introduce the policy context around zero carbon housing in the UK and the various claims and calls for action that characterise the recent debate. We provide an overview of the zero carbon housing sector to give a sense of the scale and diversity of recent development activity and foreground the ascribed role of district heating in the policy discourse, as well as emerging mainstream developments. In doing so, we contrast the clarity of the policy goal with the complexity of the development context as the housing sector attempts to come to terms with the new policy landscape. Second, we briefly revisit the debate about district heating in the UK to give an historical sense of the role (or lack thereof) of district heating in the British energy system. We identify the importance of international exemplars of district heating as influential in the current policy debate, and present the arguments of district heating advocates. Third, we explain the background and methodological approach to the case studies presented here. We introduce the case study developments and explain how and why district heating became part of the housing strategy. In each case we explore the contingency and complexity of attempts to reconfigure and rescale the energy system, in particular exploring the dynamics of development scale, design, and planning, and how this issue shapes the adoption of district heating. Finally we make links back to Summerton's work to reconsider the dynamics of district heating as a sociotechnical system. We assess the likelihood of district heating to fulfil the policy aspirations around zero carbon and the associated risks to existing systems of such a rollout. We conclude by exploring future research questions and the contribution of this work to wider studies around STS and urban technologies.

Mainstreaming zero carbon housing

About one quarter of the total UK carbon footprint is attributed to the construction, operation, and demolition of the British housing stock, and thus it provides an attractive and logical target for climate change mitigation policies (DECC 2011a). With aims to significantly reduce the carbon footprint of the UK housing stock, the Government introduced the Code for Sustainable Homes (CSH) in December 2006 with a gradually tightening specification from Level 1 compliance in 2007 to the highest Level 6 compliance in 2016, which is considered to be 'zero carbon'.[5] The definition of 'zero carbon' was vociferously debated in subsequent years, with the homebuilding industry jockeying with the Government and third sector organisations to determine the carbon footprint boundaries and to decide what types of carbon to regulate in the creation of a stringent yet viable regulatory goal.[6] Giving the homebuilding industry only ten years to completely reform the delivery of new housing was a significant challenge for building scientists, the building trades, economists and marketers, and it produced significant contestation over how zero carbon could be achieved, particularly in a commercial context. In 2014 the Government began to move away from the CSH, and as of this writing it is in the process of developing alternative approaches to reducing the carbon footprint of the UK housing stock. However, the debates over CSH from 2006 to 2014 produced new configurations between housing and realising a low carbon future for Britain.

Through the debates over the 'zero carbon' imperative and the alternative pathways of design and development that resulted, regulators and the homebuilding industry agreed upon a hierarchy of approaches to reducing the carbon footprint of new houses. In particular, there was a shared understanding that carbon reductions should start with creating an efficient building fabric (the 'fabric first' approach) and then reducing demand inside the household and providing onsite energy generation to the greatest extent possible. These two tiers of the carbon reduction hierarchy are referred to as 'Carbon Compliance' and involve onsite solutions. The original CSH called for all carbon emissions to be mitigated via Carbon Compliance measures, but as the regulatory requirements were tested on real world houses, the housing industry argued that this was unrealistic to deliver an affordable housing product in the long term (ZCH 2011).

To remedy the situation, in 2008 DCLG introduced the notion of 'Allowable Solutions' that go beyond the Carbon Compliance requirements to include strategies such as smart appliances, communal heat storage, electric vehicle charging stations, retrofit of neighbouring buildings, local micro-hydro power generation systems, investment in low carbon electricity generation, and so on. The Zero Carbon Hub (ZCH) – a public/private partnership established in 2008 to facilitate the delivery of low and zero carbon new housing – was a central actor in translating and interpreting the CSH for the homebuilding industry. Their 2011 report, *Allowable Solutions for Tomorrow's New Homes: Towards a Workable Framework,* states that: 'Allowable Solutions offer flexibility to developers, providing them with an outlet to resolve remaining emissions, where all other technically and commercially feasible on-site options have been exhausted' (2011: 12). These Allowable Solutions account for a significant portion of carbon emissions savings depending on the housing type – estimated to be 56% for flats and 40% for detached houses – and comprised some of the most highly contested elements of the zero carbon housing debate.

Allowable Solutions introduced significant opportunities for the development of new forms of energy delivery. As the ZCH (2011: 15) notes: 'The right Allowable Solutions framework could be hugely important in stimulating the fledgling (but largely unsupported) "mid-stream" sustainable-energy market, which delivers, for example, community scale sustainable energy solutions.' It is here where meso-scale infrastructures such as district heating (DH) and combined heat and power (CHP) networks emerged as important components of zero carbon housing. The Renewables Advisory Board made an explicit connection between the CSH and these meso-scale infrastructure networks, arguing that it is 'important that larger scale CHP and district heating is culturally and commercially attractive in the UK, since it offers a relatively cheap route to compliance' (RAB 2007: 4). Likewise, the Department for Business Enterprise and Regulatory Reform notes that the CSH is 'potentially a significant stimulus for renewable district heating' (BERR 2008: 75).

Allowable Solutions introduced a new spatial dimension of energy provision for zero carbon housing that resides between the individual house and the universal grid. Rather than simply hanging off the end of the conventional electricity and gas central grids or escaping altogether as an island of micro-generation, new assemblies of housing, heating, and powering communities are being imagined and designed, and in certain contexts are starting to be constructed. This is reflected in the Department of Energy and Climate Change's *Carbon Plan* of 2011, wherein district heating is part of the medium-term strategy to wean the UK building stock (both domestic and non-domestic) off high carbon fuel sources. As the report states, 'The 2020s will be a key transitional decade in delivering mainstream low carbon heat from heating networks and in buildings, and will see the expansion of low carbon heat at scale into residential areas' (DECC 2011a: 33). This new mode of infrastructure provision is evident in recent developments being proposed and built by large housebuilders to meet varying levels of the CSH, which we will discuss in more detail below. Before we do so, we will pause to consider why district heating is being promoted as key to zero carbon, and review some of virtues and challenges of this systemic change.

The importance of the meso-scale

Various terms are used to describe energy provision that lies somewhere between the universal grid and the individual house: decentralised, distributed, dispersed, embedded, and micro-generation (Alanne and Saari 2006). CHP networks involve the simultaneous production of electricity and heat while DH networks involve a complex of pipes to transport heat to a collection of buildings for space heating and hot water. These systems can range from the individual household to a small collection of buildings to a block, neighbourhood, district, or even city (e.g. Vienna), and can function as a single system or a web of interconnected smaller systems (e.g. Copenhagen). What these networks have in common is that heat and/or energy generation is located in close proximity to the point of use. In this chapter we are particularly interested in those networks that serve several different owners but are smaller than universal networks (between 10 KWe and 50 MWe). We refer to these as 'meso-scale' networks of energy provision that reside between the conventional grid and the individual household (Foresight 2008, Watson *et al.* 2010).

Meso-scale heat and power networks are a common strategy in many Nordic countries (e.g. Denmark, Sweden and Finland) as well as Eastern Europe (Poland, Estonia, Latvia and Lithuania) (Euroheat & Power 2011). The UK currently has an underdeveloped heat market; only 2% of the UK's current heat demand is met with district heating (DECC/DCLG 2009). There are several reasons for the lack of district heat networks in the UK.[7] The liberalised energy markets create separate silos of generation, transmission, distribution, and supply (Toke and Fragaki 2008, Euroheat & Power 2011) and the current arrangement of domestic energy use for space heating and hot water involves boilers fed with gas, oil or solid fuels, or using electric heating systems. These technologies are locked into national infrastructure networks as well as high carbon fuel sources (Unruh 2000, 2002).

However, the UK does have a long history of heat initiatives stretching back to World War I (see Russell 1993, Babus'Haq and Probert 1996, TCPA/CHPA 2008, Hawkey 2009, Russell 2010), and today there are a handful of projects with established track records including in Woking, Sheffield, Southampton and Aberdeen. However, as Russell (2010: 1) summarises in a review of heat networks in the UK, it is a 'long and mostly sorry history' with 'no simple reason for their neglect'. While the neglect is multi-causal, Russell sees the exclusion as essentially systematic, characteristic of a particular sociotechnical formation in which heat networks have been *locked-out*, much as in a parallel way heat networks have been *locked-in* to the national networks of other European countries. British energy providers have historically seen bulk heat supply (both pre- and post-nationalisation) as incidental, with engineers often hostile and politicians disinterested. However, Russell does note that at times of major reorganisation of the sector (e.g. nationalisation and privatisation), there may be a 'fluid phase' that allows 'options to be aired and alliances to be built around the idea' (2010: 5). While historically these moments have passed without resolution or much momentum for heat networks, it does raise the question of whether the current pursuit of low carbon Britain may present another fluid phase in which heat networks might finally emerge as a key element in British energy planning.

Certainly we are witnessing the emergence of a new coalition of interested actors and institutions supporting innovative projects, with Government agencies such as the Department of Energy and Climate Change (DECC), the Department for Communities and Local Government (DCLG), and the Homes and Communities Agency (HCA) collaborating with third sector organisations such as the Carbon Trust and the Energy Saving Trust, as well as trade organisations such as the Combined Heat and Power Association and the UK District Energy Association, to realise the benefits of meso-scale networks. They cite three primary reasons for promoting meso-scale heat and power networks: to reduce CO_2 emissions, to tackle fuel poverty, and to develop diverse and secure energy supplies (DECC/DCLG 2009, HCA 2011b). District networks provide more energy efficient forms of generation and distribution. They use technologies that make more efficient use of primary energy, and with CHP they utilise heat that would otherwise be wasted. Beyond their efficiency, district heat and power are more flexible than large grid systems. Energy generation technologies can be upgraded over time, with fossil fuels being replaced with biomass or other renewable feedstocks. As such, district heating networks are agnostic with respect to fuel source and can be leveraged in the transition from fossil fuels to renewables

(EST 2009, HCA 2011b). For example, boilers that use renewable fuel sources such as pellets and woodchips are becoming increasingly common for district heating networks (EST 2009). Further, heat can be stored quite easily in an insulated vessel, effectively decoupling generation from demand and avoiding the storage issues associated with electricity supply and demand.

Of course, there are also challenges in introducing meso-scale infrastructure networks. District heat systems have a long lifespan (forty to fifty years) compared to home heating systems, but they are capital intensive and involve significant first costs (Hawkey 2012). To justify the high first costs, these networks require high and consistent heat demands. As a rule of thumb, CHP plants need to operate for a minimum of 4000 to 5000 hours per year (the equivalent of 13 to 14 hours per day) (EST 2009). This high demand is easier to achieve in mixed-use developments where heat demand is more consistent, as opposed to domestic developments with high morning and evening peak demands (TCPA/CHPA 2008, EST 2009). Thus, it is common to design a district heat network with an anchor load such as a hospital, university, or office tower, and then add on dense residential development (i.e. flats) (Summerton 1992). Likewise, district heat networks tend to be sited in towns and city centres due to their higher heat density as well as mixed building types (TCPA/CHPA 2008).

Table 4.1 provides a comparison of densities and costs for different building types, demonstrating that the cost per dwelling increases significantly with lower housing densities and longer pipe lengths. The particular qualities of heat generation, storage and distribution shape urban form and reinforce notions of compact cities and dense, mixed-use developments, and sustainable urban development more generally.

Table 4.1 Comparison of building type, density, pipe length, and cost for various district heating configurations (Source: summarises data from TCPA/CHPA 2008)

Building Type	*Form*	*Housing density [dwellings per ha]*	*Pipe length per dwelling [m]*	*Cost per dwelling [£]*
High-rise apartment block	Corridor access, 10 to 15 storeys	240	6.75	2,500
Medium-rise apartment block	Corridor access, 5 to 6 storeys	120	8.0	2,800
Perimeter block of flats and townhouses	Stairwell or street-level access, 3 to 4 stories	80	11	4,100
Terraced street of row houses	Street-level access, 2 to 3 storeys	80	13	5,300
Detached/semi-detached houses	Street level access, compact street layout	40	19 to 24	7,700 to 9,550

Meso-scale infrastructure networks are often neglected in policy debates, falling between the micro strategies that are addressed with feed-in-tariffs, demand-side efficiency, and micro-generation technologies, and the macro strategies that legislate and regulate large-scale utility investments and upgrades. Watson *et al.* (2010: 8) note that there has been a small amount of attention to the meso scale through the Community Renewables Initiative, but conclude that 'there is no established tradition in the UK of either energy technology deployment or energy system governance and regulation at meso (regional or local) scales.' The meso scale introduces a new set of technologies and institutions that differ significantly from the micro and macro scales and require novel regulations and incentives to be viable (see table 4.2). The abolition of regional government bodies in the UK in 2010 could make the delivery of renewable energy and district heating networks more difficult at the local scale (Watson *et al.* 2010).

Despite the aforementioned challenges associated with sunk costs, minimum heat densities and demands, governance gaps, and financial disincentives, many national policy documents identify district heating and CHP networks as central components in the modernisation of the UK's energy supply networks. Examples of this include the *UK Renewable Energy Strategy* (HM Government 2009), *The Carbon Plan: Delivering Our Low Carbon Future* (DECC 2011a), *2011 National Infrastructure Plan* (HM Treasury and Infrastructure UK 2011), the *Overarching National Policy Statement for Energy* (DECC 2011b), the *UK Renewable Energy Roadmap* (DECCC 2011c), *The Future of Heating: A Strategic Framework for Low Carbon Heat in the UK* (DECC 2012), and *The Future of Heating: Meeting the Challenge* (DECC 2013). In London, all new major developments are required to connect to an existing DH network or establish a new DH network (EST 2009)

Table 4.2 Energy systems from the macro to the micro scale (adapted from Foresight 2008, Watson et al. 2010)

Scale	*Geography*	*Technologies*	*Institutions*	*Regulations and Incentives*
Macro	International	Large solar PV or wind farms	European Union, International Energy Agency	EU Emissions Trading Schemes
	National	Centralised electricity and gas grids	Central government, national energy regulators	Energy market rules, national incentives
Meso	Regional/City	City scale heat and power networks	Municipal utilities	Regional energy strategies
	Neighbourhood/ District	District heat and power networks	Local authorities	Local planning rules, local grants
Micro	Building	Micro-generation, insulation, efficient appliances	Homeowners, building managers, community organisations	Homeowner grants and incentives

and various programmes by government bodies (for example, the HCA's Carbon Challenge and Low Carbon Infrastructure Fund) feature meso-scale infrastructure networks as the centrepiece of exemplary pilot projects. Reflecting on the experiences of the Low Carbon Infrastructure Fund, the HCA cautiously concludes that 'community-level energy systems can, in the right conditions, enable developments to achieve higher levels of CSH in a more cost effective way than individual technologies' (HCA 2011b: 13). Likewise, district heat networks are increasingly being promoted as a promising option or even a necessity for achieving CSH Level 5 and 6 (RAB 2007, TCPA/CHPA 2008, Roberts 2008, Williams 2010, HCA 2011b, Williams 2012, CHPA 2014).

Configuring zero carbon housing

Given this familiar policy conundrum of technical potential and multiple 'barriers' to district heating, how might we expect the homebuilding industry to respond? In exploring the emerging landscape of zero carbon housing, we surveyed a range of recent zero carbon housing projects. Table 4.3 provides examples of low and zero carbon housing developments that include district heating and CHP networks. There is wide variation in the scope of these developments, particularly in relation to the scale of the projects. Moreover, while some are restricted to domestic buildings, others use DH and CHP to power non-domestic buildings or to supply the grid with electricity. While this is not a definitive survey, it offers a representative perspective of the low and zero carbon housing landscape and suggests that the reconfiguration of the housing/network relationship is a key element in the realisation of zero carbon housing. On the surface this survey appears to suggest a new trend, a shift towards a meso-level power system that creates a new context for the assembly of housing and networks to complete the carbon reduction hierarchy. Is it the case, then, that the UK building industry has finally woken up to the 'good sense' of international experience and home-grown demonstration projects? Does this wave of experimentation suggest a new era of district heating rolling out across our towns and cities, as a growing number of Government agencies and trade organisations suggest?

Two case studies provide insight on the motivations, challenges and opportunities related to meso-scale and zero carbon housing. In each case study we interviewed design team actors – developers, architects, housebuilders and local authorities – to understand how the CSH is being interpreted and implemented. Here, we focus on the implications of district heating networks on two projects – one that successfully implemented a district network and another where it was initially considered but ultimately dropped. The first case study is Park Dale, located in Airedale, West Yorkshire and built by Wakefield and District Housing (WDH) in partnership with the housing developer Keepmoat, the architect NPS, and the builder Bramall Construction. The project opened in November 2011 and includes 91 CSH Level 6 houses. It is touted as the 'largest zero-carbon social housing scheme in the UK' (WDH 2011) with WDH serving as landlord for low-income renters. The project includes an energy centre with a biomass district heating system (Figure 4.1). The second case study, Vista in Peterborough, Cambridgeshire, is one of two projects delivered under the HCA's Carbon Challenge Programme (the other project is Hanham Hall in South Gloucestershire). The project team included the homebuilder Morris Homes, the architect Barnett North LLP, and Peterborough City Council. The project was completed in 2013 and includes 295 houses at CSH Level 6 (HCA 2011a). The project team initially planned to include a biomass-fuelled CHP system for the project, but after the 2008 economic downturn they decided to pursue a design strategy with individual biomass boilers in each house.

In the following paragraphs we highlight some of the key challenges faced by the design teams in designing and building meso-scale infrastructure networks. We supplement the insights from the design team members of these two case studies with opinions from housebuilders, consultants and trade industry representatives involved in DH and CHP promotion and development. The findings suggests that there are multiple tensions and tactics involved in designing, constructing, and managing DH and CHP systems at the meso-scale, creating an additional layer of complexity and uncertainty in the realisation of a low carbon Britain.

Figure 4.1 The energy centre at the Park Dale zero carbon housing development in Airedale, West Yorkshire (note the asymmetrical roof and the three chimneys for the biomass boiler).

Source: authors.

Decision-making and energy provision

The pursuit of low carbon housing involves nuances in decision-making. As a design team member of the Vista project notes:

> The energy issue is clearly the big question mark and the big thing that holds back developments of this type. You can do high insulation, you can tick biodiversity boxes easily, you can do Lifetime Homes [an additional regulatory requirement] without too much trouble, you can meet the daylighting requirements. But the heat and power requirement is obviously the big thing.

A shared opinion of the respondents was that DH and CHP are the best and perhaps the only option for achieving a low carbon housing stock in the long-term future. As a trade industry representative succinctly notes, 'I've now after four years realised it's the only way forward, it really is the only solution'. This is borne out by the growing number of projects promoting decentralised schemes (see Table 4.3). Moreover, there is a growing sense within the industry that a rescaling of energy provision somehow symbolises an ecological turn in and of itself, as a builder of Park Dale suggests: 'We did a decision tree to look at the best way forward, of achieving Code 6, and ultimately the one that raised its head most was district heating, supplemented by solar PV.'

However, the implementation of this strategy is rarely straightforward; reconfiguring the system of energy provision has multiple implications that need to be considered. One of the Park Dale design team members notes:

> We're looking at a district heating system, so you'll need fuel pellets in an energy centre, in a boiler house, which will impact upon deliveries. We have to consider the physical aspect of it all, they're delivering every two weeks or four weeks. How do you want to charge your occupiers for using this fuel? We needed the PVs to supplement the district heating, roughly 36 square metres of PVs. And on a normal symmetrical roof or on a balanced roof like that you can't fit it on. So it meant having asymmetric, which is like an offset pitch.

What is clear from this account is both the often unfamiliar design and planning challenges associated with heat networks, and in response the pragmatic creativity involved in assembling these systems in the face of standardised solutions and taken-for-granted development routines. The necessity of hybrid systems, the risk involved in abandoning gas supplies, the logistical challenges of fuel pellet delivery, the complexities of financial charges, the reshaping of design priorities and the feelings of risk and uncertainty that accompany the new system all bring into question the optimism and certainty that pervade the policy literature and its promotion of district systems as the preferred solution.

Table 4.3 Examples of recent low and zero carbon housing developments in the UK with meso-scale energy networks (Source: authors)

Project	*Housing units*	*District system*
Adelaide Wharf East Ditch, London	143	Gas-fired DH
Centenary Quay Southampton, Hampshire	1620	Two CHP plants
Graylingwell Park Chichester, West Sussex	750	CHP plant
Green Space, Mendip Place Chelmsford, Essex	10	Biomass DH network
Greenwatt Way Slough, Berkshire	10	DH network powered by a biomass boiler, air source heat pump, and two ground source heat pumps
Hanham Hall South Gloucestershire	186	Biomass-fuelled CHP plant
One Brighton Brighton, East Sussex	172	Biomass DH network
Park Dale Airedale, West Yorkshire	91	Biomass DH system
Shine-ZC Derby, East Midlands	9	CHP plant with underground thermal store

The scalar implications of zero carbon

The district systems also reinforced the scalar dimensions of each development. At Park Dale, the energy centre that houses the biomass boilers for the district heating system also serves as an educational facility for green living for the residents as well as schoolchildren. In the Vista project the district heat network was integrated with other site amenities, and when the network was abandoned the amenities were also removed from the project. A Vista design team member states:

> We originally had a small shop, a café, this community hall and this sort of spa, because the energy side was going to produce so much waste heat and so we were going to use it to power the hot tub arrangement, but district heating's gone, the spa's gone, the only thing we're actually left with is a shop on the corner, so we've managed to get a small shop, there's going to be a community shop there, not central to the site but it's still there.

What emerges clearly from these case studies is the strong relationality of the technical systems with the design and planning imperatives of each development. In particular, the issue of development form and spatial configuration arises time and again as key to the optimisation of the system. In explaining the emergence of heat planning in London, a local authority engineer has a clear spatial explanation: 'the thing that does it for London is the densification … there's more people packed into a square metre, which is great because you only have to do a little investment to get a lot of people'. Moreover, micro-generation is often adversely impacted by density, as a local authority engineer confirmed: 'From a point of view of renewables, a dense urban landscape's no good for PVs, there's not enough area, you get shading. It's useless for micro wind, you just don't get the airflows sort of around buildings'. Similarly, a trade group representative confirmed that 'micro-generation gets more and more problematic the more dense the size. There aren't big enough roof spaces for PV… It makes more and more sense to have a communal solution of some kind, so district heating is that'.

There are a number of inter-related levels linking urban form and system efficiency across thermodynamics, demand dynamics, cost, and design, as a consulting engineer illustrated: 'In terms of load, the bigger the scheme the more efficient your CHP becomes, the larger CHP you can get the more efficient it generally is, the longer you can run it for'; and also financial cost: 'if you drop below probably 90 to 100 units, CHP doesn't become particularly viable from a financial point of view'. Moreover, the range and diversity of users included in the network can be critical: 'the more diversified users you can get on it, the better really. Baseload then just rockets and if your baseload is high, your low carbon generator can be high'. In other words, 'the bigger that is, the more you're going to save'. Therefore, as a recent DECC report makes clear, heat networks are more likely to be cost-effective in urban settings, where there are many buildings like blocks of flats where individual gas boilers may not be an option, and also commercial buildings such as hospitals and leisure centres that provide high and predictable heat demand (DECC 2013).

Economic viability

The consequences of this density equation in turn shape development calculations. As one homebuilder told us, 'Seemingly anything below about 350 dwellings, it just doesn't generate the revenues that they need to justify an initial infrastructure outlay'. And it's not simply a matter of the number of housing units, as the network itself is very space intensive and fussy about angles and obstructions. A local authority engineer told us: 'You have things like obstructions, bridge crossings, rail crossings … putting them in later across major junctions and things like that is a pain in the arse… Pipework is a linear development, roads and rail systems are linear developments'. The space challenges do not end with the pipe, as a consulting engineer suggests:

> People just don't understand [heat storage], they don't want to put it in. It's a pain. The first thing the clients say is, 'Where am I going to put that?' And a lot of consultants will just say, 'Well, chop that one in half and we'll put in three of a smaller size.' But thermal stores should be tall and skinny, the taller and skinnier the better for stratification.

There are similar challenges with house typologies, as a low carbon housing expert makes clear:

> With detached houses, it gets really expensive to start feeding from a central district heating centre, and the pipework has to go down all these different roads and individual roads. I think it would be something like £700 per metre, so if you've got a house down a cul-de-sac, little bit tucked away, you're suddenly up to £5,000 or £10,000, just to get to that one particular plot that's slightly offset.

In particular, the case study respondents highlight the critical importance of the economic cycle to the viability of the system, which was instrumental in the abandonment of district heat network for the Vista project. As a design team member notes:

> We went out to market for a scheme of four hundred and fifty homes, and it might have worked like more easily at that density to get the energy thing sorted out. Clearly you'd be talking more flatted developments which is easier anyway for the district heating stuff. But that was 2006 and that was before the housing bubble went bang and you get to 2009 and it was more and more apparent that that kind of density of development would not be probably desirable and anyway wouldn't be viable because it wouldn't be financed, and wouldn't be sold even if it was financed.

What becomes very clear is the requirement for some key stabilising factors to be in place if a transition to decentralisation is to be undertaken successfully. More challengingly, these stabilising factors span the full spread of issues including finance, base load, density, form, use, public/private collaboration, and so on. A Vista design team member notes:

> The infrastructural costs associated with biomass power stations, district heating schemes, just do not stack up without public sector pump priming. So unless you've got a significant heat load that is provided by a swimming pool, public sector building, [or] large office block, district heating schemes for residential uses alone are really just not economically viable. And certainly where you've got mains gas, it really just doesn't compete. Without public intervention, particularly fiscal intervention, you are not going to get waste heat, whether it's biomass or coal or gas power stations, as part of the contribution towards sustainable housing and certainly not zero carbon housing.

The Vista development faced a series of intractable issues including: the necessity of standing charges to pay for the infrastructure, which meant that customers would pay about the same for their energy as if the house was a normal Code 3 or normal standard building regulation property; the question of security of supply, as the energy companies would only look at a twenty-five- or thirty-year concession agreement to supply energy; issues of system reliability if the biomass went down, which led to suggestions of standard house boilers acting as a back-up system; and so on. In order to engineer this transition a very experimental orientation is evident both technologically and institutionally, but the final development evaluation and decision is primarily financial, with the familiar risk aversion of the homebuilding industry influencing the final design choices. As a Vista design team member states:

> We tried it at Peterborough, we explored all the options possible, on-site generation biomass, glycol, pellets. We tried co-locating it with the football stadium, but the standing charge that was going to be levied was just unacceptable to [the homebuilder] and they saw it as an impediment to actually selling the houses, and as such having gone through a full evaluation of a district heating scheme to provide what is a relatively low heat demand, [the homebuilder] took the decision to go back to HCA and argue that district heating scheme wasn't viable. And yeah, from a commercial perspective I think that's a sort of sound move.

Overall, the reasons for the success or failure of DH in the Park Dale and Vista projects (and more generally in all low carbon housing developments) are not easy to pinpoint. They involve protracted debates and deliberations over energy provision, balancing of low carbon design strategies, scalar characteristics of the particular development, economic modelling and financial viability, and long-term maintenance and upkeep.

Conclusions

As we have seen, meso-scale infrastructure networks for heat and power are increasingly being used in the UK to realise low carbon housing targets. Central provision of heat and/or power can provide energy efficiency gains beyond individual households while also utilising non-fossil fuels and taking advantage of waste heat from power generation. However, these efficiency gains are offset by challenges including significant fixed costs (specifically for the infrastructure network) as well as a new management structure for delivering heat and/or energy to houses. Beyond the technical and economic challenges of district heating networks, there are a number of inter-related challenges that arise with the rescaling of domestic heat and power services. At a national scale, the UK energy industry continues to be locked into fossil fuel and nuclear generation technologies (Foresight 2008). This is exacerbated by the nationalisation of energy provision in the 1940s, which encouraged a supply-logic, followed by privatisation in the 1980s, which created a profit logic, both arguably weakening any momentum towards energy efficiency and marginalising interest in alternatives to fossil fuels (Wilson and Game 2002, Winskel 2002, BERR 2008). The six main energy companies in the UK are likely champions of district heating. but they have a significant stake in the gas supply market, a direct competitor with district heating (Hawkey 2009). Furthermore, the energy networks have been designed for centralised generation and one-way flow; introducing distributed generation requires significant network changes (Allen *et al.* 2008).

At the institutional level, there is a lack of knowledge about (and almost a fear of) district heating. Local authorities and housebuilders have a limited understanding of the energy services market. Homebuilders have concerns about the marketability of houses that are served by district heating and the acceptance by consumers (HCA 2011b). Meanwhile, energy providers are often unfamiliar with local planning policies and decision-making processes as well as homebuilding economics. Joining up the homebuilders, local utilities, and energy providers is a significant challenge. This is similar issue with other local responses to climate change that cut across conventional divisions within local authorities and other governance structures (Bulkeley and Kern 2006). Due to the lack of in-house expertise, it is commonplace to contract the design, construction, and management of energy services to a commercial energy service company (HCA 2011b) of which there are only a handful in the UK (see Vine 2003, Williams 2010).

There are also competing energy efficiency aims that problematise the introduction of district heating network. Späth (2006: 4) notes that 'with decreasing energy demand of buildings due to improved insulation and "passive house design", the load density of urban areas is becoming critically low for such systems to be economically feasible.' In effect, the realisation of lower heat demand in houses due to higher levels of insulation makes efficient energy supply networks redundant or uneconomically unattractive. This was experienced in exemplary eco-neighbourhood projects of Hanover and Freiburg and was remedied with higher standing charge to cover capital costs (TCPA/CHPA 2008). This suggests that the strategies to reduce carbon emissions from housing may compete with one another. Producing super-efficient houses could make meso-scale heat and power networks an unaffordable and unnecessary luxury for eco-developments. Also, these new district heating networks could lock in specific levels of energy consumption and not allow for future aspirations of energy efficiency (Späth 2006). Meso-scale infrastructure networks would effectively replace the high carbon universal grid with a lower carbon local grid but prevent the possibility of further carbon savings.

The rollout of district heating networks in the UK faces the classic infrastructure issues of cost, governance and management that Jane Summerton described so vividly in the early 1990s. It is clear that a new momentum is being provided by the pursuit of lower carbon performance of the housing stock, but that challenges of cost, scale, design, logistics, coordination and so on remain and are only likely to be solved through context-specific negotiation and compromise. The case studies demonstrate that this can be done with proper funding, planning and commitment from the design team. Additionally, the emergence of decentralised systems potentially revives a role for local government in energy planning as intermediaries between the various agendas of the actors involved (Vaze and Mayo 2009). The case studies highlight the importance of local actors in the realisation of meso-scale infrastructure networks. A focus on 'heat' as a means to rescale carbon reduction efforts and to influence urban form suggests that a city with heat infrastructure will not only be governed differently, but it will also have different physical attributes in terms of density and mixtures of use. This not only involves scaling down energy provision from the national to the local level, but also scaling up energy generation from the household to a neighbourhood energy centre. As a local authority engineer told us, 'Now we're going into those heat dense areas and coming out with energy masterplanning, which is really establishing where the network is, how far the network could realistically run and be economic or market competitive, given where the demand is and given what's available as sources of heat'.

Looking forward, it remains to be seen if this is yet another discrete phase of district heating in the UK or if it is building up to a sea change in energy provision. In thirty to forty years' time, will the UK be similar to Denmark where district heating is the norm? Or will these district heating networks and zero carbon housing projects that are currently being designed and built be a short-lived and ultimately failed experiment? In a period of austerity with the building industry very much in the midst of a downturn this is difficult to predict, but it is clear that the UK is currently experiencing a crucial testing period for district networks. DH schemes that come online before April 2015 are eligible for a Government subsidy, but there are no guarantees that this programme will be extended or repeated. The energy consultant at the Vista project notes that without further fiscal support from Government, 'I think you can see the demise of district heating schemes'.

We may find a final clue by looking back to Jane Summerton's work to note the enduring power of the 'invisible grid' of inter-dependent associations that underpin the physical network and the necessity of a 'compatibility of concerns that help such systems emerge and that shape their rate and direction of system expansion' (Summerton 1992: 258). Seen this way, we might argue that until the concerns of policymakers, consumers, communities, developers, and energy suppliers are better aligned, district heating systems will struggle to find a welcome home in the UK's energy portfolio. This will involve a more systemic transition than simply 'educating' network providers and users about the virtues of DH through best practice exemplars. Instead, the viability of DH in the UK will require a more fundamental 'unlocking' of the national system and a challenge to the prevailing scalar logics of supply that continue to drive our networks of power.

Notes

1 See Hughes 1983, Mayntz and Hughes 1988, Summerton 1994, Aibar and Bijker 1997, Graham and Marvin 2001, Guy et al 2001, Coutard 2002, Graham and Marvin 2002, Coutard *et al.* 2004, Hommels 2005, Farias and Bender 2010, Graham 2010, Guy and Karvonen 2011, Guy *et al.* 2011, and Karvonen 2011.

2 Examples of recent sociotechnical studies of energy funded by the UK's Research Councils include 'Urban Transitions: Climate Change, Global Cities and the Transformation of Socio-Technical Systems' (ESRC, 2008–2012, Bulkeley), 'Governance of the Discontinuation of Socio-Technical Systems' (ESRC, 2013–2016, Sterling), 'Re-Engineering the City 2020–2050: Urban Foresight and Transition Management' (EPSRC, 2010–2014, Eames), 'Interdisciplinary Cluster on Energy

Systems, Equity and Vulnerability' (EPSRC, 2009–2012, Bickerstaff), and 'Community Innovation in Sustainable Energy (EPSRC, 2010–2014, Smith).

3 See the Sustainable Practices Research Group website, www.sprg.ac.uk.

4 See Smith 2007, Rutherford 2008, Coutard and Rutherford 2010, Hodson and Marvin 2010, Coutard and Rutherford 2011, Hodson and Marvin 2011, Loorbach and Verbong 2012, Bridge et al 2013, Rydin et al 2013, and Rutherford and Coutard 2014.

5 Note that the CSH covered more than energy efficiency and carbon footprints; it was a holistic regulatory framework that addressed the standard elements of green design, construction, and operation. In this chapter, we focus specifically on energy and carbon. For more information on the CSH, see DCLG 2007, 2008, 2010 and Williams 2012.

6 For different perspectives on the debates surrounding the rollout of CSH, see NHBC 2008, UKGBC 2008, ZCH 2009, McManus *et al.* 2010, Rydin 2010, Goodchild and Walshaw 2011, Greenwood 2012, and Reid and Houston 2013.

7 For a history of DH and CHP in the UK, see Russell 1993.

References

Aibar, E. and W. E. Bijker (1997) Constructing a city: the Cerdà plan for the extension of Barcelona. *Science, Technology & Human Values* **22**(1): 3–30.

Alanne, K. and A. Saari (2006) Distributed energy generation and sustainable development. *Renewable and Sustainable Energy Reviews* **10**: 539–558.

Allen, S. R., G. P. Hammond, and M. C. McManus (2008) Prospects for and barriers to domestic micro-generation: a United Kingdom perspective. *Applied Energy* **85**: 528–544.

Babus'Haq, R. F. and S. D. Probert (1996) Combined heat-and-power implementation in the UK: past, present and prospective developments, *Applied Energy* ***53***: 47–76.

Bolton, R. and T. J. Foxon (2013) Urban infrastructure dynamics: market regulation and the shaping of district energy in UK cities, *Environment and Planning A* **45**(9): 2194–2211.

Bolton, R. and T. J. Foxon (2014) Infrastructure transformation as a socio-technical process – implications for the governance of energy distribution networks in the UK. *Technological Forecasting and Social Change* **90**(B): 538–550..

Bridge, G., S. Bouzarovski, M. Bradshaw, and N. Eyre (2013) Geographies of energy transition: space, place and low-carbon economy, *Energy Policy* **53**: 331–340.

Bulkeley, H. and K. Kern (2006) Local government and the governing of climate change in Germany and the UK. *Urban Studies* **43**(12): 2237–2259.

Combined Heating and Power Association [CHPA] (2014) CHPA website, http://www.chpa.co.uk, last accessed 28 July 2014.

Coutard, O. (Ed.) (2002) *Governance of Large Technical Systems*. London: Routledge.

Coutard, O., R. Hanley, and R. Zimmerman (Eds.) (2004) *Sustaining Urban Networks: The Social Diffusion of Large Technical Systems*. London: Routledge.

Coutard, O. and J. Rutherford (2010) Energy transition and city-region planning: understanding the spatial politics of systemic change. *Technology Analysis & Strategic Management* **22**(6): 711–727.

Coutard, O. and J. Rutherford (2011) The rise of post-networked cities in Europe? recombining infrastructural, ecological and urban transformations in low carbon transitions, In H. Bulkeley, V. Castán Broto, M. Hodson and S. Marvin (Eds.), *Cities and Low Carbon Transitions*. London: Routledge, 106–125.

Department for Business Enterprise & Regulatory Reform [BERR] (2008) *Heat Call for Evidence.* January 2008. London: BERR.

Department for Communities & Local Government [DCLG] (2007) *Building a Greener Future*. London: DCLG.

Department for Communities & Local Government [DCLG] (2008) *The Code for Sustainable Homes: Setting the Standard in Sustainability for New Homes.* London: DCLG.

Department for Communities & Local Government [DCLG] (2010) *The Code for Sustainable Homes: Technical Guide, November 2010*. London: DCLG.

Department of Energy & Climate Change [DECC] (2011a) *The Carbon Plan: Delivering Our Low Carbon Future.* London: DECC.

Department of Energy & Climate Change [DECC] (2011b) *Overarching National Policy Statement for Energy (EN-1).* July 2011. London: DECC.

Department of Energy & Climate Change [DECC] (2011c) *UK Renewable Energy Roadmap.* July 2011. London: DECC.

Department of Energy & Climate Change [DECC] (2012) *The Future of Heating: A Strategic Framework for Low Carbon Heat in the UK.* March 2012. London: DECC.

Department of Energy & Climate Change [DECC] (2013) *The Future of Heating: Meeting the Challenge.* March 2013. London: DECC.

Department of Energy and Climate Change and Department for Communities and Local Government [DECC/DCLG] (2009) *Heat and Energy Saving Strategy Consultation.* London: DECC and DCLG.

Energy Saving Trust [EST] (2009) *The Applicability of District Heating to New Dwellings*. London: Energy Saving Trust.

Euroheat & Power (2011) *District Heating and Cooling: Country by Country/2011 Survey.* Brussels: Euroheat & Power.

Farías, I. and T. Bender (Eds.) (2010) *Urban Assemblages: How Actor-Network Theory Changes Urban Studies*. London: Routledge.

Fischer, J. and S. Guy (2009) Re-interpreting regulations: architects as intermediaries for low-carbon buildings. *Urban Studies* **46**(12): 2577–2594.

Foresight (2008) *Powering Our Lives: Sustainable Energy Management and the Built Environment.* London: Government Office for Science.

Goodchild, B. and A. Walshaw (2011) Towards zero carbon homes in England? From inception to partial implementation. *Housing Studies* **26**(6): 933–949.

Government Office for Science (2008) *Powering Our Lives.* London: BIS.

Graham, S. and S. Marvin (2001) *Splintering Urbanism: Networked Infrastructures, Technological Mobilities and the Urban Condition.* London: Routledge.

Graham, S. and S. Marvin (2002) *Telecommunications and the City: Electronic Spaces, Urban Places*. London: Routledge.

Graham, S. (Ed.) (2010) *Disrupted Cities: When Infrastructure Fails*. London: Routledge.

Greenwood, D. (2012) The challenge of policy coordination for sustainable sociotechnical transitions: the case of the zero-carbon homes agenda in England. *Environment and Planning C: Government and Policy* **30**(1): 162–179.

Guy, S. (2006) Technological convergence, cultural diversity: socio-technical perspectives on energy and building. *Environment and Planning C* **24**: 645–659.

Guy, S. and A. Karvonen. 2011. Using sociotechnical methods: researching human-technological dynamics in the city. In J. Mason and A. Dale (Eds.), *Understanding Social Research: Thinking Creatively about Method*. London: Sage Publications.

Guy, S., S. Marvin, and T. Moss (Eds.) (2001) *Urban Infrastructure in Transition: Networks, Buildings, Plans*. London: Earthscan.

Guy, S., S. Marvin, W. Medd, and T. Moss (Eds.) (2011) *Shaping Urban Infrastructures: Intermediaries and the Governance of Socio-Technical Networks*. London: Earthscan.

Hawkey, D. (2009) *Will 'District Heating Come to Town'? Analysis of Current Opportunities and Challenges in the UK.* Master's dissertation, University of Edinburgh.

Hawkey, D. J .C. (2012) District heating in the UK: a technological innovation systems analysis. *Environmental Innovation and Societal Transitions* **5**: 19–32.

Hawkey, D., J. Webb, and M. Winskel (2013) Organisation and governance of urban energy systems: district heating and cooling in the UK. *Journal of Cleaner Production* **50**, 22–31.

HM Government (2009) *The UK Renewable Energy Strategy*. July 2009. London: HM Government.

HM Treasury and Infrastructure UK (2011) *National Infrastructure Plan 2011*. November 2011. London: HM Treasury.

Hodson, M. and S. Marvin (2010) Can cities shape socio-technical transitions and how would we know if they were? *Research Policy* **39**: 477–485.

Hodson, M. and S. Marvin (2011) Making low-carbon England and Wales. *Town and Country Planning* **79**(9): 388–392.

Hodson, M. and S. Marvin (2013) *Low Carbon Nation?* London: Routledge.

Homes and Communities Agency [HCA] (2011a) A carbon challenge for Peterborough, online article, 26 July 2011, http://www.homesandcommunities.co.uk/article/carbon-challenge-peterborough, last accessed 28 July 2014.

Homes and Communities Agency [HCA] (2011b) *District Heating Good Practice: Learning from the Low Carbon Infrastructure Fund*. London: Homes and Communities Agency.

Hommels, A. (2005) *Unbuilding Cities: Obduracy in Urban Socio-Technical Change*. London: MIT Press.

Hughes, T. P. (1983) *Networks of Power: Electrification in Western Society, 1880–1930*. Baltimore: Johns Hopkins University Press.

Karvonen, A. (2011) Politics of Urban Runoff: Nature, Technology and the Sustainable City. London: MIT Press.

Karvonen, A. (2013) Towards systemic domestic retrofit: a social practices approach. *Building Research & Information* **41**(5): 563–574.

Loorbach, D. A. and G. Verbong (Eds.) (2012) *Governing the Energy Transition: Reality, Illusion or Necessity*. London: Routledge.

Lund, H., S. Werner, R. Wiltshire, S. Svendsen, J. E. Thorsen, F. Hvelplund, and B. V. Mathiesen (2014) 4th generation district heating (4GDH): integrating smart thermal grids into future sustainable energy systems. Energy **68**: 1–11.

Mayntz, R. and T. Hughes (1988) *The Development of Large Technical Systems*. Frankfurt: Verlag.

McManus, A., M. R. Gaterell, and L. E. Coates (2010) The potential of the Code for Sustainable Homes to deliver genuine 'sustainable energy' in the UK social housing sector. *Energy Policy* **38**(4): 2013–2019.

NHBC Foundation (2008) *Zero Carbon: What Does it Mean to Homeowners and Housebuilders?* Amersham: NHBC Foundation.

Reid, L.A. and D. Houston (2013) Low carbon housing: a 'green' wolf in sheep's clothing? *Housing Studies* **28**(1): 1–9.

Renewables Advisory Board [RAB] (2007) *The Role of Onsite Energy Generation in Delivering Zero Carbon Homes.* A report from the Renewables Advisory Board by Element Energy with Energy Saving Trust. URN number 07/1555.

Roberts, S (2008) Infrastructure challenges for the built environment. *Energy Policy* **36**: 4563–4567.

Russell, S. (1993) Writing energy history: explaining the neglect of CHP/DH in Britain. *The British Journal for the History of Science* **26**(1): 33–54.

Russell, S. (2010) *CHP and DH to the mid-1990s*. Heat and the City working paper, version 1, November, ISSTI, University of Edinburgh, working paper presented at international roundtable seminar, Gif-sur-Yvette, Paris, 16–18 September. http://www.heatandthecity.org.uk/_data/assets/pdf_file/0004/62419/HatC_history_paper_SR.pdf (last accessed 28 July 14).

Rutherford, J. (2008) Unbundling Stockholm: the networks, planning and social welfare nexus beyond the unitary city. *Geoforum* **39**: 1871–1883.

Rutherford, J, and O. Coutard (2014) Urban energy transitions: places, processes and politics of socio-technical change. *Urban Studies* **51**(7): 1353–1377.

Rydin, Y. (2010) *Governing for Sustainable Urban Development.* London: Earthscan.

Rydin, Y., C. Turcu, S. Guy, and P. Austin. 2013. Mapping the coevolution of urban energy systems: pathways of change. *Environment and Planning A* **45**(3): 634–649.

Smith, A. (2007) Emerging in between: the multi-level governance of renewable energy in the English regions. *Energy Policy* **35**: 6266–6280.

Späth, P. (2006) *District Heating and Passive Houses – Interfering Strategies Towards Sustainable Energy Systems.* IFZ – Inter-University Research Centre for Technology, Work and Culture.

Summerton, J. (1992) *District Heating Comes to Town: The Social Shaping of an Energy System.* Linköping University, Dept. of Technology and Social Change.

Summerton, J. (Ed.) (1994) *Changing Large Technical Systems.* Boulder, CO: Westview Press.

Toke, D. and A. Fragaki (2008) Do liberalised electricity markets help or hinder CHP and district heating? The case of the UK. *Energy Policy* **36**: 1448–1456

Town and Country Planning Association and Combined Heat and Power Association [TCPA/CHPA] (2008) *Community Energy: Urban Planning for a Low Carbon Future.* London: TCPA and CHPA.

UK Green Building Council [UKGBC] (2008) *The Definition of Zero Carbon.* London: UKGBC.

Unruh, G. C. (2000) Understanding carbon lock-in. *Energy Policy* **28**(12): 817–830.

Unruh, G. C. (2002) Escaping carbon lock-in. *Energy Policy* **30**(4): 317–325.

Vaze, P. and E. Mayo (2009) *A New Energy Infrastructure.* Discussion paper for Consumer Focus, published November 2009, available from www.consumerfocus.org.uk (last accessed 10 July 2012).

Vine, E. (2005) An international survey of the energy service company (ESCO) industry. *Energy Policy* **33**(5): 691–704.

Wakefield and District Housing [WDH] (2011) *The UK's largest eco homes project opens up at Park Dale*, 21 October press release, http://www.wdh.co.uk/documents/NewBuild/ParkDalePressRelease.doc, last accessed 29 June 2012.

Watson, J., I. Scrase, and L. Stapleton (2010) *Transforming the UK's Energy System: Policies for the 2020 Renewables Target and Beyond.* SPRU, University of Sussex.

Webb, J. (2013) If it hasn't been done in Aberdeen, it's not worth doing: governing change in urban energy in a northern UK city, http://www.heatandthecity.org.uk/_data/assets/

pdf_file/0009/128799/Webb_Urban_Energy_Govs_Abd.pdf (last accessed 28 July 2014).

Williams, J. (2010) The deployment of decentralised energy systems as part of the housing growth programme in the UK. *Energy Policy* **38**: 7604–7613.

Williams, J. (2012) *Zero Carbon Homes: A Road Map.* London: Routledge.

Wilson, D. and C. Game (2002) *Local Government in the United Kingdom, Third Edition.* Hampshire: Palgrave Macmillan.

Winskel, M (2002) When systems are overthrown: the dash for gas in the British electricity supply industry. *Social Studies of Science* **32**: 565–599.

Zero Carbon Hub [ZCH] (2009) *Defining Zero Carbon Homes: Have Your Say Report 2009.* London: Zero Carbon Hub.

Zero Carbon Hub [ZCH] (2011) *Allowable Solutions for Tomorrow's New Homes: Towards a Workable Framework.* London: Zero Carbon Hub.

5 Rethinking universality and disrepair

Seeking infrastructure coexistence in Quibdó, Colombia

Kathryn Furlong

Introduction

Improving access to potable water and sanitation is recognised as one of the key challenges of our time. Still, success continues to be limited (WHO/UNICEF 2010). The reasons for this are manifold and include issues such as precarious land tenure, rapid urbanisation, poorly-funded utilities, weak institutions, and inequality (Budds and McGranahan 2003). However, important challenges also stem from contradictions between uniform infrastructure networks, derived from experiences in the North, and diverse realities in the South (Jaglin and Zérah 2010). Instead of privileging a single universal infrastructure network, greater attention is now given to hybrid and disaggregated systems. In the North, where the "modern infrastructure ideal" of universal and uniform coverage was largely met (Graham and Marvin 2001), mounting pressures on the environment and utility capacity have encouraged the adoption of alternative infrastructure configurations (Coutard 2010). In the South, water supply has long involved multiple systems in varying degrees of coexistence. What remains controversial is whether to ignore what is external to the piped network, to regulate it, or to subsume it under the management of a single water provider (Jaglin 2004).

At the same time, many local water systems are undergoing change. For Bakker, efforts in the South have focused on a transition from "artisanal" to "industrial" water supply (Bakker 2003). Yet in many cases, such efforts have given way to accepting diversity and co-existence among water supply alternatives (Jaglin 2008). Many prominent approaches to cities and infrastructure networks do not reflect this diversity, given their tendency to be derived from circumstances in the North (Graham and Thrift 2007: 11; Robinson 2005). However, even these are shifting. As such, certain approaches could be strengthened though engagement with coexistence in sociotechnical systems, which is more obvious in the South (Graham and Thrift 2007). For the particular case of water supply, widespread concepts that are challenged by experiences common in the South include "natural monopolies" (see Budds and McGranahan 2003: 93–94), universal networks versus sociotechnical co-existence, black-boxing versus user co-production, and the conflation of stability with dominance and uniformity.

In Quibdó, Colombia the local utility is seeking to improve the piped water system in the context of extensive reliance on rainwater collection. There, networked water supply has consistently been spatially limited, unreliable, and non-potable. Residents rely mainly on rainwater, in a region with among the highest levels of rainfall in the world. The "modern infrastructure ideal" has never been reached. Instead of seeking to achieve it, however, the local utility aims to develop a functioning piped water system in co-existence with extensive rainwater collection. Rather than supplanting artisanal systems with the forward march of "modern" networked infrastructure, the goal is to integrate the two in ways that enable reliable and potable water for consumptive uses. This is a complicated process, not only because of the usual impediments to network extension in the South (e.g. land tenure, discrimination, weak institutions, and "governance failure"[1]), but because the current sociotechnical system is "embedded", despite being neither universal nor "invisible". While the sociotechnical system for water supply in Quibdó is far from "black boxed", it is stabilised through custom derived from adaptation to disrepair.

This situation compels us to push further our ideas of "stability", "transition" and even "system", which need not be monopolistic. This paper mobilises the Quibdó case to help examine these issues. It does not present an exhaustive case study; but instead draws on key elements to open widely accepted ideas about large technical systems (LTS) and associated transitions. Thus, while it does not deal with a range of issues key to understanding service provision in the South, it does examine aspects that shed new light on standard premises in LTS. In this way, the paper follows Geel's call to "accept some loss of empirical complexity" in case study research in an effort to draw out potential links that might enable the introduction of new questions and conclusions into STS theory (Geels 2007a: 635). The case study material includes documentary research from both Quibdó and Medellín, five key informant interviews in Quibdó and another two in Medellín, and participant observation during a one-day stakeholder workshop held in Quibdó on May 13, 2011.

Networks, transitions, and alternative infrastructure ideals

The evolution of large technical systems has been an important topic in STS. Studies have mostly focused on Western contexts where universal infrastructure networks had become an unconscious backdrop to everyday life (e.g. Coutard 1999; Tarr and Dupuy 1988), or our collective "exoskeleton" (Gandy 2005). Infrastructure, in this context, was considered "stabilised", "obdurate" and "black-boxed" – that is, as essentially immune to change or external influence (e.g. Hughes 1987). Change in LTS was possible, but rare. Many studies focused on a "battle of the systems" whereby one technology achieves the total displacement of the other, obtaining a virtual "natural monopoly" (David 1992: 138). Stability was not absolute, but major external shocks to a system were considered necessary for reconfiguration to occur, resulting in new stable configurations as opposed to dynamism (Hughes 1987; Mayntz and Hughes 1988).

To understand such reconfiguration, analyses in STS are often framed in terms of "multi-level transitions" (MLT). MLT posits three "levels" of interdependent activity through which transitions occur. Briefly, these include the level of technological development (the niche), the institutions, norms and practices that surround a technology (the regime), as well as broader societal trends (the landscape) (e.g. Geels 2005a; Kemp *et al.* 1998). However, the MLT approach presents a key limitation to understanding sociotechnical transitions in the South in that it assumes a universal system as the end point, i.e. one system is replaced by another (Genus and Coles 2008: 1438). Thus, while the multi-level approach recognises a range of pathways to change (Van Vliet *et al.* 2005: 36), ultimately a single, universal outcome is settled on. Consequently, the approach has rarely been applied to cases where the outcome it assumes is not already in place.

Authors in disciplines like geography and planning have also been concerned with trends in infrastructure. Typically, research has been preoccupied with the influence of political-economic change, notably neoliberalism (e.g.Swyngedouw and Kaïka 2002). From an MLT perspective, such research privileges the landscape to the virtual neglect of the niche or the regime. Most prominent is the work of Graham and Marvin, which argues that policies of economic liberalisation have led to widespread network "unbundling" and the "splintering" of the urban experience (Graham and Marvin 2001). Unlike MLT, this work does not assume a universal outcome to splintering, but it does assume universal networks as the starting point. For this reason, the "splintering urbanism" thesis is argued to be of limited applicability to experiences in the South (Coutard 2008). There, infrastructure existing in "archipelagoes" and not "networks" (Bakker 2003) could not splinter; it was not universalised (Coutard 2008).

When thinking through sociotechnical transitions in the South, not only is it problematic to assume universal networks as either the origin or the outcome of change, but universality can also be problematic as a goal. In some cases, the drive to universalisation has compounded exclusion by creating service islands and encouraging sprawl (Pflieger and Matthieussent 2008), or by limiting alternatives in contexts of diversity (Jaglin 2008). For water, important public health issues can arise from efforts to displace existing options with a single piped network supply. Such networks are often under-resourced and experience problems with maintenance, disrepair and service interruption (Jaglin 2002). Although interruption may be "normal" (Graham and Thrift 2007) with water supply in the South, it can have important consequences. Public health research finds that even a few days of service interruption can "destroy the health benefit from the provision of clean drinking water" and increase the annual risk of illness by 100% (Hunter *et al.* 2009: 2624, 2623). As such, some advocate the improvement of existing water supply systems as opposed to their replacement (Hunter *et al.* 2009: 2624).

The reasons for this are derived from both technical and biological issues as well as the practices that are necessary to cope with intermittent supply. In piped water systems microbiological growth can become a factor, especially when using surface water. This growth is dislodged from pipe surfaces in conditions of intermittent flow (Ayoub and Malaeb 2006). Intermittency can also lead to the infiltration of contaminants due to negative pressures, which can further encourage bacteriological growth, consume chlorine residuals, and include fecal contaminants from the soil or nearby sewerage systems (LeChevallier et al. 2003; Tokajian and Hashwa 2003). Further, to cope with intermittent supply, water storage becomes essential to bridge service gaps. Studies of storage practices in various countries consistently find contamination through long storage times or mixing with bacteria already present in receptacles (Ayoub and Malaeb 2006; Coelho et al. 2003). Finally, intermittency can imply increased health risks because of the shock of bacteriological exposure when a sudden return to raw water consumption becomes necessary (Hunter *et al.* 2009: 2642).

Recently, disrepair and malfunction have occupied new interest in the literature on urban infrastructure. Graham and Thrift's "out of order" piece stands as a critique of the LTS literature as being focused on "utterly fixed hard technologies … characterized by perfect order, completeness, immanence and internal homogeneity rather than leaky, partial and heterogeneous" (Graham and Thrift 2007: 10). This focus, as they note, has led much of the LTS literature to ignore systems in the South where "infrastructural improvisation and repair is too overwhelming and visible to be ignored" (Graham and Thrift 2007: 11). Still, they are concerned with infrastructure in the North and have a rather positive view of disrepair. For these authors, disconnection is not only about "catastrophe"; it is also about "learning, adaptation, and improvisation" (Graham and Thrift 2007: 5). Drawing on Summerton's idea of technical transitions through incremental change (Summerton 1994), they argue that disruption can engender important system change through manifold incremental experiments focused on adaptation (Graham and Thrift 2007: 5).

This cumulative adaption and learning can also be about custom and coping. When disrepair is normal and everyday, as opposed to aberrant, it can generate its own "momentum". Momentum is a concept developed by Hughes to describe how the links between society (including users and a variety of institutions) and technology evolve to create a stabilised sociotechnical system that is resistant to change (Hughes 1994). Once an LTS achieves momentum, it is considered to uniformly dominate a geographical space and to be black-boxed, whereby its behaviour is independent of context (Callon 1991: 152). In Quibdó, although a single universal LTS does not dominate, the local sociotechnical system – based on the coexistence of multiple water supplies, constant malfunction, and high levels of user interaction – is characterised by momentum. Sociotechnical systems characterised by dysfunction, co-existence, and co-production can likewise become stabilised. How can this be understood?

For Trentmann, like Graham and Thrift, disrepair and malfunction are normal elements of user relationships to infrastructure; they "are not freak accidents or aberrations but natural, constitutive features of lived normality" (Trentmann 2009: 68). In the North, the normalisation of breakdown and service interruption is derived from the adaptability of people's habits in the expectation that interruption is temporary. Trentmann interprets this as demonstrating flexibility as opposed to obduracy in people's habits relative to sociotechnical systems. Yet, once service returns, so too do the regular customs of use; the alternate habits are only sustained through long-term shifts in the sociotechnical system (Trentmann 2009: 74). This speaks to Star's assertion that infrastructure constitutes and is constitutive of communities of practice (Star 1999: 381): habits and customs are resistant to change, as they are bound up in the wider social order. Such communities of practice need not be restricted to the "ideal" case. They can reasonably be expected to develop around coping with permanent disrepair. In this case, such customs would be unlikely to shift in response to occasional functionality or to promises of future repair.

Disruption and its assimilation by users are revealing in terms of the mutual construction of customs and performance in sociotechnical systems. First, users "are not just recipients [of services] but co-producers, whose behavior is vital for managing shortages"; they play active roles in absorbing, coordinating, and even orchestrating disruption (Trentmann 2009: 80). Further, user reactions to system function and breakdown are defined by perceptions of what is normal or aberrant for the sociotechnical system in question (Trentmann 2009: 76). For example, the sometimes disproportionately expectations of high-income groups can place undue strain on a network, engendering interruption (Trentmann 2009). Similarly, in the South, middle-class users may demand a level of service that is unaffordable to other groups, limiting service extension (Jaglin 2004: 3). What the Quibdó case shows instead, however, is that the extremely low expectations of system performance can likewise result in customs that place strain on the system and inhibit its extension to unconnected neighbourhoods.

These issues reflect Shove's concerns with habits and custom in relation to utility services. For Shove, the focus on *efficiency* in terms of resource use is misplaced and "overshadows prior questions about the framing and formulation of 'normal' practice". It misses "the point that much consumption is customary, governed by collective norms and undertaken in a world of things and sociotechnical systems that have stabilising effects on routines and habits" (Shove 2003: 9). To understand and alter service use, we have to understand the "collective dynamic of normalization" (Shove 2003: 199). The neglect of custom, moreover, engenders a focus on "educating" users (Shove 2003: 7). Assumptions that the right information will induce behavioural change neglects the complexity of consumption (Barnett *et al.* 2005) as well as the role of artifacts in constructing social practice (Robbins 2007).

Notably, convenience is a strong aspect of constructing the ways in which sociotechnical systems develop so that users can meet their needs. Convenience is not only about accessible services, it is about interactions with the sociotechnical system that allow users to manage their time (Shove 2003: 175, 198). Thus, when a service is highly unreliable, users may seek strategies to improve reliability, either by opting out of the system, through clandestine connections, or through storage (McFarlane 2010: 135–137). Understanding users as seeking system knowledge and reliability, not only in terms of service timing and quality but also in terms of costs and billing, helps to comprehend the potential "momentum" of sociotechnical systems of malfunction.

Sociotechnical dysfunction and coexistence in Quibdó

Quibdó, nestled in the confluence of the Atrato and Cabí Rivers, is the capital of the Department of Choco (Figure 5.1). Choco is located in the northwest of Colombia between the Pacific Ocean and the westernmost cordillera of the Colombian Andes, and is considered to be among the "rainiest places on earth" (Poveda and Mesa 2000). The Department was settled in pre-Colombian times by indigenous people and by African descendents escaping slavery in the mid-1600s. Today its population is just over 388,000, with 74% identifying as Afro-Colombian and 11% as indigenous (DANE 2005). Given the historic marginalisation of the region (cf. Asher 2009), its quality of life indicators are among the lowest in the country (DANE 2005).

Issues of poverty and marginalisation are likewise important in Quibdó. According to the national bureau of statistics (DANE), 90.46% of people living in the city suffer from unmet basic needs[2] (DANE 2010). Its current population of 115 000 includes a significant number of displaced people. Of the 37,000 people that live in the city's northern barrios, Villa España, Samper, El Futuro and El Reposo, it is estimated that 90% are displaced (Vanguardia 2014). Persons living in these barrios have no access to piped water or other basic services. As of 2008,

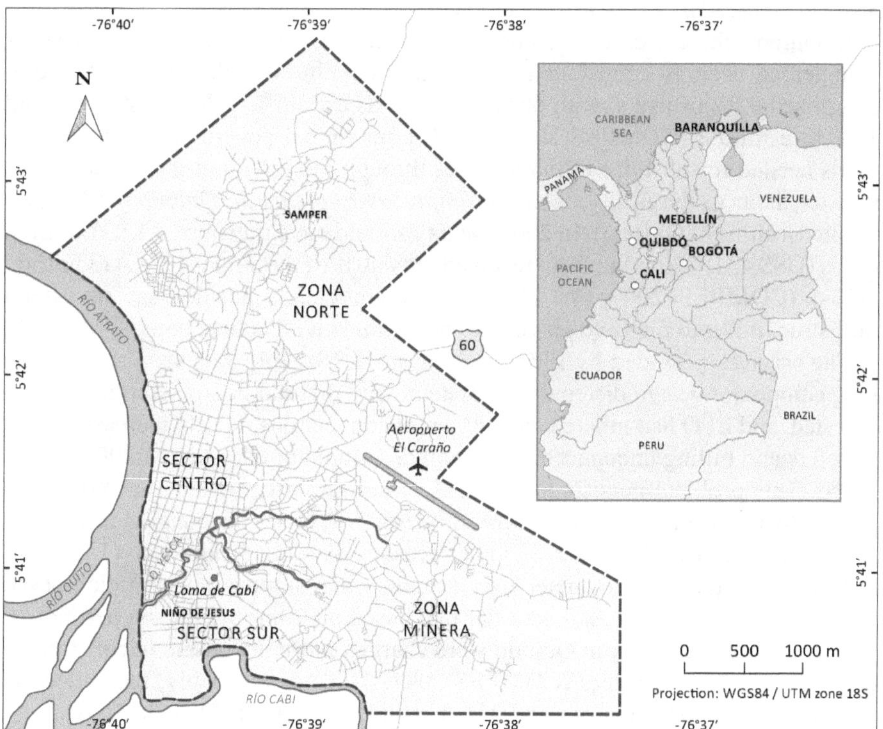

Figure 5.1 Map of Quibdó, Colombia
Source: © Rodrigo Amado, 2014.

less than 20% of the city's population had access to piped water and for those that did it was neither potable nor reliable. Sanitation coverage was around 13%. Some neighbourhoods had septic systems, but many were dependent on streams and alleyways to dispose of human and other waste (Municipio de Quibdó 2003). This situation still affects a majority of the population today (Aguas Nacionales 2014).

Many municipalities in Colombia have suffered inadequate service provision. In the early 1990s, severe problems even in major cities like Barranquilla prompted a series of water governance reforms. The main pillar was the Law 142 of 1994. Law 142 focused on the creation of new federal regulatory agencies, the liberalisation of the water supply, and incentivising private sector participation, as well as new measures to enforce rate collection and limit cross-subsidisation between high and low-income consumers. In Colombia, a system of cross-subsidisation for water (and other public services) developed gradually over the twentieth century (Acevedo et al. 2014). Since the 1980s it has been organised by dividing municipalities into six *estratos* (i.e. tiers) based on socio-economic criteria. Estratos 1, 2 and 3 pay below cost for utility services, estrato 4 pays at cost, and estratos 5 and 6 pay above cost, subsidising estratos 1 to 3. When there are not enough people in estratos 5 and 6 to fully subsidise the other users, the state transfers money to municipalities to cover the remaining costs through the *Sistema General de Participaciones* (SGP).

In Quibdó there are no households in estratos 5 and 6 and few industrial or commercial users to contribute to cross-subsidies. In 2007 the residential sector comprised 92% of users, with 66% classified as estrato 1, 21% as estrato 2, and 13% as estrato 3 (COMPES 2007: 4).[3] The municipal government received the funds needed to subsidise utility services through the SPG, but it did not always transfer them to the utility, then the *Empresas Públicas de Quibdó* – EPQ (the Public Utilities of Quibdó). In 2005 the city should have transferred 1,031 million pesos (US$ 431, 750),[4] but EPQ received only 40% of the amount (i.e. 413 million pesos) (COMPES 2007: 4). These funds are clearly insufficient to extend infrastructure or to improve services to households with connections.

The services provided by EPQ were abysmal. Not only was service coverage low, equipment was in disrepair, the state money that did get transferred was not invested, and EPQ had inflated the number of connections, falsely increasing state transfers and billing unconnected households (Municipio de Quibdó 2003; SSPD 2008). This problem persists today; 19% of consumer complaints in 2013 were related to being billed for non-existent services (Aguas Nacionales 2014: 66). Billing has been problematic for other reasons as well. Under Law 142/1994, utilities were required to meter user accounts and impose consumption-based tariffs. In Quibdó, EPQ charged a flat rate based on 20m^3 per person per month. The result was that users in Quibdó were charged more than users in comparable cities with reliable services (COMPES 2007: 5). Understandably, fee collection rates were low: 50% in water, and 23% in solid waste (COMPES 2007: 5).

The state finally intervened in 2005. It availed itself of new powers under Law 142/1994 to assume the management of and liquidate EPQ.[5] The federal utility regulator, the SSPD (the Superintendence of Public Services), took possession of

EPQ,[6] which now exists as "EPQ-in-liquidation". Then, in 2008 the SSPD arranged a seven-year contract between EPQ-in-liquidation and the *Empresas Públicas de Medellín* – EPM (the Public Utilities of Medellín) – to manage, improve, and increase coverage in solid waste, water and sanitation. EPM is a multi-utility company owned by the City of Medellín. It operates utility services in Quibdó through its subsidiary *Aguas Nacionales* under the name Aguas del Atrato.

When Aguas del Atrato took responsibility for water, sewer and solid waste in 2008, they found that EPQ's low statistics for water and sewer coverage were actually inflated. EPQ reported sufficient infrastructure for a potential coverage of 79%, but put actual coverage at 47% due to poor pressure, insufficient pumping, and missing connections (COMPES 2007: 3). On sewerage, the potential coverage was reported to be 41%, with only 21% having connections (COMPES 2007: 3). In reality, connections were 23% lower than reported (Aguas de Atrato 2013). Instead of 47% coverage was closer to 20% for water, and infrastructure was in worse condition than expected. At a public meeting on the progress of the project in May 2011, representatives from EPM noted that, while the Ministry responsible had originally allocated $37.2 billion pesos (US$ 17.9 million) (MAVDT 2007: 5), at least three times that, or more than $100 billion pesos (US$ 48.2 million), was needed. By 2013, the estimated investment required had risen to $163.8 billion pesos (US$ 88.7 million). The timeframe for the planned improvements to the network has been regularly scaled-back, and improvements have evolved much more gradually than expected (Table 5.1).

Coexistence and disrepair: the role of rainwater collection

Given the poor service quality and the abundant rainfall, residents have largely depended on rainwater collection to meet their water needs. Chocó has among the highest levels of rainfall in the world, averaging 8,000 to 13,000 mm annually (Poveda and Mesa 2000). Reliance on rainwater collection for cooking and drinking, and river water for other uses, has a long and widespread history throughout the department (Ruiz Serna 2008). However, the contamination of the

Table 5.1 Reported, projected and actual water and sewer coverage 2005–2015

	WATER		*SEWER*	
YEAR	*Projected*	*Actual*	*Projected*	*Actual*
2005	47% (Reported)	20,3%	21% (Reported)	11,3%
2008	29% (Reported)	20,3%	16,9% (Reported)	11,3%
2010	37,5%	26%	25%	17%
2012	56,9%	29,9%	46,7%	12,9%
2014	76%	28,6%	73,6%	12,2%
2015	84%		40%	

Source: Compiled by the author from COMPES (2007) and Aguas de Atrato (2014)[7]

Atrato River from illegal mining has made the population even more dependent on rainwater collection for bathing and even washing clothes (Caracol 2014). However, dependence on rainwater collection can be highly inequitable. Some households can afford large rainwater tanks or even underground collection systems, and thus are able to weather periods of little rainfall. Other households only have access to buckets, limiting their storage to three to four days of water (Córdoba 2014). In 2013 the city went nine days without rain leading to significant hardship, which was partially addressed by sending water trucks into unserviced barrios (CONtextoganadero 2013).

Thus, dependence on rainwater collection cannot relieve the state of its responsibility in the provision of potable water supply. Rainwater collection is increasingly discussed as a feasible solution for communities with abundant rainfall, but there is unreliable or absent water supply across Colombia (Ballén Suárez *et al.* 2006) as well as in other parts of the world (Ghisi *et al.* 2006). As the situation described above attests, rainwater collection does not constitute a simple and independent means of water access that enables residents to be liberated from the inadequacies of state service provision. An equitable rainwater collection

Figure 5.2 Water tanks in the centre of Quibdó.
Source: author

system necessarily involves state investment and support, not only in the provision of adequate capitation and treatment systems, but also in their maintenance, as well as the control of external contaminants.[8]

In Quibdó, rainwater collection has been supported by central government subsidies in certain barrios. In the northern barrio of Samper, whose legalisation began in 1995, funding was provided directly to the local *Junta de Acción Comunal* (community action board – JAC)[9] by the central government for rainwater tanks and septic systems. The systems were built by the individual homeowners and are thus privately owned and maintained. Still, a local representative argued that extending the piped water network to their barrio is important for reasons of hygiene, disease, and seasonal water scarcity. During the hot summer months, people may be forced to collect water from a nearby contaminated stream.[10]

In order to extend services to unconnected neighbourhoods and to improve the reliability and quality of the water supply to households that do have connections, the ongoing service extension project seeks to integrate an improved piped water network with modified rainwater collection. There are several reasons for this approach. First, in a region with a high degree of poverty, it is important to retain low-cost water supply options. Many highly consumptive water needs can be met with rainwater, while treated water can be reserved for drinking, cooking, and personal hygiene. Attempting to displace rainwater could have important social justice implications in terms of household costs and water access, even in cases of households that already have connections. Second, to extend the network, the utility needs the support of alternative water sources. That is, it does not have sufficient capacity to treat enough water to meet all the water use needs of the entire population. Third, there is a strong culture of rainwater reliance in the city and a learned suspicion of the promises of piped water. For these reasons, the integration of rainwater is understood to be necessary if services are to be technically and socially viable.

This vision reflects experiences elsewhere. Rainwater collection systems can provide a range of benefits when integrated with piped water networks. In many cities in the North and the South, efforts are being made to develop rainwater collection systems to relieve piped water networks under conditions of population growth and water scarcity. Although integration is not simple, energy and environmental efficiencies are found to increase over time (Mithraratne and Vale 2007: 7). Studies conducted across 62 cities in Brazil demonstrate that savings on piped water can range from 34% to 92% with the introduction of rainwater harvesting. Savings are especially important in areas with high rainfall and low demand for piped water (Ghisi *et al.* 2006). These savings can reduce system peaks (Vaes and Berlamont 2001), mitigate pressure on water sources, reduce pollution, flooding and pressure on sewer systems, provide a free source for non potable use (Angrill *et al.* 2012), and "could be especially important for system extension" (Ghisi *et al.* 2006: 208).

This is how utility staff in Quibdó view rainwater, as an alternative water source that can help to reduce bills and pressure on the network, enabling its extension.

Nevertheless, a number of challenges exist. In Quibdó, the project of extending the piped water network is dependent on the practices of those that already have household connections as well as the utility and local government. Given the historic unreliability of piped water in the city, the abundant rainfall, and the high instance of poverty (even where water connections exist), for those households with access to piped water, the sociotechnical system that has developed is one based primarily on rainwater collection in coexistence with deficient and unreliable piped water. Local water use customs are tailored to continuous, as opposed to occasional, disrepair. The associated practices, while perfectly rational in the face of a dysfunctional system, inhibit the ability to transition from dysfunction to function. Whereas the users in Trentmann's study easily make minor adjustments during temporary service interruptions, in Quibdó customs of disrepair – like disrepair itself – dominate.

Customs of disrepair relate to issues of quality and reliability. First, in terms of quality, in Quidbó, even where piped water is available, it is generally not trusted. This is explained by the fact that it has not been potable. In a survey commissioned by EPM in 2011, it was found that 20% to 40% of respondents believe that rainwater is of better quality than tap water, even if very few believe that rainwater itself is either clean or reliable (Table 5.2). This is true even of households that are the most likely to be connected to piped water (i.e. those in estrato 3). This indicates that not all residents are convinced that piped water is either necessary or an improvement over rainwater and, given the low expectation of water quality, available tap water is usually stored with rainwater. However, rainwater typically picks up a range of contaminants, including heavy metals and pathogens, from rooftop capitation and storage. A review of studies on harvested rainwater found that it typically does not meet drinking water standards, and is "heavily contaminated microbiologically by a variety of indicator and pathogenic organisms" (Meera and Ahammed 2006: 257).

Second, in terms of reliability, those connected to the system receive water only sporadically. Thus, in households with a connection, the appropriate custom is to fill household tanks with piped water whenever it is available. Not knowing when this might be, taps need to be left open. This also means that taps would not

Table 5.2 Perceptions of rainwater in Quibdó, n=241

Rainwater	*Estrato 1*	*Estrato 2*	*Estrato 3*	*All respondents*
Compared to piped water from the utility, rainwater is…				
Higher quality	39	3	14	22%
More clean	15	43	40	40%
Higher purity	12	27	17	23%
Rainwater itself is…				
Already treated	13	-	1	6%
Reliable	6	-	2	3%
Chemical-free	3	2	8	5%

Source: Public workshop on Aguas del Atrato's activities in Quibdó, June 2011, n=241

necessarily be turned off once tanks were filled. In many households, moreover, there is only a hose; the water cannot be turned off. As such, a significant amount of water can be lost to overflow. According to utility staff, water storage practices mean that a household can draw four to five times its actual monthly consumption through tank filling and overflow.

Water storage enables residents to ensure that there is water available when it is needed, and not simply when it happens to be provided. However, the practice, while adapted for conditions of intermittent service, makes it difficult to provide either continuous or potable water. It has several consequences in terms of service quality, cost recovery, and affordability. First, in terms of service quality, it means that the utility is unable to treat enough water to maintain continuous flow in the network. This stems from both the amount of water being drawn and the fact that everyone draws simultaneously. Yet, as discussed above, continuous flow is needed to retain sufficient pressure without which negative pressures, infiltration, and contamination ensue. A twenty-four-hour system, necessary to ensure potability, is dependent upon both the utility and the users. However, the transition to such a sociotechnical system demands that users shift their practices before the promised service quality can materialise. That is, users need to develop habits consistent with a functioning system before the system actually begins to function. This poses evident problems given one's dependence on water.

Second, in terms of affordability and cost recovery, under current water collection and storage practices, water bills would become unmanageable if service were continuous. In 2011 the utility was able to provide water for ten to eighteen hours a day in certain sectors of the city. However, they estimated that if they went to ten hours a day without preparing the user, a household water bill could jump from 38,000 to 500,000 pesos (US$ 18 to US$ 241). In 2014, the utility finally managed to achieve twenty-four-hour potable supply in one barrio, that of Niño Jesus, which is adjacent to the treatment plant (Loma de Cabí, Figure 5.1), and is currently moving to twenty-four-hour service in the city centre. Beyond affordability, the lack of credibility of the city's institutions – in terms of poor services and billing unconnected households – has important impacts on revenue collection. According to the utility, given past experiences, people are often reluctant to deal with local functionaries, making them more likely to seek a "clandestine" connection from a neighbour. These issues have also been identified in other contexts (Bakker *et al.* 2008; Merrett 2002).

Strategies for transition: from dysfunction to co-existence?

In attempting to understand the transition underway in Quibdó and the difficulty in achieving it, there are limits to LTS and MLT approaches. Beyond the assumption of the replacement of one universal system with another (see above), the approaches have been critiqued for their lack of attention to the roles of users and consumption in transitions (Van Vliet *et al.* 2005: 36). Indeed, according to Geels, the literature posits two basic groups of human actors and associated roles. The first has the ability to affect transitions and includes "system builders" or "heroes", such as scientists, engineers and policy makers. The second includes the "faceless automata following iron rules, institutions or given roles/functions" that they cannot change (Geels 2005b: 12); these are the users. Still, other work in STS underscores the role of users in enabling sociotechnical change, due to the importance of public trust (Ormerod and Scott 2012) or their influence on networks through their use of household mediating technologies (Furlong 2011).

In the case of Quibdó, users and customs are evidently central to achieving the desired form of coexistence between the piped network and rainwater collection. As such, Aguas del Atrato emphasises efforts to improve the level of public trust in the water utility. Community education is an important focus. Programmes include youth activities as well as door-to-door canvassing and a range of workshops. These activities reached over 2000 people a year between 2010 and 2012 (Aguas de Atrato 2011; Aguas de Atrato 2012). The utility also donated libraries to two different schools, in an effort to reach "the next generation of users".[11] Further, they focus on convincing their own workers (97% of whom are local) of the project's importance, such that the utility sends a united message.

In keeping with Shove, however, the results of education initiatives have been beneath the utility's expectations. Correspondingly, in an interview, a community leader stressed the importance of demonstrating project results over public education. Experiences elsewhere in Colombia echo this, finding that reducing the time-lag before concrete improvements in living conditions become apparent is crucial to garnering project credibility (Trejos Gómez *et al.* 2008: 231). This can also help with cost-recovery. Utility representatives underscored that between 2008 and 2011 their collection rate rose from 24% to 44% through improvements to existing infrastructure and management practices. Still, the repeated extension of the expected time to generalised service coverage (see Table 5.1) is a subject of criticism (OPT 2013).

In Quibdó, the limited results of public education encouraged the utility to add other programmes which integrate education, scheduled service provision, small household technologies and community involvement. In terms of scheduled service provision, initially the utility tried to provide water for as many hours a day as possible, up to eighteen hours a day for three months in one barrio. However, finding that they were losing the water to storage and overflow and not recovering costs, they elected to reduce the number of service hours. They began to provide water all day by sectors, each sector getting one to two hours a day of service. This follows Trentmann's assertion that planned disruption, through rolling blackouts or rota water disconnections, can usefully serve "to reduce stress on crisis-prone systems" (Trentmann 2009: 70–71). In Quibdó, the controlled interruption – or rather, sporadic but planned provision – is likewise a strategy to shift customs and norms, but here the goal is to transition from perpetual malfunction and exclusion to reliability, potability and service extension.

In some ways the strategy of maintaining scheduled if intermittent services also addresses Shove's problem of convenience. Although water cannot be obtained at all times, users know that at particular times it will be available. Mediating technologies, which help users to regulate their consumption, can also be useful in improving convenience while reducing demand on the network. For example, the utility installs floaters in household water tanks such that the valve automatically shuts when the tank is full. The floaters are given to households in estrato 1, while households in estratos 2 and 3 are charged a fee. Utility personnel also offer home visits to ensure that the floater is functioning. The installation of floaters, combined with the improvement of household fixtures (including the replacement of hoses with taps) and measures to address leakage, enabled the utility to move from two to twelve hours of service in the Barrio San Martin (also adjacent to the treatment plant) in 2011, affecting 191 users (Aguas de Atrato 2012).

Such measures to enable the reduction of household consumption are also important in insuring that users are given time to adjust to regular, if intermittent, service in order to avoid sudden increases in utility bills. Aguas del Atrato planned to install meters, as required by law,[12] only once the services were continuous and potable. Unable to reduce storage, they decided instead to install micrometers prior to shifting to a twenty-four-hour service. Metering is coupled with measures to help people adjust their consumption by clarifying the eventual costs of metered water. Workshops called "understanding my bill" are held and garnered 834 participants in 2011 (Aguas de Atrato 2012: 2). Utility personnel also go door-to-door prior to meter installation to inform people about ways to minimise their water bills. They use the first bill to inform the household what the cost would have been given their consumption, but charge only a fraction. If the first bill is still too high, communications personnel visit the home again; the bill is cancelled if the user cannot pay. This has similarities with meter installation programmes in other parts of the world (Furlong and Bakker 2011). At the time of the research, 3250 connections were metered. In 2014, the utility expressed plans to install another 6000 micrometers for connected residents in the South and Central Zones of the city (Aguas Nacionales 2014).

The community and its leaders are integral to managing the challenges described above. On a practical level, in areas where physical access is difficult the utility is dependent on the locally elected barrio leadership (the JACs) to organise aspects of the service. The JACs also help to keep the utility informed of needed infrastructural repairs, which helps to avoid additional service interruption. JACs can also bring information about project advancements directly to the community. Building on these links, the utility is planning a Social Contracting (Contratación Social) programme. There, the utility retains the responsibility for service extension, but contracts to the local JACs to build the infrastructure in their barrios. The JACs then hire local residents to do the construction. In Medellín the programme dates back to 1996. There, they find that contracting to the community instead of to a construction company results in a variety of benefits for the newly-built systems, their maintenance, and relationships between community, the utility, and local government (Furlong 2013).

Conclusions

In Quibdó, the sociotechnical *transition* sought by the utility is to a system that fosters coexistence between a functioning piped water network and extensive rainwater collection. Although such coexistence is rarely considered in studies of LTS, it is not unheard of. Moss, for example, finds a "battle of the systems" approach inadequate for understanding sewage systems in Berlin, which are better characterised by fluid and flexible coexistence (Moss 2001). Van Vliet and coauthors, moreover, argue that utility services are more often characterised by "complex combinations of multiple modes" than monopolistic forms of provision (Van Vliet *et al.* 2005: 21).

Still, co-existence, while not anomalous, does not simply emerge in the way it is desired. Like the more often discussed sociotechnical transition based on replacement, achieving sociotechnical coexistence in which "multiple modes" function in mutually supportive ways is complex. For the particular case of rainwater and piped water, actualising a sociotechnical system that results in sufficient and accessible potable and non-potable water can run counter to established sociotechnical relationships. Indeed, while there are technical issues related to the fact that the two types of systems "differ widely in terms of scale, useful life and level of service" (Mithraratne and Vale 2007: 1), it is the sociotechnical barriers that result from long-term disrepair that were of interest here.

In Quibdó, as in many cities in the global South, large numbers of people have lived with disrepair as the norm. They may have no service, sporadic service, and/or poor water quality (McFarlane 2010). Disrepair is the essential character of the network. The Quibdó case evinces that the continuous experience of unreliability can result in a sociotechnical system that perpetuates dysfunction. System unreliability can itself be embedded. Recalling Hughes, momentum is not inherent in a technology but develops through the variety of social institutions and practices that have co-evolved with it (Hughes 1994). Indeed, "[p]eople adapt their lifestyles to artifacts" (Geels 2007b: 124) and infrastructure "shapes and is shaped by the conventions of a community of practice" (Star 1999: 381). These factors can help to produce momentum irrespective of uniformity and universality. Following this reasoning, the actual functionality or dominance of the LTS are not the sine qua none of its momentum; rather, it is the fact that society has adapted itself to it, whatever its configuration.

Overcoming these barriers requires changes in collective cultures of water use in anticipation of changes in service quality. However, such cultures and the disrepair they negotiate are bound up in a variety of historical, political, and institutional factors that, while beyond the scope of this paper, have been treated in detail elsewhere (e.g. Bakker *et al.* 2008; Jaglin and Zérah 2010). Rather than a full analysis of piped water supply in Quibdó, the case was used to open a discussion about sociotechnical coexistence, stability, and transitions beyond the "modern infrastructure ideal". Further, users are not the only actors with roles in creating momentum around dysfunction. In the case of India, for example, research has shown that the dual roles of many actors as local councillors and private water vendors inhibits the improvement of pipe water networks, given their conflicting economic and political interests (Angueletou-Marteau 2010: 155). As such, future research could examine the roles of those who benefit economically from disrepair in producing momentum. Indeed, malfunction often provides fertile ground for the creation of new economies to fill service gaps. Long-term malfunction means that such economies can become entrenched.

This study sought to identify possible openings in the LTS and transitions literatures to integrate a wider variety of contexts and configurations into the study of infrastructure. It has shown the transferability of certain concepts like momentum, while underscoring the need to decouple them from assumptions of uniformity, monopoly, invisibility, and black boxes. At the same time, the study underscored the potential importance of system coexistence instead of replacement for sociotechnical systems in transition. As stated in the introduction, this is of relevance in the North as well as in the South. Infrastructure monopolies – even for water supply – are produced as opposed to "natural" (Budds and McGranahan 2003). Finally, users constitute an important element of sociotechnical transitions. Rather than "automata" comfortably ignorant of their "exoskeleton", they are co-determinants of its momentum and its potential trajectories of change.

Notes

1 See Bakker *et al.* (2008).
2 The indicator for unmet basic needs (Necesidades básicas insatifechas – NBI) is based on levels of overcrowding (more than three people per room), absence of water supply or sanitation, high levels of economic dependency (more than three people per working member of the household) or on a member of the household with up to two years of primary education, and children between the ages of six and twelve that do not attend school.
3 CONPES – *Consejo Nacional de Política Económica y Social* – is Colombia's highest national coordinating authority for political-economic and social planning.
4 All monetary data is converted to American dollars using the Bank of Colombia conversion rates for 2010 (http://www.banrep.gov.co/series-estadisticas/see_ts_trm.htm#cotizaci%C3%B3n – last accessed 6 March 2014).
5 Law 192/1994, Article 34.1.
6 Resolución SSPD No. 40455 of October 24, 2006.
7 Projected values for 2014 differ slightly in Aguas Nacionales (2014).
8 For example, the policy of aerial spraying to destroy coca leaf production is introducing chemical contaminants onto people's roofs and therefore into the rainwater that they collect (Acevedo 2014).
9 First created in 1958, the JACs are locally elected councils at the barrio level that exist across Colombia. These are often tied to broader political and clientelist networks (Leal Buitrago and Davila 1990).
10 Interview, May 2011.
11 Interview, Quibdó, May 2011.
12 Decree 951/1989.

References:

Acevedo, T. (2014) Aguas Lluvias. *El Espectador*. Bogotá.

Acevedo, T., J. A. Castaño and K. Furlong (2014) *Regulación del Agua en el Contexto del Sur: Colombia 1909–2013 – Presentación de la Base de Datos*. Canada Research Chair in Urban Water and Utility Governance, Montreal.

Aguas de Atrato (2011) *Balance Social 2010*. Aguas del Atrato, Quibdó, Colombia.

Aguas de Atrato (2012) *Balance Social 2011*. Aguas del Atrato, Quibdó, Colombia.

Aguas de Atrato (2013) *Indicadores Servicios – Enero-2013*. Aguas del Atrato. *http://www.aguasdelatrato.com/web/files/INDICADORES ENERO 2013a.pdf* (accessed 14 December 2014).

Aguas de Atrato (2014) *Indicadores Servicios – Enero-2014*. Aguas del Atrato. *http://www.aguasdelatrato.com/web/images/enero_indicadores.pdf* (accessed 14 December 2014).

Aguas Nacionales (2014) *Informe de Sostenibilidad 2013*. Grupo – EPM, Medellín.

Angrill, S., R. Farreny, C. M. Gasol, X. Gabarrell, B. Viñolas, A. Josa and J. Rieradevall (2012) Environmental Analysis of Rainwater Harvesting Infrastructures in Diffuse and Compact Urban Models of Mediterranean Climate. *International Journal of Life Cycle Assessment* **17**: 25–42.

Angueletou-Marteau, A. (2010) Les Petits Opérateurs Privés Dans la Chaîne D'approvisionnement d'eau Potable Dans Les Petites et Moyennes Villes Indiennes. *Revue Tiers Monde* **3**(203): 141–158.

Asher, K. (2009) *Black and Green: Afro-Colombians, Development, and Nature in the Pacific Lowlands*. Durham and London: Duke University Press.

Ayoub, G. M. and L. Malaeb (2006) Impact of Intermittent Water Supply on Water Quality in Lebanon. *International Journal of Environment and Pollution* **26**(4): 379–397.

Bakker, K. (2003) Archipelagos and Networks: Urbanization and Water Privatization in the South. *The Geographical Journal* **169**(4): 328–341.

Bakker, K., M. Kooy, N. Endah Shofiani, and E.-J. Martijn (2008) Governance Failure: Rethinking the Institutional Dimensions of Urban Water Supply to Poor Households. *World Development* **36**(10): 1891–1915.

Ballén Suárez, J. A., M. Á. Galarza García, and R. O. Ortiz Mosquera (2006) Sistemas de Aprovechamiento de Agua Lluvia para Vivienda Urbana. *Seminário Iberoamerican sobre Sistemas de Abastecimento Urbano de Água*. João Pessoa, Brasil.

Barnett, C., P. Cloke, N. Clarke and A. Malpass (2005) Consuming Ethics: Articulating the Subjects and Spaces of Ethical Consumption. *Antipode* **37**(1): 23–45.

Budds, J. and G. McGranahan (2003) Are the Debates on Water Privatization Missing the Point? Experiences from Africa, Asia and Latin America. *Environment and Urbanization* **15**(2): 87–113.

Callon, M. (1991) Techno-Economic Networks and Irreversibility. In J. Law (Ed.), *A Sociology of Monsters: Essays on Power, Technology and Domination*. London and New York: Routledge.

Caracol (2014) Chocoanos Construyen Sus Propios Acueductos Ante Falta de Agua Potable. in *Caracol.com.co*. Bogotá.

Coelho, S. T., S. James, N. Sunna, J. Abu Jaish and J. Chatila (2003) Controlling Water Quality in Intermittent Supply Systems. *Water Science and Technology: Water Supply* **3**(1–2): 119–125.

COMPES (2007) *Documento COMPES 3470: Importancia Estrategica del Plan de Inversiones para Garantizar la Prestación de los Servicios de Acueducto, Alcantarillado y Aseo en el Municipio de Quibdó*. Consejo Nacional de Política Económica y Social, República de Colombia y Departamento Nacional de Planeación, Bogotá.

CONtextoganadero (2013) Grave Crisis Sanitaria en Chocó por Falta de Agua. *CONtextoganadero*. Bogotá: FEDEGAN.

Córdoba, D. L. (2014) *Diagnostico Ambiental de los Barrio Paraiso y Caraño de Quibdo-Chocó*. Universidad Tecnológica del Chocó Quibdo, Colombia.

Coutard, O. (1999) The Governance of Large Technical Systems. London: Routledge.

Coutard, O. (2008) Placing Splintering Urbanism: Introduction. *Geoforum* **39**: 1815–1820.

Coutard, O.. (2010) Services Urbains : la fin des Grands Réseaux? In O. Coutard and J.-P. Lévy (Eds.), *Écologies Urbaines, Collections Villes*. Paris: Économica: Anthropos.

DANE (2005) *Censo General 2005*. DANE, Bogotá.

DANE (2010) *Boletín: Censo General 2005 Perfil Quibdó Choco*. DANE, Bogotá.

David, P. A. (1992) Heros, Herds and Hysteresis in Technological History – Thomas Edison and 'the Battle of the Systems' Reconsidered. *Industrial and Corporate Change* **1**(1): 129–180.

Furlong, K. (2011) Small Technologies, Big Change: Rethinking Infrastructure through STS and Geography. *Progress in Human Geography* **34**(4): 460–482.

Furlong, K. (2013) The Dialectics of Equity: Consumer Citizenship and the Extension of Water Supply in Medellín, Colombia. *Annals of the Association of American Geographers* **103**(5): 1176–1192.

Furlong, K. and K. Bakker (2011) Governance and Sustainability at a Municipal Scale: The Challenge of Water Conservation. *Canadian Public Policy – Analyse de Politiques* **37**(2): 219–237.

Gandy, M. (2005) Cyborg Urbanization: Complexity and Monstrosity in the Contemporary City. *International Journal of Urban and Regional Research* **29**(1): 26-+.

Geels, F. W. (2005a) Co-Evolution of Technology and Society: The Transition in Water Supply and Personal Hygiene in the Netherlands (1850–1930) – a Case Study in a Multi-Level Perspective. *Technology in Society* **27**: 363–397.

Geels, F. W. (2005b) *Technological Transitions and System Innovation: A Co-Evolutionary and Socio-Technical Analysis*. Cheltenham, UK: Edward Elgar.

Geels, F. W. (2007a) Feelings of Discontent and the Promise of Middle Range Theory for STS – Examples from Technology Dynamics. *Science Technology & Human Values* **32**(6): 627–651.

Geels, F. W. (2007b) Transformations of Large Technical Systems – A Multilevel Analysis of the Dutch Highway System (1950–2000). *Science Technology & Human Values* **32**(2): 123–149.

Genus, A. and A. M. Coles (2008) Rethinking the Multi-Level Perspective of Technological Transitions. *Research Policy* **37**(9): 1436–1445.

Ghisi, E., A. Montibeller and R. W. Schmidt (2006) Potential for Potable Water Savings by Using Rainwater: An Analysis over 62 Cities in Southern Brazil. *Building and Environment* **41**: 204–210.

Graham, S. and S. Marvin (2001) *Splintering Urbanism: Networked Infrastructures, Technological Mobilities, and the Urban Condition*. New York: Routledge.

Graham, S. and N. Thrift (2007) Out of Order – Understanding Repair and Maintenance. *Theory Culture & Society* **24**(3): 1–25.

Hughes, T. P. (1987) The Evolution of Large Technological Systems. In W. E. Bijker, T. P. Hughes, and T. J. Pinch (Eds.), *The Social Construction of Technological Systems*.

New Directions in the Sociology and History of Technology. Cambridge and London: Cambridge University Press.

Hughes, T. P. (1994) Technological Momentum. In M. R. Smith and L. Marx (Eds.), *Does Technology Drive History? The Dilemma of Technological Determinism*. Cambridge, MA: MIT Press.

Hunter, P. R., D. Zmirou-Navier and P. Hartemann (2009) Estimating the Impact on Health of Poor Reliability of Drinking Water Interventions in Developing Countries. *Science of the Total Environment* **407**: 2621–2624.

Jaglin, S. (2002) The Right to Water Versus Cost Recovery: Participation, Urban Water Supply and the Poor in Sub-Saharan Africa. *Environment and Urbanization* **14**(1): 231–245.

Jaglin, S. (2004) Vingt Ans de Réformes Dans Les Services d'eau Urbains d'Afrique Subsaharienne: Une Géographie de la Diversité. *Cybergeo: European Journal of Geography, revues.org* (last accessed 6 March 2014).

Jaglin, S. (2008) Differentiating Networked Services in Cape Town: Echos of Splintering Urbanism? *Geoforum* **39**: 1897–1906.

Jaglin, S. and M.-H. Zérah (2010) Eau et Ville : Repenser des Services en Mutation. Introduction. *Revue Tiers Monde* **3**(203): 7–22.

Kemp, R., J. Schot and R. Hoogma (1998) Regime Shifts to Sustainability through Processes of Niche Formation. The Approach of Strategic Niche Management. *Technology Analysis and Strategic Management* **10**(2): 175–195.

Leal Buitrago, R. and L. Adres Davila (1990) *Clientelismo: el Sistema Políttico y su Expresión Regional*. Bogotá: Tercer Mundo Editores and Instituto de Estudios Políticos y Relaciones Internacionales.

LeChevallier, M., R. Gullick, M. Karim, M. Friedman and J. Funk (2003) The Potential for Health Risks from Intrusion of Contaminants into the Distribution System from Pressure Transients. *Journal of Water and Health* **1**(1): 3–14.

MAVDT (2007) Convenio Interadministrativo de Apoyo Financiero Suscrito Entre el Ministerio de Ambiente, Vivienda y Desarrolo Territorial, el Municipio de Quibdó y Empresas Públicas de Quibdó ESP en Liquidación. In *Convenio Interadministrativo 124*. Bogotá: Mnisterio de Ambiente, Vivienda y Desarrollo Territorial, República de Colombia.

Mayntz, R. and T. P. Hughes (1988) *The Development of Large Technical Systems*. Boulder, CO: Westview.

McFarlane, C. (2010) Infrastructure, Interruption and Inequality: Urban Life in the Global South. In S. Graham (Ed.), *Disrupted Cities: When Infrastructure Fails*. New York: Routledge.

Meera, V. and M. Mansoor Ahammed (2006) Water Quality of Rooftop Rainwater Harvesting Systems: A Review. *Journal of Water Supply: Research and Technology – Aqua* **55**(4): 257–268.

Merrett, S. (2002) Deconstructing Households' Willingness-to-Pay for Water in Low-Income Countries. *Water Policy* **4**: 157–172.

Mithraratne, N. and R. Vale (2007) Rain Tanks or Reticulated Water Supply? In *Proceedings of Rainwater and Urban Design*. Sydney, Australia.

Moss, T. (2001) Battle of the Systems? Changing Styles of Water Recycling in Berlin. In S. Guy, S. Marvin, and T. Moss (Eds.), *Urban Infrastructure in Transition: Networks, Buildings, Plans*. London: Earthscan.

Municipio de Quibdó (2003) *Plan de Ordenamiento Territorial – Municipio de Quibdó*. Alcadia, Quibdó, Colombia.

OPT (2013) Ocho Años de Intervención y Quibdó Sin Agua. *Quibdó, Colombia*. Quibdó, Colombia: Observatorio pacífico y territorio.

Ormerod, K. J. and C. A. Scott (2012) Drinking Wastewater: Public Trust in Potable Reuse. *Science Technology & Human Values* **00** (online first) (0): 1–23.

Pflieger, G. and S. Matthieussent (2008) Water and Power in Santiago de Chile: Socio-Spatial Segregation through Network Integration. *Geoforum* **39**: 1907–1921.

Poveda, G. and O. J. Mesa (2000) On the Existence of Lloró (the Rainiest Locality on Earth): Enhanced Ocean-Land-Atmosphere Interaction by a Low-Level Jet. *Geophysical Research Letters* **27**(11): 1675–1678.

Robbins, P. (2007) *Lawn People: How Grasses, Weeds, and Chemicals Make Us who We Are*. Philadelphia: Temple University Press.

Robinson, J. (2005) *Ordinary Cities: Between Modernity and Development*. Abingdon, UK and New York: Routledge.

Ruiz Serna, D. (2008) Gente de Agua: Comunidades Negras en el Bajo Atrato (Water People: Afro-Colombian Communities in the Lower Atrato River). *Maguaré* **22**: 339–359.

Shove, E. (2003) *Comfort, Cleanliness + Convenience: The Social Origins of Normality*, Edited by D. Slater. Oxford and New York: Berg.

SSPD (2008) *Acta de Reunión – Vigilancia Itinerante, Quibdó-Choco*. Superintendencia de Servicios Públicos Domiciliarios, República de Colombia, Bogotá.

Star, S. L.(1999) The Ethnography of Infrastructure. *American Behavioural Scientist* **43**(3): 377–391.

Summerton, Jane (1994) Introductory Essay: The Systems Approach to Technological Change. In J. Summerton (Ed.), *Changing Large Technical Systems*. Boulder, CO: Westview Press.

Swyngedouw, E. and M. Kaïka (2002) Urban Water: A Political Ecology Perspective. *Built Environment* **28** (2).

Tarr, J. A. and G. Dupuy (1988) Technology and the Rise of the Networked City in Europe and America. Philadelphia, PA: Temple University Press.

Tokajian, S. and F. Hashwa (2003) Water Quality Problems Associated with Intermittent Water Supply. *Water Science and Technology* **47**(3): 229–234.

Trejos Gómez, C. L., J.-C. Garcia-Zamor and J. M. Castaño Rojas (2008) Domestic Wastewater Management in a Rural Community in Colombia. *Comparative Technology Transfer and Society* **6**(3): 212–235.

Trentmann, F. (2009) Disruption is Normal: Blackouts, Breakdowns and the Elasticity of Everyday Life. In E. Shove, F. Trentmann, and R. Wik (Eds.), *Time, Consumption and Everyday Life: Practice, Materiality and Culture*. Oxford & New York: Berg.

Vaes, G. and J. Berlamont (2001) The Effect of Rainwater Storage Tanks on Design Storms. *Urban Water* **3**(4): 303–307.

Van Vliet, B., H. Chappells and E. Shove (2005) *Infrastructures of Consumption: Environmental Innovation in the Utility Industries*. London: Earthscan.

Vanguardia (2014) Defensoría Advierte Precaria Situación de Niños en Quibdó. *Vanguardia.com*. Bucaramanga, Colombia: Galvis Ramírez y Cía. S.A.

WHO/UNICEF (2010) *Progress on Sanitation and Drinking-Water 2010 Update*. WHO/ UNICEF, Geneva.

6 De-networking the poor

Revanchist urbanism and hydrological apartheid in Mumbai

Stephen Graham, Renu Desai and Colin McFarlane

Beyond the Pale: Water, 'Worlding' Mumbai and the War on the Poor

The *Mumbai Mirror*, Friday, 8th January 2010. A photograph shows a lineup of proud Mumbai police officers standing behind row upon row of what appear at first sight to be rusted machine guns. However, this is not one of the arms caches regularly unearthed to demonstrate the police force's effectiveness against the myriad of terrorist networks that regularly target urban sites in contemporary India. Rather, the objects are water booster pumps, pulled from the ground in Mumbai slums in a new campaign of dawn raids against 'illegal' water theft within the city's burgeoning informal settlements in the city's Shivaji Nagar and Govandi districts (see Figure 6.1).

"Stealing water to earn a few bucks?" the article's headline reads. "Pay a hefty price!" It details how the raids are being backed up by new legal moves to criminalise people 'stealing' water. Hundreds of people, arrested for installing and using the pumps, are to be prosecuted under draconian and non-bailable laws such as the Damages to Public Property Act. All this is portrayed unproblematically as a heroic response to the threat that water theft in slums poses to the wider, formal, legitimate and law-abiding city. "Pilferages, if not controlled," writes the author Abhijit Sathe (2010), "could exhaust the potable water reserves before the next monsoon."

Such statements tap into a mainstream discourse where recent monsoons have led to a major "water crisis" in Mumbai which necessitates radical, emergency measures to address widespread "water theft" or "water pilferage" – especially by the urban poor. What such discourses occlude, however, are the ways in which current systems of urban water provision work to systematically dehydrate and profit from urban slum communities while water wastage by the affluent and their preferred urban facilities goes unchecked.

On a rapidly urbanising and increasingly thirsty planet, contemporary Mumbai is one possible harbinger of a stark global future. Already the majority of humans are urbanites; some estimate that 75% will be by 2050. The inevitable result is that water, like everything else, is being urbanised: larger and larger swathes of the earth's 'natural' hydrological systems are being organised to sustain burgeoning metropolitan areas (see Swyngedouw 2004; Kaika 2005). This transformation challenges the easy separation of nature and the city that is a legacy of modernist thinking, because the earth's water is increasingly metabolised through the pipes, channels and sewers of urban areas (and, of course, the bodies of the people that live in them).

With urban residents largely unable to provide for their own water needs, access to technological systems providing water of adequate quality and quantity is now arguably the most basic element of the right to a decent urban life. This is especially so for the world's billion or more slum dwellers, who are usually denied access to formal water supplies because their claims to space are deemed illegitimate or illegal. For such communities, the challenge of even very basic hydration, sanitation or washing often involves the negotiation of complex circuits of predation, corruption and patronage, arrangements which seek to fully exploit both the nature of urban water as the ultimate, inelastic, life-giving commodity, and the distance of such communities from adequate formal water infrastructures.

Crucially, slums and slum dwellers are demonised and criminalised in many global South megacities. Indeed, often they are often portrayed as being not actually 'of the city' at all. Another *Mumbai Mirror* article, for example, published three days later than the previous one (Singh 2010), recounts the systematic destruction of another network of 'illegal' pipes improvised by slum dwellers in the Dahisar district, twelve miles to the northwest of Shivaji Nagar and Govandi. Here a punitive act of state violence against one of the poorest and most vulnerable communities in Mumbai, undertaken after protracted lobbying by nearby affluent communities, is sanitised as a heroic act of water conservation, 'saving water' for 'the city'.

In this case, 'the city' signifies, crucially, not the whole of Mumbai but only the formal city of apartment blocks, malls, corporate towers, technology centres and leisure parks organised to meet the needs of elites, the middle classes, foreign investors, and tourists. All else – the majority of the population living in self-built slums on a mere twentieth of the city's land – is here discursively cast out from the modern rights, entitlements and promises of city life. The urban poor are rendered instead as a parasitic threat requiring an increasingly violent response and draconian control. Given the most essential nature of urban water, Georgio Agamben's (1998) notion of 'bare life' – human existence reduced to a biological process, that can be extinguished through sovereign power with impunity – has rarely seemed so apposite.

In the essay that follows, we draw on an in-depth ethnographic field study into the water and sanitation practices within informal settlements in Mumbai (see note on methods) to examine further the complex politics of water in contemporary Mumbai.[1] We demonstrate in particular how the contemporary situation in Mumbai reveals the costs of marginalising the majority urban poor in global megacities: in public health, death rates and communicable and water-borne diseases; in the burden of waiting for and carrying water (especially for women and children); in the burden of incurring high expenditure on water; in water extortion against the poor by predatory water rackets; and in the problems of systematic dehydration. Beyond this, we also show how these costs are often obfuscated in dominant discourses of the urban water crisis emanating from governing elites, security and police forces, and mainstream media. In Mumbai, as we shall see, these discourses often tend to demonise informal settlements and their residents as the ultimate *cause* of the wider water crisis afflicting the city. They portray informal settlements and their inhabitants as spaces that must be reclaimed and reconstructed in the production of Mumbai as a new 'global city' to rival Shanghai.

For example, the influential and controversial 2003 report by McKinsey and Company and the organisation Bombay First, entitled *Vision Mumbai: Transforming Mumbai into a World-Class City*, which was subsequently taken up by the Maharashtra state government, argues that Mumbai needs to transform its infrastructure and governance in order to become a 'world class city' (McKinsey/ Bombay First 2003). It draws on examples from Shanghai, Singapore, New York, and London, and emphasises high-impact projects with public-private partnerships, largely ignoring informal settlements and sanitation. When slums do feature in the report, they are portrayed as sites of proliferation that mark Mumbai's 'slippage' down the rankings of 'top' Asian cities, and as spaces that must be removed because they do not fit with the image of the world-class city. The report argues that 'the percentage of the population living in slums must fall to 10–20 percent' (2003: 20), and says very little about the causes of slum formation beyond housing prices being too high. It does suggest building more low-income housing, creating 'Special Housing Zones' for 300,000 people, reforming existing schemes such as the controversial market-led Slum Rehabilitation Authority (SRA) scheme by increasing contributions from slum residents to the cost of construction, ensuring that housing prices should be no more than three to four times the annual household income, and insisting that no post-1995 slums should be allowed (i.e. they should be demolished). Not only are these minor interventions that are nowhere near significant enough, they also insist in slums as spaces outside the project of 'worlding' Asian cities to meet global aspirations of market competitiveness and aesthetics (Roy and Ong 2011). The political agendas around the promotion of a new Shanghai continue to reverberate in the city, not just from elements of politico-corporate Mumbai but from public NGOs like Citispace and mainstream journalists. For example, one *Times of India* journalist celebrated Shanghai over Mumbai for its expensive hotels and shopping malls that beat 'even Singapore' for the 'new overpasses, metro rail links, a dream subway system', and for the fact that the 'unruliness visible on Indian streets is almost absent' (Joseph 2005).

Finally, dominant discourses surrounding Mumbai's contemporary water crisis tend to simplistically equate the city's poor citizens with the interests of a powerful and shadowy 'water mafia' in Mumbai. This term, which is very vaguely and loosely deployed by elites and mainstream media outlets, refers to a set of loose alliances that exploit water scarcity among Mumbai's population. It involves corrupt municipal officials, police officials, and middlemen (some of whom are better-off residents of informal settlements). This complex set of arrangements, practices, and rackets works to profit from Mumbai's water shortages by organising, for high levels of kickbacks and payment, both legal and illegal water connections and private water tanker deliveries.

By failing to reveal how such corrupt water practices impact negatively on Mumbai's poor, the 'water mafia' discourse further compounds the vulnerabilities of informal city dwellers to them, especially in the context of increasingly extreme water scarcity. In the process, the exploitative power relations linking the police, Brihanmumbai Municipal Corporation (the City's municipal Government), politicians, private water tanker companies, influential middlemen, and residents of informal settlements are rendered all but invisible. Our analytic strategy here is thus to deliberately juxtapose elite, discursive invocations of 'water crises' in Mumbai with ethnographic research exposing some of the lived materialities surrounding access to water among Mumbai's most marginalised inhabitants. We adopt such an approach in order to illustrate some of the jarring collisions that characterise both the research process and the nature of hydropolitics in Mumbai.

Parched City

In 2008, some 19.5 million people were squeezed into the 438 square kilometres of Mumbai's tiny island-peninsular and its immediate hinterland (Figure 6.1). The city was thus the fifth biggest in the world by population, having quadrupled in size in a mere four decades.

Dominating this growth overwhelmingly has been Mumbai's burgeoning population of people living in informal settlements. Indeed, one estimate suggests that while in 1861 only 12% of Mumbai's population could be classified as 'poor,' by 1991 this figure had risen to 51% (Kumar et al. 2003). (Such definitions are contested, however, and are complicated by the fact that Mumbai's extremely high real estate prices force many lower middle class people to live in informal settlements.)

Mumbai has a fragmented and polarised metropolitan landscape (see Gandy 2008; Appadurai 2000). The World Bank (2009) estimates that 54% of Mumbai's population – around 9 million people – are now concentrated into the city's archipelago of dense informal settlements or *Zopadpatti*, forced to occupy just the 5% of the city's land (see Figure 6.1). Such places have among the highest population densities on Earth: the biggest and most famous – Dharavi – squeezes over a million people into just one square mile. Mumbai's slum-dwellers are the city's most vulnerable population: they are the prime victims of environmental pollution, a whole spectrum of water-borne and other diseases, and suffer inadequate housing conditions, infrastructure services, and waste disposal routes.

Within thirty years, moreover, most projections suggest that Mumbai's extraordinary rate of continued growth will lead it to become the world's second largest city (after Tokyo), with over twenty-six million people. The majority of this new population is expected to squeeze into informal cities.

Mumbai's polarised social landscape is reflected in the politics of water provision in a city which, despite having copious monsoon rains, has never been able to provide decent, piped water for all. Instead, Mumbai's water politics are

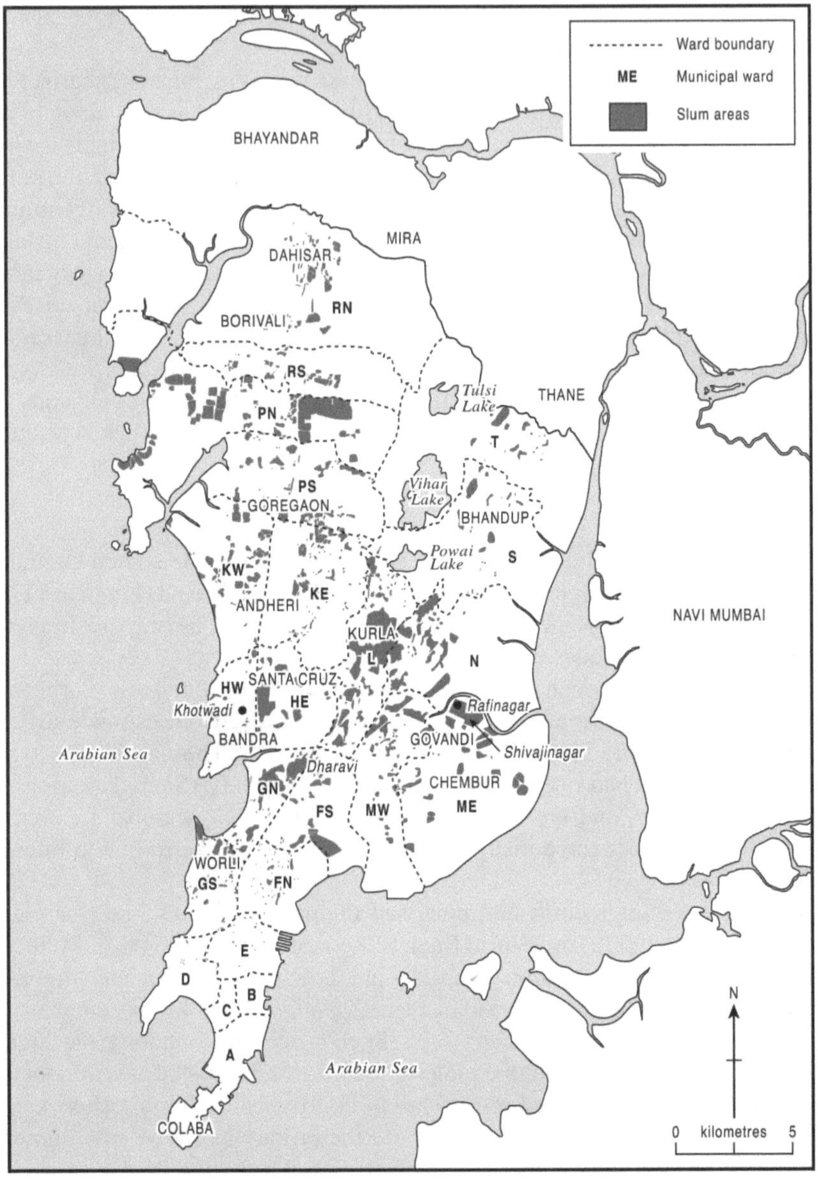

Figure 6.1 The city of Mumbai, 2011, showing municipal wards and the main *Zopadpatti* or 'slum' areas mentioned in the text

Source: authors.

labyrinthine. A complex colonial and planning legacy of half-built improvements laces a shaky and partial piped system. Based on a crumbling, century-old network of fragile pipes, Mumbai's water system is already at crisis point. Ongoing efforts to build new reservoirs north of the city have faced delays and funding shortages and have so far failed to increase supply to match Mumbai's burgeoning population. Such a situation has been compounded by a series of failed monsoons as well by continuously high levels of water leakage and regular pipe ruptures. All this means that while Mumbai, with a notional per capita water availability of 180 litres, is theoretically endowed with more water per person than London, which has 150 litres, it fails even to offer its most affluent citizens twenty-four-hour piped water services (Bjorkman n.d., p.1).

Mumbai currently receives about 2,500 million litres of water per day from the jungles, lakes, and mountains in the north of the Maharashtra state within which Mumbai sits. This has been estimated to represent only around 65% of demand (Jadhav 2005). In 2009 this deficit led the Brihanmumbai Municipal Corporation (BMC), the city's municipal government, to limit the hours of water supply to all users in the city.

Mumbai's water crisis means that even many wealthy residences are now supplied with piped water for only two or three hours a day. Around 20% of all the city's piped water escapes into the ground through leaks. Another 20% or so is tapped illegally by water tanker providers, the construction industry, or informal and formal settlements. Periodic ruptures in the mains supply add to the sense of a pivotal urban resources being wasted before reaching its rightful users. Complete bans and rationed quotas are regularly introduced during relatively dry monsoon seasons.

The presence of the burgeoning *Zopadpatti* – informal settlements -- dominates Mumbai's water politics. Many such places have for generations been denied adequate linkages to formal, piped water supplies, and are thus the most vulnerable to the current water crisis. The BMC offers inhabitants of informal settlements water connections with a daily notional amount of 45 litres per person per day, 30% of what is allocated to a resident living in formal residential areas (135 litres per person per day). However, even this amount is often unavailable. Informal settlements rely on unmetered municipal standposts, metered municipal group connections, wells, boreholes, tanks, tankers, filtration systems and improvised pipes, as well as many other efforts to seek water of adequate quality and quantity. Meanwhile, a complex world of organised crime, which overlaps with the world of municipal officials, works to build high profits from illegal piped supplies and water tanker deliveries, focusing especially on the poorest slums. Their practices also delay efforts to extend adequate formal piped water supplies to slum communities.

Recent proposals to plan and improve Mumbai haven't even attempted to universalise decent water for the city. Instead they tend to disguise or camouflage the problem. Strategies like *Vision Mumbai*, commissioned by the elite citizens' group Bombay First from McKinsey management consultants, seek instead to superficially 'clean up' or 're-brand' Mumbai, improving quality of life and services for affluent and middle-class citizens yet doing little or nothing for the impoverished communities that are in such dire need. Planners and local government officials hope that such changes will help Mumbai reproduce the success of 'global' cities such as Shanghai or Singapore. The result is an increasingly authoritarian stance – including mass demolitions (particularly intensive in the winter of 2004/5), declarations that newer slums will not receive any water service, and efforts to remove the poor from public spaces – against Mumbai's slum dwellers.

The growth of revanchist urban politics in Mumbai relates closely to the broader construction of a xenophobic, nativist, Marathi-centred, regional political party – the Shiv Sena – which has dominated both municipal and State government in recent years (see Appadurai 2000). As Matthew Gandy points out (2008, p.122), the embourgeoisement of public policy has also been fed by an increasingly moralistic discourse amongst Mumbai's media and public institutions. This tends to demonise the poor migrants from elsewhere in India as imposters within Mumbai and to legitimise authoritarian and punitive policies against them. It therefore tends to weaken already feeble concepts of civil society, citizenship and human rights within the City. The Shiv Sena has long argued that immigrants in informal settlements are the main reason for Maharashtrians being unemployed, and has consistently invoked the figures of the migrant and the Muslim as a source of crime and social disorder. The Sena's grip over the municipal corporation since the mid-1990s has entrenched these politics, and closely linked them to the election of councillors across the city. As a political movement, the municipal success of the Sena has been part of a wider marginalisation and disproportionate targeting of Muslim groups in the city, evidenced not just in the pogroms of the early 1990s (which helped the Sena's rise to power in the city and state) but the ongoing demolition or infrastructure denial and removal to Muslim informal settlements, for example in Rafinagar (a neighbourhood we discuss in detail later) and surrounding areas in northeast Mumbai (on the pogroms, see Hansen 2001; Appadurai 2000; Varma 2004).

In the context of water, this politics of denial, violence and disconnection have also been made possible because of perceptions that the municipal water shortages faced by many of Mumbai's middle-class residents are caused by slum dwellers. Moreover, the water-borne diseases that increasingly afflict slum dwellers – Surat plague, malaria – tend not to directly endanger middle-class residents who can access a range of clean water alternatives far more easily. As a result, rather than the emergence of a broad-based campaign for citizenship and human rights for all urban residents in the context of water, there has been a further fragmentation of civil society in Mumbai. Not only are there competing claims on the city's water resources, but the embourgeoisement of public policy has supported the view that affluent groups have more legitimate claims on the city's water than the urban poor (see Baviskar, 2002 on bourgeois environmentalism). As with the example of the New Link Road Residents' Forum detailed below, organisations of middle-class residents in Mumbai increasingly mobilise to protect and further their interests at the expense of the urban poor, while organisations working for the rights of the poor find it ever more challenging to make themselves heard and realise systemic change in the city's water policies.

Mumbai's water crisis means that water protests increasingly convulse the city; water scarcity dominates political mobilisation and debate; new legal sanctions are mobilised against transgressions to cope with the new sense of emergency. In addition, water engineers are regularly attacked and frequently complain that they get no protection from the police. Political parties often mobilise their loyal mobs on water demonstration against water engineering departments and officials, and local groups have beaten up neighbours accused of locking collective taps, parading them around streets with their hair shorn in humiliation. 'The city's water crisis,' warned the *Mumbai Mirror* in December 2009 at the height of a particularly acute shortage, 'seems to be turning Mumbaikars into monsters' (Sadhwani 2010).

This politics of water scarcity and the systematic denial of water to the poor is quickly translating into a militarised struggle to control and protect Mumbai's water supplies. Thus the status quo of effective hydrological apartheid, forged through lesser water entitlements for residents of informal settlements as well as predation by them, is sustained rather than undermined or overturned.

This is being done through a campaign to blame and criminalise *Zopadpatti* dwellers, and their efforts to gain even minimal access to water, for the city's water crisis as a justification to 'take back' or 'reclaim' the city's water for elite and middle class 'tax payers' and real estate interests. Such a campaign works to obscure the origins of the current water crisis and deny the stratification of supply which starves the majority of the urban poor of water.

In making this argument, we are not suggesting that this eviscerated hydropolitics is itself new. Mumbai's water and sanitation infrastructures have always been deeply fragmented and politicised. While colonial investments made clear their biopolitical choices about which neighbourhoods were metabolised through water and sanitation and which were not, nationalist discourses emphasised an egalitarian focus that envisaged a modern urbanite in need of civic transformation (see McFarlane 2008a; Dossal 1991; Gandy 2008). The Nehruvian view of the city as an important site for the expression and negotiation of modernity was in practice always heavily prescribed both by the inequities of India's capitalist urbanisation that consistently drove surpluses away from infrastructure investment for the poor, and the piecemeal will of India's political elites. The city has never, for example, come close to providing full and adequate sanitation coverage. The colonial legacies of disinvestment in the extension of sewer coverage to poorer areas structure the possibilities of contemporary sanitation intervention, where public toilets are provided to informal settlements off the sewer grid and at nowhere near sufficient number for most neighbourhoods.

A 2001 survey, for example, found that 63% of those living in informal settlements depended on often defunct toilet blocks, and a ration of people to toilet seats of 273:1 in A ward to 56:1 in F/S and S wards, often resulting in queues lasting two hours or more and an untold range of minor and serious health complaints (see MW-YUVA 2001; McFarlane 2008b). There is a complex story to be told around the changing logics and imaginations which have shaped this deeply unequal metabolic geography, which exceed the scope of this paper, but it is important to be clear that the present moment represents a particular intensification of unequal water politics for certain neighbourhoods (often 'unauthorised', often predominantly Muslim) over others.

What is new, however, is the intensification of a bourgeois urbanism that is both revanchist in its aims – seeking to seize urban space from the poor, often to realise property prices – and splintering in its form, as elites increasingly shift to infrastructure-rich gated tower blocks and as political imaginations all but abandon more progressive urban projects in favour of investments for the wealthy.

Indian cities increasingly reflect the logic of what Amita Baviskar calls 'bourgeois environmentalism' – disparate efforts to remove informal settlements, street hawkers, and (often Muslim) immigrants from spaces across the Indian city, which depend partly on notions of urban contamination and beautification, and the often violent removal of the poor (Baviskar 2002; see also Chatterjee 2004). As Baviskar argues in relation to Delhi, a lack of public toilets means that any open space with sufficient shelter becomes a potential place to defecate. She uses the example of the public park. To the expanding middle classes the park embodies a sense of 'gracious urban living', a place devoted to leisure and recreation; to the poor it may be the only environment where defecation with relative safety and dignity is possible. Baviskar uses this conflict to point to the increasingly powerful presence of bourgeois environmentalism as an ideology shaping urban landscapes in India, a notion that combines political, economic, social and ecological dimensions. Bourgeois environmentalism discriminates between 'good' and 'bad' natures, such as between the park and the 'unsanitary slum', and privileges commodified socionatures, and if it is a discourse with a long history, it is nonetheless intensifying. It is clear that there is now a growing impetus to 'cleanse' the streets of India's major cities, whether through violence or through regulation, and to focus infrastructure resources on high-end residential and industrial secessionary network enclaves that disproportionately benefit the wealthy, including roads, fly-overs, air-conditioned malls, and telecommunications and media infrastructures.[2]

Water Wars: De-Networking the Poor

As with the *Mumbai Mirror* articles detailed above, the official discourse surrounding Mumbai's water crisis offers a simple tale of a malign and burgeoning informal population threatening the order, security and public health of the formal city of legitimate, tax-paying citizens through their criminal theft of water. One report by the Mumbai High Court, published in July 2009, notes that the number of encroachments on the mains pipelines in Mumbai has grown from 6,687 in 1995 to 15,780 in 2009 (Vyas 2009). It sees these 'encroachments' as a 'health and security hazard' which can 'puncture the entire system' – what they term 'the real lifeline of the city' (Vyas 2009).

A powerful revanchist logic is at play here: if only 'we' – i.e. middle class and elite consumers and corporate investors – could reclaim the City's hydrological – and urban – commons from the mass of illegitimate slum dwellers, Mumbai might attain civil order, a high quality of life for elites and the middle classes, and its aspiration, solidified in 2003 with the 'Bombay First' and 'Vision Mumbai' strategies for the city, to emerge as a truly 'world class' or 'global city' to compete with the likes of Singapore and Shanghai. Bombay First, a powerful corporate coalition which commissioned the influential *Vision Mumbai* report from

McKinsey and Company in 2003 (McKinsey/Bombay First 2003), have since lobbied hard for the wholesale privatisation of Mumbai's water industry, part of what Matthew Gandy (2008 p.125) has called its 'Neo-Haussmannite' agenda of forcibly reengineering Mumbai through slum clearances into its vision of a 'global' city.

These polices powerfully reinforce the message that slum-dwellers are not actually of Mumbai; rather, their presence acts as a pathology and an obstacle to the legitimate aspirations of Mumbai's elites. Thus, no political effort is made to address the root causes. Crucially, for example, no adequate alternative, public or formal water supplies are made available for the communities whose water supplies are destroyed in the raids. Thus, acute water shortages and extreme reliance of already vulnerable communities on water rackets are further accentuated.

Powerful residents' groups representing affluent areas have been especially vocal about the need to destroy the 'illegal' taps and end the complicity of local municipal and water officials in the so-called 'water mafia' operations that construct them at huge profit. The New Link Road Residents' Forum, for example, which represents affluent communities in an area of Mumbai's north western suburbs (Figure 6.1), mentions these efforts to provide basic hydration in the same breath as what they term 'terror threats to Mumbai's water network' (Desai 2010a). Criticising a huge slum by the name of Ganpat Patil Nagar which has been built recently on the New Link Road, the group has filed many legal complaints about thefts of the mains water that previously flowed unencumbered to their communities. The group has been 'consistently filing complaints of alleged water theft by the slumlords and unsocial elements who indulge in puncturing the MCGM [Municipal Corporation of Greater Mumbai] lines laid under the link road'. The result of the 'water theft', they argue, 'is being faced by the honest tax paying citizen of Borivali and Dahisar who often face water cuts and shortages and inconsistent supplies'.

Fuelled by such allegations of ineffectiveness, corruption or complicity, the BMC have recently joined forces with the police to launch draconian raids against slum settlements during which improvised water pumps, tapping into the city's formal water system, have been destroyed (the results have been the startling press conferences discussed at the start of this essay) and supposedly illegal pipes have been disconnected (Figure 6.2).[3] Certain slum settlements, portrayed as especially important centres of 'water theft', have been the particular focus of these activities. Raids in January 2010 against the Dahisar slums identified by the New Link Road Residents, for example, were portrayed heroically in the *Mumbai Mirror* as being part of a 'save water' drive which successfully saved the city 400,000 litres of water (Singh 2010). Harish Pandey, secretary of the New Link Road Residents' Forum, was 'thankful that the BMC has finally taken some action against this water theft'. He argued, however, that .the civic body should look at a permanent solution as slums end up getting water in full pressure, while taps at the taxpayers' home run dry'.

The most notorious slum for municipal raids, however, has been the predominantly Muslim settlement of Shivaji Nagar in Govandi in Mumbai's

Figure 6.2 De-networking the poor: water pipes cut in Rafinagar during the water raids in December 2009.

(Photograph by Renu Desai)

M-East ward, the poorest of Mumbai's twenty-four municipal wards (and a ward which has gradually become a concentration of poverty as people have been resettled here over thirty years of continuous slum clearance in central Mumbai). Here the BMC deems the destruction to be so 'successful' as to consider it a model to be followed around Mumbai's other informal settlements. After one raid in Shivaji Nagar in early December 2009, Senior Police Inspector Dundapa Jodgujari reported that his forces raided and destroyed 156 illegal water pipes which had been connected to the mains illegally. The police also seized 96 booster pumps. In the process they 'detained around 40 people. Of these, 21 were arrested, as they did not have valid documents to use the pumps. Many of those arrested sold water to the slum dwellers' (Dalvi 2009). Moreover, Jodgujari argued that local residents were relieved that the 'menace' of the water mafia was no longer hanging over their community. However, this argument masked the fact that because of the denial of adequate formal water connections in the area by the BMC and the high costs of acquiring even a legal water connection from the BMC, many residents were in fact forced to rely on the corrupt practices of water rackets to obtain water for their daily needs. Thus, while many local residents of Shivaji Nagar got better water pressure in their taps after the municipal raids, many others spiraled into a deepening water crisis since the BMC cracked down on their only source of water while not providing them with alternate adequate water supplies. And yet, the Shivaji Nagar experience was deemed to be so successful by Rahul Shewale, standing committee chairman at BMC, that they began to consider implementing similar programmes of continuous raids across the whole of Mumbai (Desai 2010b).

Our research in Rafinagar, a predominantly Muslim informal settlement located at the edge of Shivaji Nagar (see Figure 6.1), suggests that water raids by the BMC have been a regular occurrence in some informal settlements. Almost every year the BMC has cut some 'illegal' water pipes supplying water to Rafinagar's residents. However, this time around these raids were not only carried out at a larger scale, with more intensity, but also for the first time they involved the police. The water wars in Mumbai have thus evolved from routinised practices of a BMC crackdown on 'illegal' connections to militarised operations. Initially, basic criminal law surrounding damage to public property or inhibiting public services was used to prosecute those deemed to be guilty of setting up illegal water connections. The discourse of 'water emergency', however, soon allowed political leaders to invoke the more draconian Maharashtra Control of Organised Crime Act (MCOCA) (Suryawanshi 2010) against offenders. Not only were arrests made under these laws, but the militarised crackdown also translated into police harassment on the streets of ordinary residents trying to obtain water from surrounding areas by foot or cycle. There is a sense among NGOs working in Shivaji Nagar, where the population is 70% Muslim and is increasingly comprised of north-Indian migrants (both Hindu and Muslim), that the water raids in the area reflect the religious and ethnic biases of Mumbai's municipal officials and the police.[4] The vehement anti-Muslim and anti-north-Indian politics of the Shiv Sena, which have penetrated the work cultures of the municipal government and the police, thus also shape how the broader revanchist politics against Mumbai's poor unfold on the ground.

Beyond the use of draconian laws, further efforts to securitise Mumbai's water system are emerging. The Brihanmumbai Municipal Corporation has drawn up plans to physically demolish and erase informal settlements located along the city's water mains (Suryawanshi 2009). Even if residents are relocated under the city's controversial slum demolition and resettlement programme – which was launched by the Hindu nativist Shiv Sena party when it was in power in the State of Maharashtra in the mid-1990s and which continues to be implemented under the Congress Party which is currently in power – it is likely that many will be excluded, since only residents able to prove pre-1995 residency in the city are eligible for resettlement under the programme.

Ambitious, although still unimplemented, plans were also announced in late 2009 to place protective walls around every major exposed water main in the city, supported by a complex network of closed circuit television cameras and security guards. One BMC official revealed: "We will float a tender for construction of a protective wall. At some places where this is difficult, we will put up barbed wire. We have even planned to deploy security guards to keep a vigil and protect the pipelines, as we have received terrorist threats stating that the pipelines could be blown up. Besides, there is the peril of someone poisoning the city's water supply through the pipelines. We will take all possible measures in due time to prevent a disaster" (Suryawanshi 2009). The CCTV system, costing Rs. 2.3 million (around $50,000 US), was put out to tender in April 2010 (Vyas 2010).

Hydrological Apartheid

Crucially, at no point does Mumbai's water revanchism allow for the provision of adequate alternative sources of formal, public or 'legal' water supplies to the majority of the city's population who live in slums. What such discourses and actions mask is the inevitability of mass water theft by *Zopadpatti* dwellers in Mumbai, in a context where political elites have long worked to use the systematic dehydration of such places as a deliberate strategy to discourage the formation, or force the abandonment, of slums. Since 1996, the Maharashtra State Government – first under the Hindu nativist Shiv Sena and then under the Congress Party – has deliberately linked the right to water to the geography of tenancy in Mumbai. Slums deemed to have been formed after January 1st 1995 are considered unrecognised and have consequently been totally denied rights to access formal water mains.

Official, recognised slum-dwellers are entitled to water connections on production of a photo-pass, a name on the voters' list or a ration card with the requisite pre-1995 date. However, these comprise of metered group connections, shared between five to fifteen households, which are supposed to provide 45 litres per person per day (as opposed to 135 litres per person per day in residences in the formal city). Moreover, these connections don't always materialise, and when they do they don't always provide adequate water. Meanwhile, the 1.2 million or more people living in post-1995 settlements (Tatke 2010) remain unrecognised and consequently live under the permanent threat of the bulldozer, bereft of legal entitlements to water and other basic services.

Of course, this ruling has done nothing to inhibit the continued growth and formation of slums, as India's population booms and ever-more migrants move to Mumbai from impoverished areas and violent rural states in search of a better life. Indeed, in the absence of affordable housing in the formal city, many slums are brought into being through corrupt public officials working in liaison with elements of organised crime, the 'slumlords', and the 'vote-bank' politics whereby politicians seek to gain the votes of the city's majority-slum dwellers in exchange for electoral promises.

In such a context, the demands among slum-dwellers for even minimal hydration have provided a motivation for the growth of the water racketeering carried on by the so-called water mafia. Both pipeline connections and tanker supplies are accelerated and made possible through large payments organised through middlemen, in liaison with municipal officials, police officers and politicians. In Rafinagar, our research found that obtaining a legal water connection from the BMC usually involves having to resort to a middleman, and residents therefore have to pay between Rs. 20,000 and Rs. 60,000 (and often more) to organise a legal water connection from the public mains. Thus, not all residents can afford to get a legal connection and are forced to remain dependent on those who can incur such costs. The middleman, who is sometimes a plumber, organises the installation of the water pipe, giving a cut to municipal officials and even the police and local politicians. Even after this, there is no guarantee of how long water will flow through the pipe. Nonetheless, having incurred these high expenses, these "tap-owner" residents then sell water at high prices to others in their neighbourhood in order to recover their investment, pay the BMC's water bill, and make some cash for themselves. Our research in Rafinagar revealed that many residents paid between Rs. 5 and Rs. 20 for a 35-litre jerry-can of water. This is between 30 and 200 times more than the official municipal water tariff for slums. Others paid for water based on the number of minutes during which they drew water from the tap-owner's connection.

Along with such formal connections, there are also "illegal" connections organised by middlemen for some residents, who similarly sell water at prices that rise and fall with the seasons and the extremities of the water crises. Geographies of legal and illegal water provision thus become extremely blurred and materially entangled During the raids in Rafinagar, for example, many legal connections were also cut by the BMC. Meanwhile, the raids created a deeper water crisis for those who had been forced to depend on these legal/illegal connections for water, and they increasingly had to pay more money to obtain water from inside Rafinagar. An increasing number of residents also had to go to neighbouring areas on foot or bicycle to obtain cheaper water.

Mumbai's predatory water rackets have also grown through the involvement of municipal officials. An undercover investigation by the *Mumbai Mirror* in April 2010, for example, showed how Rs. 6,500 that was paid to hydraulic department and municipal-ward officials, ostensibly to film a "rain scene" in a Bollywood movie, was able to provide instant water tanker deliveries, filled illegally from the public water system and delivered by one of the 10,000 or so private water tankers in Mumbai. (Obtained through official BMC means, such tankers take at least a week to arrive, cost only Rs. 600, and come via the BMC's over-stretched fleet of only 24 water tankers; see Raina 2009.) The newspaper's investigations revealed a classic racket which forced slum dwellers – and, indeed, more affluent communities who often resort to private tankers as well – to pay extremely high prices to the illicit tanker companies for water; these prices rise and fall with the levels of supply and the depth of the city's water crisis. Such arrangements are organised by geographic neighbourhood, with kickbacks distributed to key political, municipal and hydraulic department officials to keep everyone happy.

In such a context it is no surprise that city officials attempting to crack down on the water rackets surrounding pipeline or tanker provision have struggled to cope with the complex vested interests involved. By destroying pipes built illegally in slum settlements, BMC official were, of course, further adding to levels of demand for highly profitable tanker supplies.

On a broader scale, and to compound the geographies of hydrological apartheid across Mumbai, the BMC has presided over the construction of a whole archipelago of massive water-hungry architectural edifices. These are geared towards Mumbai's powerful middle class and social and political elites, and are endlessly invoked as evidence that Mumbai is 'going global'. Most obvious here has been the complicity of the BMC in licensing the proliferation of private swimming pools within Mumbai's burgeoning range of upscale gated communities and apartment complexesSingh 2010). However, large-scale water theme parks are also being built across the city's suburbs. As ordinary Mumbaikers reeled under the 10% supply cuts during the 2003 water crisis, Charubala Annuncio (2003) observed that, at the same time, the periphery of Mumbai was being ringed by complexes of water parks, theme parks, leisure parks, elite high-rise complexes, clubs, and bungalows with swimming pools rented out for Bollywood film shootings. "Just the two dozen existing water parks in Mumbai and adjoining areas like Thane and Raigad use over 50 billion litres of water every day. Of this, Esselworld's Water Kingdom, which is spread over 24 acres, gets over 7 billion litres." Indeed, the water used by these water parks is twice the water supply to the whole of Mumbai.

India's burgeoning bottled water industry only adds to the city's deepening hydrological apartheid. As elsewhere, in Mumbai those who can afford it increasingly buy bottled water to try and insulate themselves from the perceived health risks of piped municipal water, which has often been found to be contaminated. Such strategies reflect a broader tendency among more affluent Mumbaikers to attempt to secede from the public city through complexes involving gated communities, new raised highways, private malls and the 'capsular' mobilities afforded by private cars.

The Indian market for bottled water, valued at Rs. 10 billion in 2010, is growing at an astonishing 40% per annum (gits4u.com 2010). The industry, which tripled in size between 1999 and 2004, is one of India's fastest growing. Over two hundred brands jostle for the market, their utopian advertising looming large above the city's streets (see Figure 6.3). For those who can buy-in, the latest "premium" bottled brands offer the fantasy of a pure, "natural" escapism amid the dense urbanisation of Mumbai. "I was born in the Shivalik range of the Himalayas," proclaims the label on the Tata corporation's "Himalayan" bottled water. "In a place most of you visit only in an Atlas. In a time that wasn't measured by cuckoos that sprang out of clocks. In a silence that was sometimes punctuated by howling winds and gushing streams. In a world that had northing to do with yours." This mythologised "journey" of Mumbai's premium, elite bottled water, a Tata press release argues, "seeks to make the consumer one with its source – the Himalayas – with its inherent and pristine goodness that sets it apart from ordinary water" (Tata Group 2008). The strap-line seems especially startling given the wider context of Mumbai's water wars: "Hydration with wellness."

The middle classes and elites tackle the water crisis by a combination of groundwater from their borewells, calling in private water tankers, and, more recently, installing rainwater harvesting and grey-water recycling systems, thus seceding from the public city when required.

Figure 6.3 A Bisleri bottled water advert looms large over Mumbai, north of the Fort District
(Photograph by Colin McFarlane)

"Can Shanghai be made on the graves of the poor?"

Mumbai's water raids have further accentuated the public health, economic and social impacts of systematic dehydration among already vulnerable communities. Paradoxically, the reliance of such communities on extremely expensive water provision organised through water rackets and, ironically, further "illegal" piped supplies delivered through BMC corruption, has also been deepened still further.

The limited evidence available already paints a bleak picture of the health and social impacts of systematic water shortages among Mumbai's slum dwellers. Before the raids, for example, a study of 1070 households in four slums in the year 2000 led by Hideki Harada of Nagaoka University of Technology, published in India's prestigious *Economic and Political Weekly* (Kumar et al. 2003), showed that where one water tap was provided between thirteen to thirty households, it was woefully inadequate both quantitatively and in terms of quality. The median per capita daily water consumption per person in the four slums was found to be as low as 26, 27, 33 and 25 litres per day. Not surprisingly, slum-dwellers ranked access to water and toilet facilities as being the most important issues they faced. Among slum-dwellers living with such extreme water scarcity, the annual diarrhoea, typhoid and malaria cases were estimated to be 614, 68 and 126 per thousand people respectively. The study found that "all types of water-borne disease are occurring with great severity" in all the four poor communities that were studied. Consequently, the authors calculated that at least 30% of all deaths in Mumbai's could be attributed to water-related infections (Kumar et al. 2003).

Dr. Ravindra Rathod,[5] of the Niramaya Health Foundation in Rafinagar, recounted to us some of the chronic health impacts from systematic dehydration that he has to treat daily among his patients. Diarrhoea, dysentery, scabies and typhoid are all extremely common. As desperation takes hold, many people dig their own wells, accessing limited supplies of extremely contaminated ground water in the process. Tests by the World Health Organization have found cholera, hepatitis and, most disturbing of all, polio viruses in water samples from such wells (Sayed and Desai 2009). Nazia Sayed and Geeta Desai point out that, to date, the Indian state has spent Rs. 26 billion trying to eradicate polio, only to expose residents to the virus because of the BMC's aggressive water raids policy. Sangeeta Yadav, a Mankhurd resident, discussed a newly-dug well in her house with the journalists: "How do they expect us to survive without water?" she asked. "We have no option but to drink this dirty water. We will die if we don't drink water, and we will die if we drink this water. Whatever has to happen, will happen" (Sayed and Desai 2009). Iqbal Shaikh, a social worker in the area lobbying for more municipal pipelines, stressed that "the residents all know the water is contaminated but are helpless. For days we don't get water in the area and if we complain the BMC people demand a huge amount of money for a pipeline. That's the reason why people opt for such illegal measures."

Following the destruction of pipes serving Rafinagar in November 2009, our research demonstrated how the community was pushed more deeply into a water crisis. Two weeks after the raids the BMC did install two temporary water storage tanks just outside Rafinagar. However, beyond the burden of long hours queuing, even this completely inadequate supply was next to useless for most residents as the arrival of the BMC water tankers to fill the tanks was extremely sporadic. In addition, many physical struggles occurred as residents fought over the inadequate water supplies, especially when the tankers finally arrived (Figure 6.4). "Either beat up each other for water or die [without water] yourself," a Rafinagar woman remarked to us caustically.[6] "Those whose mouths have strength [to speak up and shout]," recounted another, "those whose bodies are strong, those who can curse, they are the ones who can fill [water from the tanks]."[7] Even after a group of residents improvised a system for regulating the queues at the water tanks, various exclusions continued.

The extreme water scarcity in Rafinagar meant that, for the first time, private water tankers, organised through corruption and offering extremely expensive water to those who could afford it, began to be regularly called in. Many of the private water tankers also obtained water from borewells, pointing to profits being made through excessive and often unregulated groundwater extraction.

Water prices within Rafinagar increased after the raids. While water delivered through a legal connection is charged at Rs. 2.25 for 1000 litres, Rafinagar residents, the poorest of Mumbai's poor, were increasingly paying up to 100 to 200 times that amount just to fill one jerry-can holding 35 to 40 litres. Some of our

Figure 6.4 Rafinagar residents waiting at one of the temporary water storage tanks for the BMC water tanker to come and fill it.

(Photograph by Renu Desai)

interviewees thus ended up spending over Rs. 600 per month on minimal water supplies, out of a total monthly income of Rs. 3000–4000. Most of our interviewees were unable to bathe on a daily basis, even in the unbearably hot summer months, leading to skin infections. Water was available for lower rates only if it was "brown" or "yellow" in colour.

After the raids, our interviews showed that more of Rafinagar's residents than before were forced to rely on sources of water outside the settlement, accessing these by time-consuming and exhausting one-to-five-kilometre journeys on foot or bicycle. Many women and young girls in particular went on foot with water pots to search for water from outside, while many men and boys went by bicycle with jerry-cans. Children often had to forego schooling or paid work because of the imperative of accessing water, or else they took on multiple burdens.

These itinerant water-carriers in turn then became subject to state violence and harassment. The BMC and police started to confiscate bicycles and jerry-cans and even punctured jerry-cans. An additional problem was that the threats of arrest against those caught "illegally" selling water meant that water-carriers found it harder and harder to gain supplies even beyond the community's boundaries. For instance, one male resident from Rafinagar (who often used to collect water from outside the settlement even before the water raids) told us: "People in Shivaji Nagar aren't giving water. They say that if we give you water then the police will catch us."[8]

The systematic state harassment, and the active denial of an essential source of urban life, led to deep feelings of outrage and alienation among Rafinagar and Govandi inhabitants. "I sold my jewellery to acquire this connection and now police have seized the motor. How can we live without water?" pleaded one Govandi woman who has a family of six to the *Mumbai Mirror*'s Nazia Sayed (2009). Many complained that they weren't even aware that their connections, which, after all, were organised with BMC complicity, were technically illegal. Others were incensed that raids destroyed pipes that were legally supplied and for which they had been paying bills. A further group of residents actually backed the raids because the resulting increases in pressure further down the pipes meant that their supplies were improved.

Govind, a seventeen-year-old boy who was visibly upset as he recounted his experience, discussed with us how as a Rafinagar resident he was violently assaulted by residents in neighbouring Gautamnagar: "Once when I was coming back from there, some five or six men from there stopped me on my bicycle and forcefully took the cans and upset all the water. They told me that if you people from Rafinagar come here to fill water then Gautamnagar's connections will also be cut, so don't come here again."

Mumbai's slum-dwellers are painfully aware of the way in which their demonisation has served the purposes of the city's political and economic elites. They are also not slow to see links between the wider political aspirations of reengineering Mumbai into a "global city" or a "new Shanghai", and the increasingly orchestrated state violence against them and their needs.

Salma, one our Rafinagar interviewees,[9] discussing the ongoing threat of demolition against the settlement as well as the water raids, put it bluntly. "The government is not listening to us," she says, "because it wants to make [Mumbai into] Shanghai [...]. We don't oppose Shanghai. But [the government] comes and crushes us and goes away, like [one might crush] ants... We would also like that Shanghai is made. We might not see it, but at least our children will. Our children's dreams will be fulfilled." But, Salma argues, if Mumbai's government and ruling elites "try to make Shanghai at the cost of the dreams and aspirations of the poor, then this Shanghai will not be successful... Can Shanghai be made on the graves of the poor? If there is a funeral opposite me and I play music, can I really enjoy it?"

Water protests by the urban poor in Mumbai have often targeted political parties whose vote-bank politics traditionally surround the politics of water provision for Mumbai's poor. Through these systems, election cycles tend to coincide with promises to address questions of water provision to slums. After the municipal water raids in and around Rafinagar too, residents mobilised to approach their political leaders, reminding them of their election promises to bring adequate piped water to Rafinagar. However, instead, inadequate temporary arrangements were made and the realisation of these promises was continually delayed. This reveals an increasingly predatory vote-bank politics in which politicians on the one hand come to power based on their electoral promises to the urban poor, while on the other they are often involved in perpetuating the dependency of the urban poor on exploitative or predatory water rackets for their own financial benefit.

Renetworking? The Right to Urban Water

In Mumbai, water "infrastructure" encompasses much more than pipes, aquifers, rivers, borewells and taps. Crucially, it also involves complex social and political arrangements and the very people of the city (see Simone 2004). The fragmented nature of civil society has paved the way for water revanchism – an attempt by the middle classes to "claim back" the City's hydrological commons from the poor through a discourse that casts themselves as tax-paying citizens who are denied their legitimate right to municipal water because this is diverted through vote-bank politics and corruption to "encroaching" slum-dwellers who have no legitimate right to the city's resources. This discourse erases the experiences of the urban poor of the hydrological apartheid, just as it does their right to urban water. This discourse further dovetails with attempts by powerful elites to "claim back" the City from the poor majority under the spurious justification of competing with aspirational cities elsewhere (Shanghai, Singapore). As a result, civil society organisations that cast the city's water crisis through the experiences of the urban poor and argue for policies that would recognise their right to water face profound challenges (for a broader discussion of such challenges, see Appadurai 2001).

Mumbai's water wars reveal in stark detail what emerges when slum-dwellers are cast out from the rights of a modern urban existence in the world's burgeoning mega-cities, in a context where, in contrast to previous periods, elites are not vulnerable to the public health disasters that befall the urban poor who are both systematically dehydrated and prone to water-borne diseases caused by inadequate sanitation provision. Mumbai's water raids reveal that revanchist politics extend to the systematic destruction of the means of hydration. Examining them shows in startling detail how the complete denial of the status of water as a public good to be organised and distributed collectively to benefit all urban residents opens the door to the hydrological apartheid which surrounds the hyper-commodification of the ultimate inelastic good. Finally, our story demonstrates that apparently benign aspirations to "clean up" global South megacities, to allow them to be reengineered into "global" metropolises imitating some shining and semi-mythical exemplar (often Shanghai, but increasingly Dubai) can camouflage extreme campaigns of violence, erasure, intimidation and plunder by political and economic elites against the vulnerable and poor.

The results of water wars – dehydration, death, disease and desperation – demonstrate the urgent need to challenge the politics of water at its foundation. Only by building a politics of water rights, addressing the legitimate needs of Mumbai's poor for clean, adequate water, will city and State authorities ever hope to eradicate the symptoms of poverty and social fragmentation that their regressive and violent mobilisations are ostensibly designed to address. We are encouraged by the range of activist groups and organisations that attempt to draw attention to Mumbai's deeply unequal urban metabolism, but we have yet to see a broader movement in the city coalesce around water and sanitation in order to continually call into question and offer robust alternatives to the nature of the city's development strategies.

It is difficult to offer normative solutions in such a context. Barring a major revolution, what political transformation is likely to have the power to overturn the deep-seated connections between the bourgeois environmentalism, revanchist and ethnocratic populism, and militarised neoliberal 'world city' posturing exposed in this chapter? How to counter a deep culture of accumulation by dispossession and elite criminality within the urban and hydrological commons of one of the world's foremost megacities? How to radically engineer a city like Mumbai based on a pro-poor ideology and a notion of universal hydrological citizenship?

Innovations involving NGOs and social movements involving slum upgrading and legalisation clearly have their place. Technologies involving rain-water harvesting and community-level innovation in water sharing are also worth exploring. However, we have the feeling that more substantial changes will only be possible through a sustained and radical change in the nature of infrastructural, urban and national governance in India, similar to the radical shifts that have emerged in Latin America over the past two decades. Then, perhaps, a democratic infrastructure politics centred on the possibilities of networking – or, in some cases, *re*networking – the urban poor into life-giving systems of hydration may emerge as more than a pipe dream.

Note: This chapter is a revised version of a paper published in the journal *Public Culture*, Volume 25, issue 1.

Note on Methods: The paper reports on a nine-month period of ethnographic fieldwork conducted in two contrasting informal settlements in Mumbai:

Rafinagar, a predominantly Muslim settlement in northeast Mumbai, and Khotwadi, a more established and predominantly Hindu settlement in western Mumbai. The authors would like to thank the UK's Economic and Social Research Council (Grant RES-155-25-0087) for the support which made the research outlined in this essay possible. In-depth ethnographic research was conducted in the two case studies; four months were spent at each of the two sites. Key methods involved observation, interviews and focus groups. Observation involved spending time in the lanes of the case study areas, talking with groups and residents and observing water and sanitation practices (e.g. cleaning of drains, delivery of water for toilet blocks etc.). Interviews were particularly important, most of which were informal and unstructured, including ad hoc conversation with residents and workers.

Notes

1 We gratefully acknowledge the support of the ESRC for the grant 'Everyday Sanitation', number RES/0360/7277, which made this research possible.

2 For an excellent discussion of these processes connected to the city's disparate but crucial film culture, see Mazumdar, R. (2007) *Bombay Cinema: An Archive of the City*. University of Minnesota Press, Minneapolis/London.

3 This process links to the ways in which particular civic groups have become part of a broader architecture of exclusion in Mumbai. See, for example, Jonathan Shapiro Anjaria's study of exclusive Advanced Locality Management (ALM) and Local Area Citizens' Groups (LACG), 'Guardians of the Bourgeois City: Citizenship, Public Space, and Middle-Class Activism in Mumbai', *City and Community*, 8:4, pp. 391–406.

4 Personal interviews with Leena Joshi, Apnalaya (a community-development NGO).

5 Personal interview on Feb 6, 2010.

6 Resident, Rafinagar Part-2, personal interview, Jan 8th, 2010.

7 Resident, Rafinagar Part-1, personal interview, Jan 12th 2010.

8 Resident, Rafinagar Part-1, personal interview, Mar 7, 2010.

9 Resident, Rafinagar Part-2, personal interview, Jan 8th, 2010

References

Agamben, G. (1998) *Homo Sacer: Sovereign Power and Bare Life*. Stanford: Stanford University Press.

Anjaria, J. S. (2005) "Guardians of the Bourgeois City: Citizenship, Public Space, and Middle-Class Activism in Mumbai." *City and Community* **8**(4): 391–406.

Annuncio, C. (2003) "Watery divisions." *Outlook India.Com*, May 26th, 2003, at *http://www.outlookindia.com/article.aspx?220265* (accessed 31 March 2012).

Appadurai, A. (2000) "Spectral Housing and Urban Cleansing: Notes on Millennial Mumbai." *Public Culture* **12**(3): 627–651.

Appadurai, A. (2001) "Deep Democracy: Urban Governmentality and the Horizon of Politics." *Environment and Urbanization* **13**(2): 23–43.

Baviskar, A. (2002) "The politics of the city." Seminar 156, available at http://www.india-seminar.com/2002/516/516%20amita%20baviskar.htm (accessed 31 March 2012).

Bjorkman, L. (n.d.) *Water and Resettlement on the Urban Periphery: the case of Mumbai's M-East Ward*. Unpublished paper.

Chatterjee, P. (2004) *The Politics of the Governed: Reflections on Popular Politics in Most of the World.* Delhi, Permanent Black.

Dalvi, V. (2009) "BMC, police begin crackdown on illegal water vendors." *Mumbai Mirror,* December 1st, 2009, p.2.

Desai, G. (2010a) "Where does Mumbai's water go?" *New Link Road Residents Forum,* March 26, 2010, at http://newlinkroad.blogspot.com/2010/03/where-does-mumbais-water-go.html (accessed 31 March 2012).

Desai, G. (2010b) "BMC wants ACB, cops to curb water thefts." *Mumbai Mirror*, June 7th, 2010, p.4.

Dossal, M. (1991) *Imperial Designs and Indian Realities: The Planning of Bombay City, 1845–1875.* New Delhi, Oxford University Press.

Gandy, M. (2008) "Landscapes of disaster: water, modernity, and urban fragmentation in Mumbai." *Environment and Planning A,* **40**: 108–130.

Gits4u.com (2010) 'Bottled water industry in India,' 2010 at http://www.gits4u.com/water/water16.htm (accessed 31 March 2012).

Hansen, T. B. (2001) *Wages of Violence: Naming and Identity in Postcolonial Bombay.* Princeton, NJ, Princeton University Press.

Jadhav, N. (2005) *Economic Renewal of Mumbai City: Opportunities and Constraints.* Paper presented at the International Conference on Urban Renewal: Learning for Mumbai, Mumbai, May 24–26, 2005 at http://www.drnarendrajadhav.info/drnjadhav_web_files/Published%20papers/Economic%20REnewal%20of%20Mumabai%20City.pdf (accessed 31 March 2012).

Joseph, J. (2005) "What will make Mumbai into a Shanghai?" *Times of India*, April 6th, http://timesofindia.indiatimes.com/business/india-business/What-will-make-Mumbai-into-a-Shanghai/articleshow/1069800.cms?flstry=1 (last accessed 31 March 2012).

Kaika, M. (2005) *City of Flows: Modernity, Nature and the City.* London, Routledge.

Kumar, S., S. Shikura and H. Harada (2003) "Living environment and health of urban poor: A study of Mumbai." *Economic and Political Weekly*, August 23 2003, 3575–3586.

Mazumdar, R. (2007) *Bombay Cinema: An Archive of the City.* Minneapolis/London, University of Minnesota Press.

McFarlane, C. (2008a) "Governing the contaminated city: Infrastructure and sanitation in colonial and postcolonial Bombay." *International Journal of Urban and Regional Research*, **32**(2): 415–35.

McFarlane, C. (2008b) "Sanitation in Mumbai's Informal Settlements: State, 'Slum', and Infrastructure." *Environment and Planning A*, **40**: 88–107.

McKinsey/Bombay First (2003) *Vision Mumbai: Transforming Mumbai into a World-Class City.* Mumbai, McKinsey/Bombay First, at http://www.maharashtra.gov.in/english/reports/taskforce.pdf (accessed 31 March 2012).

MW-YUVA (2001) *Slum Sanitation Project: Final Report.* Mumbai, Municipal Corporation of Brihan Mumbai.

Raina, A. (2009) "Water, yes, but at 10 times the cost." *Mumbai Mirror*, November 12, 2009.

Roy, A. and A. Ong (2011) (Eds.) *Worlding Cities: Asian Experiments and the Art of Being Global.* Oxford, Wiley-Blackwell.

Sadhwani, Y. (2010) "Water war!" *Mumbai Mirror*, November 17th, 2010, p. 2.

Sathe, A. (2010) "Stealing water to earn a few bucks? Pay a hefty price." *Mumbai Mirror,* Friday 8 January 2010, p. 4.

Sayed, N. (2009) "Artificial water shortage in Govandi." *Mumbai Mirror*, December 05, 2009, p. 6.

Sayed, N. and G. Desai (2009) "Thirst for water leading to polio." *Mumbai Mirror*, December 29th, 2009, p. 1.

Simone, A. M. (2004) "People as infrastructure: Intersecting fragments in Johannesburg." *Public Culture,* **13**(3): 407–408.

Singh, V. (2010) "Pools to be built despite water crunch." *Mid-Day.Com,* April, 2010, 2nd April, at http://www.mid-day.com/news/2010/apr/020410-swimming-pools-water-crunch-mumbai.htm (accessed 31 March 2012).

Singh, V. A. (2010) 'How the BMC saved the city 4 lakh litres of water." *Mumbai Mirror,* Monday 11 January 2010, p. 8.

Suryawanshi, S. (2009) "BMC to reclaim city's pipelines." *Mumbai Mirror*, September 11th, 2009, p. 2.

Suryawanshi, S. (2010) "Water thieves to be tried under MCOCA." *Mumbai Mirror,* April 10th, 2010, p. 2.

Swyngedouw, E. (2004) *Social Power and the Urbanization of Water: Flows of Power.* Oxford, Oxford University Press.

Tata Group (2008) "Mount Everest Mineral Water unveils the new Himalayan natural mineral water." May 7th, 2008 at http://www.tata.com/media/releases/inside.aspx?artid=Mw/mP8IlwTE= (accessed 31 March 2012).

Tatke, S. (2010) "Slumdwellers rely on ration shops for water." *Times of India*, December 19th, 2010, p. 9.

Varma, R. (2004) "Provincialising the global city: from Bombay to Mumbai." *Social Text,* **22**(4): 65–89.

Vyas, H. (2009) "HC pulls up BMC over state of water pipelines." *Mumbai Mirror*, July 30, 2009.

Vyas, S. (2010) "BMC kickstarts process to install CCTVs at pipelines." *The Times of India*, Mumbai edition, April 6, 2010, p. 5.

World Bank (2009) "Bombay fights the markets, and more than half of Mumbai's residents live in slums." World Bank.Org, 2009 at http://web.worldbank.org/WBSITE/EXTERNAL/HOMEPORTUGUESE/EXTMAININPOR/EXTDATAANDRESINPOR/EXTRESINPOR/EXTWDRINPOR/EXTWDR2009INPOR/0,,contentMDK:21954976~pagePK:64168445~piPK:64168309~theSitePK:5593535,00.html (accessed 31 March 2012).

7 Another urban infrastructure is possible

Contesting energy and water networks in Berlin

Ross Beveridge and Matthias Naumann

Introduction

> *A 35,000 km electricity network. Good for you that you don't have to be bothered by such things. The electricity network we operate is one of the world's safest and most advanced.*
>
> (Vattenfall GmbH 2013; translation by the authors).

The above advertisement from Berlin's electricity provider Vattenfall appeared in the city's newspapers in autumn 2013. What is striking about it is not so much the company's attempt to present itself as expert and reliable, but Vattenfall's assertion that Berlin's citizens should not be bothered about how their electricity is provided and managed. In fact, many of Berlin's citizens were bothered, and they aimed to bother Vattenfall (hence the advertisement). Citizen-led initiatives have been campaigning to return the electricity network to public hands and these even prompted a referendum on remunicipalisation, which only very narrowly failed. These developments reflect a broader awakening of public and political interest in infrastructure networks in the city as social movements and political parties have developed new, often conflicting concerns regarding private and public ownership and democratic urban governance more generally. Most notably, a long-running citizens' campaign against the partial privatisation of the Berlin Water Company (BWB) played a key role in the remunicipalisation of the water company in 2013. However, the simple return of ownership to the state has itself been criticised by social movements and left-leaning politicians on the grounds that the "old way", a return to a pre-privatisation form of state ownership, is no guarantee of democracy, equity and sustainability. Hence in Berlin we witness not simply the contestation of privatisation and the remunicipalisation of infrastructure networks, but also political contests over what remunicipalisation should actually entail. There is an ambitious, vibrant and often progressive politics of infrastructure in Berlin, with discourses of both radical and reformist change apparent, and the current and future roles of the state, civil society and the private sector are heavily contested.

What has made Berlin's water and energy infrastructures the focus of so much politics? Why has the provision of water and energy supply become so *political*? This chapter addresses these questions, arguing that the struggles against privatised forms of infrastructure provision not only illustrate the interrelationships of infrastructural and urban change (as well as their contestations). They are going further, in that they are tentatively promoting new urban futures, and can be seen, we argue, as part of a more general attempt to reclaim elements of the urban and assert the "Right to the City".

In response to these developments, the main aim of this chapter is to bring the infrastructural dimension of urban social struggles into clearer focus. Put differently, we detail the increasing embeddedness of infrastructure networks in urban social struggles in Berlin. While the political dimension of infrastructure is well-researched (McFarlane and Rutherford 2008), the inter-connections and co-evolution of contemporary urban social movements, emancipatory politics, and urban infrastructure are less so, at least in Europe.[1] This might be due to the fact that the provision of essential infrastructure services has often been taken for granted in contemporary European cities, with privatisation and liberalisation not always as controversial as they are in cities of the global South, for example. We argue that our understanding of the contemporary politics of infrastructure, where privatisation is not only being rolled out but is also being reversed and resisted, could benefit from being brought into closer relation to wider urban struggles, such as the various movements using the banner of the "Right to the City" as well as the critical urban theory which deploys the same term. At the same time, there may be some gains for our understanding of the increased materiality of urban social struggles, where rights and democracy are ever more asserted in relation to cities ("Right to the City"/"*Wem gehört die Stadt*") and infrastructures ("Water Democracy", "Energy Democracy"). Therefore, we analyse the links between the "Right to the City" and a "Right to Infrastructure" (Corsín Jiménez 2014) and explore the fundamental elements of such a right. By connecting research on urban infrastructure systems with the academic as well as activist oriented discussion on the "Right to the City", we argue that a better understanding of current struggles for public and more participatory urban infrastructures is possible. In doing this, the chapter provides insights into ongoing contestations of neoliberal logics in infrastructural governance, which hold that market-based organisation and practice and private ownership are per se superior, with the consequence that former political goals of infrastructure supply, such as social and spatial cohesion, are replaced by the aims to reduce costs and maximise revenues (Wissen and Naumann 2006).

The chapter is structured as follows. The next section discusses the ongoing urban struggles around infrastructure in Berlin. Reflecting on events in the city, the third section provides a preliminary discussion of how the "Right to the City" perspective may be brought into conversation with urban infrastructure politics. We conclude with some thoughts on the relevance of remunicipalisations for the re-appropriation of urban space and the potential of the notion of a "Right to Infrastructure".

Contesting Urban Infrastructures in Berlin

The city of Berlin provides a useful case of contested urban politics and the role of infrastructures within these contests. As we have argued elsewhere (Beveridge and Naumann 2014), Berlin can be seen as emblematic of the apparently "inevitable" rise of neoliberal urban governance after reunification (the assertion that there were no alternative paths available to the city and its politicians), as well as its contestation and the beginning of debates about post-neoliberal urban governance (for an overview of the city's development and debates, see Bernt et al. 2013). The specific importance of Berlin's infrastructural struggles is threefold: first, the city has experienced a complex transformation over recent decades – of re-unification, socio-economic change, and mostly failed global city ambitions (Cochrane and Passmore 2001; Gornig and Häußermann 2002; Krätke 2004; Beveridge 2012); second, and partly following on from this, the imperatives of austerity politics and sharp cuts in public budgets have shaped the development of the city for years; third, the city not only has a long history of social movements (Mayer 1993), but also had a so-called red-red coalition government from 2001 to 2011, in which the Left Party (*Die Linke*) was the junior partner of the Social Democrats. The general possibilities of and constraints on political agency, the gains and losses of the urban left in and outside government can thus be observed in Berlin.

Recently, the city has become an international focal point for young people, temporary 'creative' residents and tourists (Füller and Michel 2014). After more than a decade of decline Berlin's population is growing and even the city's economy is slowly recovering, with a concentration of start-ups apparent. However, these developments have been far from unproblematic. Rapidly rising rents and the gentrification/touristification of inner-city neighbourhoods as well as urban redevelopment projects, such as Berlin's waterfront initiative *Mediaspree*, have caused huge public protests and the emergence of a number of urban social movements (for an overview, see Holm 2014).

The turbulent developments in the city are closely entwined with its infrastructure networks. Indeed, Berlin's unique place in twentieth century history (the grand designs of the Third Reich, the destruction experienced in the Second World War, the Cold War division) have made infrastructure curiously "visible" (for a historical overview see Moss 2014). Unlike many cities, where the politics of infrastructure development have to be unearthed, in Berlin infrastructure has often been centre stage of urban governance. The period since reunification can be thought of in three general phases, all of which are embedded in broader socio-economic changes in the city.

1 Post-reunification privatisation during the 1990s: a first phase in the immediate post-reunification period was characterised by the systematic privatisation of infrastructure (gas, electricity and water) in the 1990s, within the broader restructuring of Berlin as a 'global city'. The city's electricity, gas and water utilities, originally owned by the state of Berlin, were sold completely or partly to private, national as well as international companies such as Vattenfall, Véolia

and RWE (see for the case of water see below). Despite the magnitude and pace of change, public opposition to privatisation was limited in this period (Fitch 2007: 144ff.). The dominance of a 'Third Way' (*Neue Mitte*) wing in the Social Democratic Party (SPD) party ensured a general consensus on the issue with the coalition government partners the CDU, while "no alternative", austerity-like arguments brought Berlin's rising financial and economic crisis in relation to state ownership and spending (Beveridge 2012).

2 Reaffirmation and extension of privatisation during the early 2000s: a second phase witnessed a move to the left with the establishment of a red-red coalition (2002–2011) of the SPD and *Die Linke*. This followed the collapse of the CDU-SPD coalition in the wake of the scandal surrounding a partly publicly-funded bank, *Berliner Bankgesellschaft*, which had been providing real estate development loans. In 2001, the so-called "banking scandal" emerged when it transpired that leading CDU politicians, as well as their associates, had profited from the bank loans (Ugarte Chacón 2012). However, these events did not lead to a new approach to privatisation in the new red-red coalition, with privatisation continuing in other sectors (e.g. housing and state property). Alongside these developments, there was a growth in social movements in Berlin contesting urban developments such as *Mediaspree*.

3 Contestation and reversion of privatisations during the late 2000s/2010s: a third phase, still ongoing, has seen infrastructure networks become the object of political contestation (along with housing and gentrification). Infrastructural urban movements became visible with the establishment of the Berlin Water Roundtable (*Berliner Wassertisch*) in 2006. A range of citizen initiatives are campaigning on infrastructure issues, contesting policy and forging a new, variegated political discourse on urban democracy related to infrastructure. The following initiatives are working on infrastructure issues:

- The group "Berliner S-Bahn-Tisch" (Roundtable on the Berlin overground train system) has been campaigning since 2011 against the impact of the commercialisation of Deutsche Bahn, Germany's rail company, and its impact on the city's overground train system, S-Bahn, which Deutsche Bahn controls. Its main demand is the return of public and city-level control.
- The coalition of groups, "Aktionsbündnis A100 stoppen", is an alliance against the planned extension of Berlin's inner-city motorway (which the group calls "Germany's most stupid motorway"), which goes directly through inner-city neighbourhoods. It emerged in the early 2000s when the first plans for the motorway extension were confirmed. They campaign under the slogan 'Berlin remains our city' ('Berlin bleibt unser Stadt'), and are linked to other campaigns against rising rental prices and the displacement of people from the inner city as a result of gentrification and urban development projects.
- Since 2012 the initiative "100% Tempelhofer Feld" has been concerned with the future use of the former inner-city Tempelhof airport, which has

been turned into a public park. It aims to ensure that the former airport is an open area available for different public uses and to prevent any privatisation or commercial development. The Berlin Senate had planned to allow a partial development of the former airfield and the former airport buildings. This was blocked via a successful public referendum in May 2014.

- There are several initiatives campaigning about the future use of undeveloped areas, such as former squats or informal settlements within the inner city, e.g. "*Mediaspree versenken*".
- The "Berliner Wassertisch" (Berlin Water Roundtable) initiated a successful referendum to publish the confidential contracts of the partial privatisation of Berlin's water company, and was a key promoter of remunicipalisation.
- Two initiatives, "Berliner Energietisch" (Berlin Energy Roundtable) and "BürgerEnergie Berlin" (Citizens' Energy Berlin), are concerned with (different models of) remunicipalising Berlin's energy supply.

This list illustrates the importance of infrastructure within current urban struggles. The demands made by these groups differ, but they share a rejection of the direction of current urban policy and development and make claims to common ownership of the city and its infrastructure, e.g. 'our city', 'water in citizens' hands' etc. These initiatives are also entwined with other social movements in Berlin, e.g. related to housing issues. The failure of neoliberal urban policies became especially obvious with the collapse of the *Berliner Bankgesellschaft* in 2001, the contested privatisation of public housing companies and the dramatically increasing rents in Berlin's inner city neighbourhoods (Holm 2013) as well as the contestation of Berlin's waterfront re-development *Mediaspree* (Scharenberg and Bader 2009). Furthermore, the social movements in the city – who traditionally worked together with the Left Party – were deeply disappointed with the *Realpolitik* of the party in coalition government, who joined the "There is no Alternative" consensus on reducing public spending (for a critical review of the Left Party's ten years in power, see Holm et al. 2011).

Figure 7.1 Main title reads: 'Berlin remains our city'.
Source: MS Versenken, 2014.

Figure 7.2 Main title reads: 'Public referendum. Vote and unplug Vattenfall'.
Source: Berliner Energietisch 2014

Figure 7.3 Main title on left-hand page reads 'Our Water', while that on the right-hand page reads 'Berliner Wassertisch – Water belongs in the hands of the citizens, Enough of the secret contracts'.
Source: Berliner Wassertisch 2011.

Figure 7.4 Placard on the right reads "Red Card for the Urban Development".
Source: Aktionsbündnis A100 stoppen.

Against this background social movements on urban and infrastructural issues have grown rapidly in size, numbers and prominence. "Right to the City" and "Reclaim your City" increasingly provide an umbrella and shared basic vocabulary for all of these quite heterogeneous movements. What all initiatives in Berlin have in common is that they are – in different ways – rejecting privatisation and other norms of neoliberal urban governance as practiced by the Berlin executive. Remunicipalisation became an admittedly vague alternative to the neoliberal vision of privatised utilities and acted as a focal point for diverse social campaigns for a more just and democratic city.

In the following sections we discuss the initiatives to remunicipalise Berlin's water and energy supply in more detail. These campaigns are not only concerned with crucial sectors of urban services, but they are also respective examples of successful and failed public referendums on the remunicipalisation of infrastructure utilities. The two cases show how infrastructure became intrinsically political in Berlin, and reflect a wider trend in Germany towards remunicipalisation in various cities and sectors (Matecki and Schulten 2013). The struggles over water and energy infrastructures are, despite their similarities, also characterised by some significant differences.

"Publish the contracts". The remunicipalisation of Berlin's Water Company

After twelve years of partial privatisation, one private partner, the German energy utility RWE, agreed to sell its 24.9% share of the Berlin Water Company (BWB) back to the city of Berlin in May 2012. After initial threats of legal action against RWE, the French utility Véolia, the second partner, sold back its 24.9% share in January 2014. The story of Berlin's Water Company, with its "inevitable" privatisation in the 1990s (Beveridge 2012), the contestation of the privatisation deal during the 2000s and the final decision for remunicipalisation in 2012, illustrates how a key policy norm of neoliberal urban governance became contested, experienced crisis and was ultimately overturned. Further, it shows how urban utilities can became a vehicle for a fundamental paradigm shift in urban politics. Although the privatisation of infrastructure and other services continues in some parts of the world (e.g. in the austerity programmes of southern Europe), Berlin is not an isolated case. Other major capitals, such as Buenos Aires and Paris, have reversed privatisations.

BWB was privatised in 1999 in the context of deep fiscal and economic crisis. After German reunification in 1990, Berlin experienced large-scale de-industrialisation and a loss of jobs in the public sector in both parts of the city (Ellger 1992: 45), as Cold War subsidies were phased out and Berlin re-entered the global economy. In the face of these challenges, the city government presented privatisation of BWB as "inevitable" (Beveridge 2012) and provoked little public opposition (Fitch 2007: 144ff.). Even the unions, usually opponents of privatisations, ultimately supported the partial privatisation. A public-private partnership was agreed with RWE, Véolia (Vivendi at this time) and the insurance and financial investment company SA/Allianz (who withdrew soon after the sale). The purchase price for the 49.9% share of the company was $1.96 billion.

However, despite the continuing and even increasing indebtedness of the city, the remunicipalisation of BWB became a mainstream objective in city politics within a few years, supported by three of the four major political parties (SPD, Greens and *Die Linke*, with the CDU dissenting) in the city election campaign of 2011. The reason for this shift was the increasing lack of legitimacy of the privatisation deal, which prompted public opposition and culminated in the public referendum in 2011.

Opposition centred on three interrelated issues. First, the Partial Privatisation Law contained a mechanism which ensured a return of around 8% for the private investors, similar to the annual profit rates of privatised water companies in England and Wales (for detail see Hüesker 2011). Second, some of the privatisation contracts were (until a Berlin newspaper published them online) secret and not accessible to the public. The agreement to ensure an annual return of around 8% was contained in one of the additional confidential contracts. Certain aspects of management and decision-making were also covered in these contracts, and hence these aspects were not transparent or subject to parliamentary or public scrutiny. Third, water prices increased by 21% from 2003 to 2006, making them the highest in Germany (Bundeskartellamt 2012).

Ultimately, public and hence political unease grew as partial privatisation led to high profits and high prices, as had occurred during the 1990s in England and Wales where the affordability of water for low-income customers became a political issue (Bakker 2003: 8).Furthermore, original plans to make BWB a "global player" failed, as well as the hope that privatisation would relieve Berlin's fiscal problems. Unlike other cases of privatisation in the water sector around the world, the maintenance, quality and efficiency of drinking water and wastewater disposal services were not the major concerns in the debate. The BWB remunicipalisation debate was also not particularly centred on the current challenges of water management in the region, such as the impact of climate change or changing demand because of socio-economic change (Hüesker et al. 2011). Nevertheless, criticism of the partial privatisation led to increasing public protest.

The main leader of the campaign to publish the privatisation contracts and remunicipalise BWB was the Berlin Water Action Group, Berliner Wassertisch, established in 2006. Disappointed by the post-communist party *Die Linke*, which was junior partner in Berlin's government from 2001–2012 and approved further privatisations in this period, Berliner Wassertisch organised opposition against the privatisation deal, mainly focussing on the publication of the confidential contracts. The initiative was not a coalition of organisations and was not part of the "classic" non-dogmatic radical movements found in the city, e.g. the squatter movement (Holm and Kuhn 2011). Still, the public referendum to publish the privatisation contracts was supported by environmental organisations (e.g. *Grüne Liga*, *Naturschutzbund*) and tenant groups (e.g. *Berliner Mietergemeinschaft*), as well as other social movements in the city and even local branches of *Die Linke*.

This initiative has to be understood within the wider context of increased campaigning and social movements within the city in the 2000s, many of which actively associated themselves with the slogan "Right to the City". These were usually focused on opposition to the local effects of economic globalisation, the 'normalisation' of the city,[2] and the government's promotion of it. During this entire period there was a red-red government in Berlin and *Die Linke* in particular became the subject of heavy criticism for their failure to outline an alternative, leftist urban politics for the city. Emblematic examples of movements included the rise of a campaign (*Mediaspree versenken!*) opposing a government programme to develop ex-industrial and residential areas along the river Spree into a focal point for media and creative industries (*Mediaspree*), the Squat Tempelhof movement which occupied the newly closed Tempelhof airport in 2009, and the proliferation of small groups (e.g. *Karla Pappel* in 2008) campaigning against the rapidly rising rental prices in the city. Some of these groups are part of the traditional left-wing milieu of the city, but many of the initiatives go far beyond the "usual suspects" of urban protest in Berlin. The initiative "Kotti und Co", fighting against dramatic increases in rents in a neighbourhood of the district Kreuzberg, is supported mainly by migrant tenants (Kotti & Co 2014).

Given its pragmatic and less fundamental demands – centred only on the water company, not on capitalism in general – the Berliner Wassertisch was also able to attract support beyond left activist groups. After collecting the signatures of around 280,000 Berliners they met the formal requirements to start a public referendum, a *Volksbegehren*. In the end more than 660,000 people, a majority of 98.2%, supported the proposal to publish the privatisation contracts. Although opposition to the privatisation of the water company was mounting in all political parties, reversing privatisation was still seen as impossible given the long-term (twenty-nine-year) contracts signed with the private partners Véolia and RWE in 1999. The city government initially claimed there were legal concerns over publishing the contracts, but finally did so after the Berlin newspaper *Die Tageszeitung* published them online in October 2010, confirming the confidential agreement to ensure annual return.

Within this context of widespread public hostility, the three shareholders of BWB – the city of Berlin and the two private investors – started negotiations about a remunicipalisation of the company in late 2010. The privatisation contract lasted formally until 2028, so an earlier cancellation required the agreement of all the partners. Having decided to withdraw from the water market to concentrate on energy, RWE sold its 24.95% share in May 2012, when the Berlin executive agreed to a price of around €650 million. In December 2013, fourteen years after privatisation, the return of the last shares of the private owner Veolia were finalised and BWB is now again a full municipal enterprise (Senatsverwaltung für Finanzen 2013). In the wake of this decision, Berliner Wassertisch published a Berlin Water Charter (Berliner Wassertisch 2013) outlining principles by which the water company should be run, such as affordability of tariffs, direct democratic participation, a not-for-profit-oriented approach, decent labour conditions and strict environmental requirements (ibid.). Though they have yet to filter into policy, they remain a clear articulation of citizen rights to participation in water governance in the city.

The case of the Berlin Water Company illustrates how quickly alternatives to neoliberal urban governance can arise. Alternatives which had previously been seen as "utopian" soon became possible (Beveridge et al. 2014). Though it should be noted that both RWE and Veolia had additional strategic reasons for wanting to withdraw from Berlin, and the remunicipalisation has not yet created a new democratic form of water governance, the seemingly "impossible" reversal of an "inevitable" privatisation (Beveridge and Naumann 2013) must be seen as a striking example of the possibilities for infrastructural and urban change.

Despite this apparent success, the return to public ownership and management raises many questions about what a remunicipalised BWB should be. Besides lowering water tariffs, which the company has done, it operates according to the same commercial logics as before. The 'new' BWB may be publicly-owned and hence ultimately more accountable to parliament and citizens, but it certainly does not operate in a more directly democratic way. Hence, what are the necessary instruments for public control and participation? Berliner Wassertisch (2013) have been campaigning for "direct democratic participation" but they no longer have the public exposure or support from urban social movements that they once enjoyed. It appears that having been highly visible and political, water infrastructure has retreated again from the public imagination of the city. Is it possible, beyond the contested issue of ownership, to achieve and sustain public interest in services that are generally taken for granted, or are the truly political manifestations of urban infrastructure fated to be as fleeting and unpredictable as Erik Swyngedouw

(2009) has argued the urban political generally is? What does this say about the influence and interests of urban social movements, and how can they become long-term players in urban politics? How do they avoid the fate of Berliner Wassertisch, which has split into three different groups after its successful referendum campaign, mainly because of personal differences?

"New energy for Berlin": The referendum to remunicipalise Berlin's energy supply network

The German energy transition (*Energiewende*) is not only a huge technological challenge; it also changes the governance – the scale, scope and institutions – of Germany's energy sector (Beveridge and Kern 2013; Moss et al. 2014). The political decision to phase out nuclear energy and to extend the use of renewable energy sources contributed to a dynamic development of new forms of energy utilities, such as energy cooperatives, so-called bio-energy villages and remunicipalised energy providers. Since 2005 more than 120 new public energy utilities have been founded in Germany (Berlo and Wagner 2013). Remunicipalisations in the energy sector have two major objectives: first, German municipalities want to control local energy transitions and see energy utilities as an important instrument for this task. Second, the German Renewable Energy Sources Act (EEG) provided reliable feed-in tariffs for electricity and heat generation based on renewable energy sources. Therefore, municipalities aim to take advantage of this funding and use revenues of local energy provision for subsidising municipal services. A common model is the traditional multi-utility provider (*Stadtwerk*), where the profitable energy branch provides resources for loss-making services such as public transport. Berlin's executive is also part of these current efforts towards increasing municipal influence on energy infrastructure configurations.

Berlin's energy system has often, but not always, been the subject of political contestation (Monstadt 2004; Moss 2014). Still, periods of relative consensus and calm are apparent. As in many other German cities (Bontrup and Marquardt 2011), Berlin's energy supplier BEWAG and its gas utility GASAG were fully privatised during the 1990s (Monstadt and von Schlippenbach 2005). Similar to and preceding the privatisation of BWB, the privatisations occurred in a context of "inevitability" and without large public protests. Indeed, a depoliticisation of private sector involvement in energy (and other sectors) characterised the late 1990s and 2000s. For a long time the privatisation of Berlin's energy utilities was not questioned or contested, even as the campaign against water privatisation got underway. This changed in 2011 when the local action group Berliner Energietisch started a campaign to buy back Berlin's energy network (on the following see Moss et al. 2014). The reasons for this new interest in the city's energy supply were threefold:

- First, the incumbent energy provider of Berlin, Vattenfall, is one of the "big four" within the German energy market and has been widely criticised for its

use of lignite as well as nuclear power. According to the Berlin remunicipalisation campaign it is not possible to realise a transition to renewable energy sources in Berlin with a utility company involved in such practices.
- Second, the experience of the BWB referendum as well as success stories in other German cities, such as Hamburg where a municipal energy utility has been re-established, illustrated that remunicipalisations are possible.
- Third, the expiry of the concession contract for Berlin's electricity network at the end of 2014, currently run by Vattenfall, created a window of opportunity for a change of ownership. The bidding for the concession creates the possibility that a municipal utility could be re-established.

The campaign to reclaim Berlin's energy infrastructure has been (and still is) promoted by two citizen-led organisations. The first initiative, Berliner Energietisch (Berlin Energy Roundtable), a platform of about forty different NGOs, has campaigned for a Bürgerstadtwerk (Citizen's Power Utility) which is municipally owned and committed to ecological orientation and social issues as well as to democratic control. The Berliner Energietisch defines ecological orientation in terms of "decentralized renewable energy plants in the Berlin-Brandenburg region" as well as "the goal (…) to provide you with 100 percent real green power" (Berliner Energietisch 2013). Social issues include "targeted energy-saving advice" as well as "energy-saving household appliances" (ibid.). According to Berliner Energietisch, democratic control would be realised through direct elections for membership on the company's Board of Directors (ibid.). Based on these objectives, Berliner Energietisch compiled a draft law which was the subject of a public referendum in 2013. The campaign was supported by the opposition parties within the city's parliament. The city government, a coalition of Social Democrats and Christian Democrats, the Berlin Chambers of Industry and Crafts as well as some of the unions, campaigned against the proposal. In what could be seen as a diversionary act, in the days running up to the referendum the Berlin Senate even announced the creation of a municipal energy supplier, BerlinEnergie. Over 600,000 people, a majority of 83% voted for the draft, but the referendum missed the necessary quorum of all eligible voters by 21,000 votes (Landeswahlleiterin Berlin 2013).

The second group, the energy cooperative BürgerEnergie Berlin (Citizen Energy Berlin), has been aiming to buy the city's electricity network and to supply Berlin with energy entirely from renewable sources. In May 2014, the cooperative had around 2,300 members with a total capital of 10.3 million Euros (BürgerEnergie Berlin 2014). The cooperative submitted its offer for the electricity network in June 2014 and is now competing with the incumbent Vattenfall and the newly established municipal energy utility BerlinEnergie for the new concession contract.

Berliner Energietisch and BürgerEnergie Berlin have thus proposed different ways for re-appropriating Berlin's energy system: Berliner Energietisch has favoured a municipal utility, where every voter in Berlin is able to participate – as a voter or as part of an initiative which should be listened to by the executive board of the utility; BürgerEnergie Berlin has aimed for a cooperative which requires a minimum deposit of 500 Euro per member to take part in decisions of the cooperative. This also allows people outside Berlin and Germany to become a member of BürgerEnergie Berlin. Despite these differences, both have questioned the limits to existing forms of energy provision and governance. Mainly because of the public referendum of the Berliner Energietisch, the Berlin Senate was put under pressure and established a municipal energy utility which is currently bidding for Berlin's electricity and gas network.[3]

The different efforts to remunicipalise Berlin's energy supply are not only emblematic of numerous and often contested energy transitions in cities and regions around the world. The public referendum and the established cooperative BürgerEnergie Berlin are also notable examples of the Germany-wide trend of remunicipalisation, with its potentials and limitations. Remunicipalisation campaigns have not always been successful or only progressive in their outcomes, but they have promoted a debate about reconfigurations of urban infrastructures.

These infrastructural struggles are embedded in wider urban struggles in Berlin around the "Right to the City". We turn now, therefore, to the "Right to the City" debate and assess its potential as a means of analysing contested urban infrastructural change.

A "Right to Infrastructure" within the "Right to the City"

The "Right to the City" (RTC) has become an increasingly visible and potent mobilising concept for urban social movements around the world struggling against, inter alia, increasing rents and displacements of the poor, the exclusion of immigrants, the privatisation of municipal services, the commercialisation and militarisation of public spaces, as well as the commodification of urban development in general. However, RTC is also an increasingly important notion for critical urban theorists. Numerous progressive and radical scholars have discussed the meaning and possible realisation of the "Right to the City" (see e.g. Brenner et al. 2011, Mitchell 2003, Harvey 2012, Purcell 2013, Mullis 2014). Consequently, the academic and activist debates are closely interwoven. Going back to Lefebvre's "Le droit à la ville" (1968) the term has been used as both a slogan and umbrella for a range of movements and approaches (for a brief summary see Grell 2014 or Mullis 2013).

Gebhardt and Holm (2011: 13ff.) distinguish between four dimensions of the "Right to the City": first, as a conceptual approach following Henri Lefebvre's work on the "Production of Space": according to Lefebvre, societal emancipation is only possible within emancipatory modes of the production of space (Mullis 2013: 59); second, as a political project for formulating utopias; third, as an agenda for political demands, such as the "World Charter for the Right to the City"; and fourth, as a form of organisation for connecting very heterogeneous urban social movements. David Harvey has argued that the "Right to the City" should be adopted "as both a working slogan and a political ideal" (cited in Attoh 2011: 676). He understands the "Right to the City" to entail the right to reclaim the city back from the Bourgeoisie (Harvey 2012: 138). "Urban commons" are Harvey's alternative to the capitalist city, dominated by privatisation. Critiques of the "Right to the City" approach, and the Marxist interpretation of it, mention the neglect of issues like gender, ethnicity etc. (Souza 2010). Other scholars criticise the vagueness and radical openness of the meaning of "right", whereby "the right to the city constitutes not a singular right, but a set of rights" (Attoh 2011: 675).

Despite these criticisms, the key promise of the RTC, as it is understood in practice, is that it is about opening up political possibilities, about laying claim to the city in all its forms. Hence there may be a lack of precision about what these rights are in general, but it is a positive open-ended notion that can be adapted to the specificities of local urban struggles. Underlying this, and following Lefebvre, RTC "is not a right that merely leads towards the participation of the urban citizen in urban social life; it is not only a right to urban space, but to a political space as well" (Dikec 2005: 48). Hence, the city in all its material and social forms is conceived as a political space which is up for grabs, in which claims for rights are made. In this perspective the city is, then, a site of struggle as global capital has also increasingly made claim to the city (ibid.: 47). However, it is also the object of struggle: rights are articulated in relation to the material and social forms of the urban – the things that constitute the city.

The contested character of the urban within the understanding of the "Right to the City" is the first starting point for connecting the RTC debate to infrastructure. Infrastructure systems are – as an essential part of the materiality of the urban and its contestations (Rutherford 2014) – highly conflictual and never politically "neutral". Second, infrastructure issues can be linked to debates around local "urban commons" to develop practical means of realising the "Right to the City". The question of new forms of ownership and management of infrastructure networks could lead us to concrete materialisations of "urban commons". Approaching water, energy, telecommunication systems etc. as "commons" can be essential elements of other possible urban worlds. The issue of infrastructure provides both a subject and object for RTC. It allows concrete demands ("Stop privatisation") to be made, rights to be articulated in relation to the urban fabric, and utopias (such as socially and ecologically just cities) to be articulated. Infrastructure systems are central to the process of urbanisation and therefore exerting collective power over them would reshape urbanisation according to Harvey's notion (cf. Harvey in Attoh 2011: 676).

While infrastructure is implicitly an integral part of the demand for RTC, as for example when Lefebvre defines RTC as a "right to centrality", infrastructure is rarely mentioned in explicit terms in the academic discussion. While socio-technical systems of energy and water supply, public transport, waste and wastewater disposal etc. are undoubtedly essential for all urban societies, they have not been a key concern of the critical urban studies engagement with RTC. A notable exception is Edward Soja (2010), who combines struggles around public transport in Los Angeles with more general questions of "spatial justice". Corsín Jiménez (2014) has developed a notion of a "Right to Infrastructure", in relation to open source urban technologies (part of "open source urbanism" more generally). However, he does so in a deliberately indeterminate fashion, which lacks the political immediacy of rights claiming as outlined above. Taking inspiration from Bruno Latour and others' understanding of displaced agency and power across human and non-human hybrid networks, RTC is presented neither as "a *right to* infrastructure, nor an infrastructure *made right*. Rather, the right to infrastructure allows us to escape the human-nonhuman and epistemology-ontology dichotomies altogether by opening up the agential work of infrastructures as a source (an open source) of possibilities *in their own right*" (2014: 343).

We believe therefore that the large social science literature on urban infrastructure systems has yet to connect to RTC debates. This is despite the literature's emphasis on the "inherently political nature and implications of networked urban infrastructure" (McFarlane and Rutherford 2008: 364). This literature has been very successful in drawing out the dynamic interplay between infrastructure developments and broader political philosophies and aspects of modernity (e.g. Gandy 1997, 2004, 2005; Kaika 2004; Heynen et al. 2005). Moreover, work on the global South in particular has highlighted the location of infrastructures within wider urban struggles. Examples include Swyngedouw's (2004) key work on Guayaquil in Ecuador which showed explicitly the urban social struggles over water, and more recently, Gandy (2006) and Graham et al. (2013) on Mumbai's water conflicts.

However, the inter-connections and co-evolution of infrastructure and contemporary urban social movements are less well understood in relation to Europe. While Graham and Marvin's (2001) "Splintering Urbanism" thesis, as well as its various empirical applications, problematise the socially and spatially uneven character of infrastructural change, struggles around RTC are not mentioned within current research on urban infrastructure. For example, it has been shown that energy transitions always have varying winners and losers and there is a need to find more socially just transitions (Rutherford and Coutard 2014: 1371), but this has yet to be linked more explicitly to a discussion of the work of urban social movements and a normative consideration of the future city. Ultimately, the extended, largely European-focused literature on STS, sustainable and socio-technical transitions understands innovation as always contested but not necessarily connected to practical, political urban struggles.

Against this background of debate, we connect our findings from Berlin with the approach of a "Right to the City". The debates about the remunicipalisation of Berlin's water and energy infrastructures reveal the ongoing contingencies of urban governance. Neoliberal thinking about the superiority of private ownership and management of infrastructure remains potent, but it has been and is being contested in Berlin and elsewhere. The city's debt level remains high and the unemployment rate is still one of the highest among the German federal states. In other words, the remunicipalisation of the water company and the broader emergence of this as a salient political process have occurred in a continuing context of austerity. Hence, the campaigns for remunicipalisation of Berlin's water and energy infrastructure are striking because they illustrate a "politics of possibility" (Beveridge et al. 2014): the ongoing struggles for new forms of political agency and practice, within the broader constraints of contemporary urban governance framed by austerity budgeting (Peck 2012), the needs of global markets and neoliberal policy and logics.

The shift away from privatisation and financial profitability can be understood as a tentative, incomplete form of post-neoliberal urban governance. The changing discourses, with demands for greater democracy and more ecologically and socially oriented forms of infrastructure management, reflect new forms of political agency constituted by infrastructure – a move towards a "Right to Infrastructure". Though the outcomes remain unclear, the fact that a SPD-CDU coalition government undertook a remunicipalisation and that, at least formally, the SPD is in favour of establishing a *Stadtwerk* is indicative of the effects citizen groups have had. All this shows the degree to which citizens are in fact – to come back to the Vattenfall quotation at the beginning of this chapter – bothered about infrastructure. Given the level and impact of activism in the city, it appears that Berlin's infrastructure will be characterised by more municipal forms of ownership in the future. If so, public ownership could provide a kind of "formalisation" of the "Right to the City". Indeed, the examples from Berlin show that municipal infrastructure utilities can provide a legal platform for political demands. The Berliner Wassertisch (2013) proposed Berlin Water Charter makes direct links between political and economic management of water in the city and the social rights of citizens living in that city. Water supply and sanitation are not defined in terms of the needs of financial profitability, but in terms of democratic control and environmental sustainability.

A particularly salient issue raised by the Berlin case is the way in which new forms of political agency and urban democratic governance are emerging and being fundamentally shaped by urban infrastructure. In a sense, infrastructure networks have acted as mediums of urban change in Berlin. Infrastructural struggles have not only made the often "invisible" energy and water networks visible again, but they have placed them centre stage in urban politics. Infrastructure has become a subject and an object of urban political agency, and the contestations show the possibilities – as well as the limitations – of urban social movements. Future research might think much more specifically about infrastructure and citizenship. With the restructuring of economies and consequent changes in political systems, the nation state's role as the provider of citizenship and democracy has been challenged, with new notions of legal rights emerging, for example, which stresses the social struggle through which they are won or lost (Dikec 2005: 47). This might involve conceptual research, but it could just as easily look at the actual practices of social movements such as the Berliner Wassertisch and Berliner Energietisch, and the politics they have helped to create.

Conclusion: The 'Right to Infrastructure' in a post-networked city

> A Right to the City means for us the right to stay in the inner city, even if we are not rich, young or creative. (…) It is about being able to participate in the development of the city we are living in.
>
> (Kotti & Co 2014: 354; translation by the authors).

This chapter has shown that the reconfiguration of infrastructure networks in Berlin has been less of an issue of technological networks, about pipes and plants, but more about socio-political governance, especially around ownership, prices/profits and environmental sustainability. While in the energy sector a re-orientation to more sustainable means of electricity and heat generation played a role, in the water sector there has not been a debate about a technological "re-making" of infrastructure networks. In both cases issues of quality of service provision were not addressed prominently. In that sense, only selected components of water and energy systems have been problematised and made "political".

Second, the idea of a *Bürgerstadtwerk* entails a vision and *Leitbild* for infrastructure and the city which aim to be socially and ecologically more just. However, labour conditions have played an ambivalent role within both struggles, despite their prominence in many other examples of anti-privatisation campaigns around the world. This is because trade unions have been supportive of many privatisations and their members have achieved secure (well-paid) jobs in privatised companies. Hence they have been very sceptical about the efforts for remunicipalisation. As issues of labour are generally an inseparable part of "Right to the City" movements, the links between remunicipalisation campaigns and trade unions remains an uncertain and open issue for future infrastructural struggles.

A third insight from Berlin concerns the stability or fluidity of urban social movements and the effects this can have on RTC campaigns. Despite its success in the public referendum the *Wassertisch* no longer acts as a single organisation. This is also the case with the "Tempelhof" initiative which, despite its success, is characterised by internal conflicts. In contrast, despite the failure of the energy referendum, *Berliner Energietisch* still exists. This might be due to the fact that the energy initiative consists of more experienced activists. Both campaigns benefitted from the fact that their issues became important to more than the "usual suspects" of urban social movements in Berlin, such as *Die Linke* and environmental organisations. They were able to build coalitions and to exit the "left niche".

It remains to be seen whether the remunicipalisation debate in Berlin is able to go beyond questions of ownership. Indeed, it is important to remember that remunicipalisations are not necessarily progressive, and that a return to the state is not enough. The crisis of municipal enterprises in the 1990s, and the commercialisations of services which followed, illustrate that public ownership is not a guarantee of fairer, democratic governance, let alone emancipatory urban infrastructure politics (Lederer and Naumann 2011). We suggest that remunicipalisation movements might promote those issues that Andrew Cumbers has identified as being crucial to furthering economic democracy, and what we see as being integral to the assertion of a right to urban infrastructure: community control, distributional justice, environmental sustainability, justice and enhanced participation (Cumbers 2012: 165).

Notes

1 In the global South this importance has already been revealed in cities such as Mumbai by Gandy (2006) and McFarlane (2008), among others.

2 One of the main proponents of a "Right to the City" within academia, Peter Marcuse, made a similar argument in 1998, stating that the reconstruction of Berlin should be seen as an ideological project to transform Berlin into a 'normal' business-friendly city, erasing its historical and political eccentricities.

3 The concession for Berlin's gas network has also ended and bidding for a new concessionaire has started (Zawatka-Gerlach 2014).

References

Attoh, K. A. (2011) 'What kind of right is the right to the city?' *Progress in Human Geography* **35**(5): 669–685.

Bakker, K. (2003) *An Uncooperative Commodity: Privatising Water in England and Wales*. Oxford: Oxford University Press.

Berliner Energietisch (2013) Seven good Reasons… for a community-owned energy supply. http://www.berliner-energietisch.net/argumente/7-gute-gruende-deutsch-und-tuerkisch (accessed 10 September 2014).

Berliner Wassertisch (2013) *Berlin Water Charter*. http://berliner-wassertisch.net/assets/docs/Berlin%20Water%20Charter.pdf (accessed 10 September 2014).

Berlo, K. And O. Wagner (2013): *Stadtwerke-Neugründungen und Rekommunalisierungen. Energieversorgung in kommunaler Verantwortung: Bewertung der 10 wichtigsten Ziele und deren Erreichbarkeit*. Wuppertal: Wuppertal Institut für Klima, Umwelt, Energie.

Bernt, M., B. Grell and A. Holm (Eds.) (2013) *The Berlin Reader. A Compendium on Urban Change and Activism*. Bielefeld: transcript.

Beveridge, R. (2012) *A Politics of Inevitability: The Privatisation of the Berlin Water Company, the Global City Discourse and Governance in the 1990s*. Wiesbaden: VS.

Beveridge, R. and K. Kern (2013) 'The Energiewende in Germany: Background, Developments and Future Challenges.' *Renewable Energy Law and Policy Review* **4**(1): 3–12.

Beveridge, R. and M. Naumann (2013): 'The Berlin Water Company. From "Inevitable" Privatization to "Impossible" Remunicipalization.' In: Bernt, M.; Grell, B.; Holm, A. (Eds..): *The Berlin Reader. A Compendium on Urban Change and Activism*. Bielefeld: transcript (Urban Studies), 189–203.

Beveridge, R. and M. Naumann (2014) Global norms, local contestation: privatisation and de/politicisation in Berlin. *Policy and Politics* **42**(2): 275–291.

Beveridge, R., F. Hüesker and M. Naumann (2014) 'From post-politics to a politics of possibility? Unravelling the privatization of the Berlin Water Company.' *Geoforum* **51**(1): 66–74.

Bontrup, H.-J. and R.-M. Marquardt (2011) *Kritisches Handbuch der deutschen Elektrizitätswirtschaft*. 2nd Edition. Berlin: editon sigma.

Bundeskartellamt (2012) *Pressemeldung – Bundeskartellamt erlässt Preissenkungsverfügung gegen Berliner Wasserbetriebe von insgesamt 254 Mio. € für die Jahre bis 2015*. Bonn: Bundeskartellamt.

BürgerEnergie Berlin (2014) *Fragen und Antworten*. http://www.buerger-energie-berlin.de/faq (accessed 7 August 2014).

Brenner, N., P. Marcuse and M. Mayer (Eds.) (2011): *Cities for People, not for Profit: Critical Urban Theory and the Right to the City*. New York: Routledge.

Cochrane, A. and A. Passmore (2001) "Building a national capital in an age of globalisation." *Area* **33**(4): 341–352.

Corsín Jíménez, A. (2014) "The right to infrastructure: a prototype for open source urbanism." *Environment and Planning D* **32**(2): 342–362.

Cumbers, A. (2012) *Reclaiming Public Ownership. Making Space for Economic Democracy*. London: Zed Books.

Dikec, M. (2005) (In)justice and the 'Right to the City': The case of French National Urban Policy. In D. Wastl-Walter, L. Staeheli and L. Dowler (Eds.), *Rights to the City*. IGU – Home of Geography Publication Series Volume III. Rome: Societa Geografica Italiana, 45–53.

Ellger, C. (1992) 'Berlin: Legacies of Division and Problems of Unification.' *Geographical Journal* **158**(1): 40–46.

Fitch, K. (2007) *Liquidating the public sector? Water privatization in France and Germany*. Dissertation available from ProQuest, http://repository.upenn.edu/dissertations/AAI3260904 (accessed 1 January 2007).

Füller, H. and B. Michel (2014) "'Stop being a tourist!' New dynamics of urban tourism in Berlin-Kreuzberg." *International Journal of Urban and Regional Research* **38**(4): 1304–1318.

Gandy, M. (1997) "The Making of a Regulatory Crisis: Restructuring of New York City's Water Supply." *Transactions of the Institute of British Geographers* **22**(3): 338–358.

Gandy, M. (2004) "Rethinking urban metabolism: Water, space and the modern city." *City* **8**(3): 363–379.

Gandy, M, (2005) "Cyborg Urbanization: Complexity and Monstrosity in the Contemporary City." *International Journal of Urban and Regional Research* **29**(1): 26–49.

Gandy, M. (2006) "Landscapes of disaster: water, modernity and urban fragmentation in Mumbai." *Environment and Planning A* **40**(1): 108–140.

Gebhardt, D. and A. Holm (2011) Initiativen für ein Recht auf Stadt. In A. Holm and D. Gebhardt (Eds.), *Initiativen für ein Recht auf Stadt. Theorie und Praxis städtischer Aneignungen*. Hamburg: VSA, 13–21.

Gornig, M. and H. Häußermann (2002) "Berlin: economic and social change." *European Urban and Regional Studies* **9**(4): 331–341.

Graham, S. and S. Marvin (2001) *Splintering Urbanism. Networked Infrastructures, Technological Mobilities and the Urban Condition*. London/New York: Routledge.

Graham, S., R. Desai R. and C. Mcfarlane (2013) "Water wars in Mumbai." *Public Culture* **25**(1 69): 115–141

Grell, B. (2014) Recht auf Stadt. In B. Belina, M. Naumann and A. Strüver (Eds.), *Handbuch Kritische Stadtgeographie*. Münster: Westfälisches Dampfboot, 238–243.

Heynen, N., M. Kaika and E. Swyngedouw (Eds.) (2005) *In the Nature of Cities*. London: Routledge.

Harvey, D. (2012): *Rebel Cities. From the Right to the City to the Urban Revolution*. London: Verso.

Holm, A. (2013) Berlin's Gentrification Mainstream. In M. Bernt, B. Grell and A. Holm, Eds., *The Berlin Reader. A Compendium on Urban Change and Activism*. Bielefeld: transcript, pp.171–187.

Holm, A. (Ed.) (2014): *Reclaim Berlin. Soziale Kämpfe in der neoliberalen Stadt*. Hamburg, Assoziation A.

Holm, A. and D. Gebhardt (Eds.) (2011) *Initiativen für ein Recht auf Stadt. Theorie und Praxis städtischer Aneignungen*. Hamburg: VSA.

Holm, A. and A. Kuhn (2011): 'Squatting and Urban Renewal: The Interaction of Squatter Movements and Strategies of Urban Restructuring in Berlin.' *International Journal of Urban and Regional Research* **35**(5): 644–658.

Holm, A., K. Lederer and M. Naumann (Eds.) (2011) Linke *Metropolenpolitik. Erfahrungen und Perspektiven am Beispiel Berlin*. Münster: Westfälisches Dampfboot.

Hüesker, F. (2011) *Kommunale Daseinsvorsorge in der Wasserwirtschaft: Auswirkungen der Privatisierung am Beispiel der Wasserbetriebe Berlins*. Munich: oekom.

Hüesker, F., T. Moss and M. Naumann (2011) 'Managing Water Infrastructures in the Berlin-Brandenburg Region between Climate Change, Economic Restructuring and Commercialisation.' *Die Erde* **142**(1–2): 187–208.

Kaika, M. (2004) *City of Flows: Modernity, Nature and the City*. London: Routledge.

Kotti & Co (2014) Kotti & Co und das Recht auf Stadt. In A. Holm, Ed., *Reclaim Berlin. Soziale Kämpfe in der neoliberalen Stadt*. Hamburg: Assoziation A, 343–354.

Krätke, S. (2004) "City of talents? Berlin's regional economy, socio-spatial fabric and 'worst practice' urban governance." *International Journal of Urban and Regional Research* **28**(3): 511–529.

Landeswahlleiterin Berlin (2013) *Volksentscheid "Neue Energie" am 3. November 2013. Ergebnis des Volksentscheides*. https://www.wahlen-berlin.de/abstimmungen/VE2013_NEnergie/ErgebnisUeberblick.asp?sel1=6052&sel2=0797 (accessed 7 August 2014).

Lederer, K. and M. Naumann (2011) Linke Metropolenpolitik und öffentliche Unternehmen. Eine Bestandsaufnahme am Beispiel Berlins. In A. Holm, K. Lederer and M. Naumann, Eds., *Linke Metropolenpolitik. Erfahrungen und Perspektiven am Beispiel Berlin*. Münster: Westfälisches Dampfboot, 128–144.

Lefebvre, H. (1996 (1968)) 'The Right to the City.' In E. Kofman and E.Lebas (Eds.), *H. Lefebvre, Writings on Cities*, Cambridge, Mass: Blackwell, 63–184.

Marcuse, P. (1998) "Reflections on Berlin: the meaning of construction and the construction of meaning." *International Journal of Urban and Regional Research* **22**(2): 331–338.

Marcuse, P. (2012) Whose Right(s) to What City? In N. Brenner, P. Marcuse and M. Mayer, Eds., *Cities for People, not for Profit: Critical Urban Theory and the Right to the City*. New York: Routledge, 24–41.

Matecki, C. and T. Schulten (Eds.) (2013) *Zurück zur öffentlichen Hand? Chancen und Erfahrungen der Rekommunalisierung*. Hamburg: VSA.

Mayer, M. (1993) The Career of Urban Social Movements in West Germany. In R. Fisher and J. Kling, Eds. *Mobilizing the Community. Local Politics in the Era of the Global City*. London, Sage, pp.149–170.

Mayer, M. (2011) The "right to the city" in urban social movements. In N. Brenner, P. Marcuse and M. Mayer, Eds., *Cities for People, not for Profit: Critical Urban Theory and the Right to the City*. New York: Routledge, 63–85.

McFarlane, C. (2008) "Sanitation in Mumbai's informal settlements: state, 'slum' and infrastructure. *Environment and Planning A* **40**(1): 88–107.

McFarlane, C. and J. Rutherford (2008) "Political infrastructures: governing and experiencing the fabric of the city." *International Journal of Urban and Regional Research* **32**(2): 363–74.

Mitchell, D. (2003) *The Right to the City: Social Justice and the Fight for Public Space*. New York: Guilford.

Monstadt, J. (2004) *Die Modernisierung der Stromversorgung. Regionale Energie- und Klimapolitik im Liberalisierungs- und Privatisierungsprozess*. Wiesbaden: VS.

Monstadt, J. and U. von Schlippenbach (2005) *Privatisierung und Kommerzialisierung als Herausforderung regionaler Infrastrukturpolitik. Eine Untersuchung der Berliner*

Strom-, Gas- und Wasserversorgung sowie Abwasserentsorgung. Berlin: German Institute of Urban Affairs (netWORKS-Paper, Nr. 20).

Moss, T. (2014) "Socio-technical change and the politics of urban infrastructure: managing energy in Berlin between dictatorship and democracy." *Urban Studies* **51**(7): 1432–1448.

Moss, T., S. Becker and M. Naumann (2014) 'Whose energy transition is it, anyway? Organisation and ownership of the Energiewende in villages, cities and regions.' *Local Environment*, DOI: 10.1080/13549839.2014.915799.

Mullis, D. (2013) 'Recht auf die Stadt. Facetten und Möglichkeiten einer Parole.' *Emanzipation* **3**(2): 57–70.

Mullis, D. (2014) *Recht auf die Stadt. Von Selbstverwaltung und radikaler Demokratie*. Münster: Unrast.

Peck, J. (2012) "Austerity Urbanism." *City* **16**(6): 626–655.

Purcell, M. (2013) 'The right to the city: the struggle for democracy in the urban public realm.' *Policy and Politics* **43**(3): 311–327.

Rutherford, J. (2014) 'The Vicissitudes of Energy and Climate Policy in Stockholm: Politics, Materiality and Transition.' *Urban Studies* **51**(7): 1449–1470.

Rutherford, J. and O. Coutard (2014) 'Urban Energy Transitions: Places, Progresses and Politics of Socio-technical Change.' *Urban Studies* **51**(7): 1353–1377.

Scharenberg, A. and I. Bader (2009) 'Berlin's waterfront site struggle.' *City* **13**(2–3): 325–335.

Senatsverwaltung für Finanzen (2013) *Rückkauf der Berliner Wasserbetriebe abgeschlossen*. http://www.berlin.de/sen/finanzen/vermoegen/nachrichten/artikel.30112.php (accessed 6 August 2014).

Soja, E. W. (2010) *Seeking Spatial Justice*. Minneapolis/London: University of Minnesota Press.

Souza, M. L. de (2010) 'Which right to which city? In defence of political-strategic clarity.' *Interface* **2**(1): 315–333.

Swyngedouw, E. (2004) *Social Power and the Urbanisation of Water: Flows of Power*. Oxford University Press.

Swyngedouw, E. (2009) 'The Antinomies of the Postpolitical City: In Search of a Democratic Politics of Environmental Production.' *International Journal of Urban and Regional Research* **33**(3): 601–620.

Ugarte Chacón, B. (2012) *Berlin Bank Skandal. Eine Studie zu den Vorgängen um die Bankgesellschaft Berlin*. Münster: Westfälisches Dampfboot.

Vattenfall GmbH (2013) *Mediakampagne* [online]. Berlin, Vattenfall. http://www.vattenfall.de/de/berliner-energiewende/mediakampagne-berlin.htm [accessed 8 October 2013].

Wissen, M. and M. Naumann (2006) "A new logic of infrastructure supply: the commercialization of water and the transformation of urban governance in Germany." *Social Justice* **33**(4): 20–38.

Zawatka-Gerlach, U. (2014) "Grüne wollen die Gasag retten." *Der Tagesspiegel*, 31 January 2014.

8 A political economy of urban solid waste management in emerging countries

Learning from Vitória (Brazil) and Coimbatore (India)

Jérémie Cavé

Introduction

Municipal solid waste (MSW) is something of an "*orphan child*" (Melosi 2000) in urban service studies. It is nevertheless acquiring major importance today because of its obvious environmental impact, particularly in emerging countries, where waste quantities are starting to soar and waste content is becoming more diverse. At the same time, industrial growth in emerging countries is playing a key role in the rise in worldwide demand for raw materials, contributing to a diminution in easily exploitable sources and to the rising prices of these materials. In these circumstances, urban waste is increasingly perceived as a potential (secondary) source of materials.

Having never been implemented in the form of a "pipe-shaped" infrastructure, solid waste management (SWM) services are more adjustable than "hard" physical networks: their "soft" forms (Debout 2012) can arguably be more easily and quickly adapted to contemporary social needs and technical constraints. Because of this flexibility, this essential urban service may help shed an early light on transformations currently taking place in all urban services.

Emerging countries appear as a particularly promising field for innovations. Being neither part of the group of "rich", long-industrialised countries, nor among the poorest ones where policies are largely shaped by foreign and international donors (Cavé 2008), cities in emerging countries are highly unequal from a socio-economic point of view, but they are also places of soaring industrial development. Emerging countries thus provide interesting contexts in which to study issues of solid waste management and to explore the implications of the "*bipolar*" nature of waste (O'Brien, 1999: 271) – sometimes "*garbage*", sometimes "*resource*" (Bertolini 1992). Indeed, this dual nature is particularly true in highly unequal societies, where what the rich throw away as garbage may have value for the poor. In addition, in many urban contexts in emerging countries today, solid waste management policies are being reformulated: dumping sites are banned, hygienic landfills are imposed, separate collections (i.e. door-to-door collection of recyclable waste) and recycling schemes are starting to be implemented by public authorities in order to avoid simply burying all urban residues.

This "modernisation" process has raised the issue of the banning of waste picking. The various kinds of eviction processes or threats aimed at informal wastepickers[1] in southern countries in general have been extensively documented (cf. among others: Coing and Montaño 1985; Furedy 1992; Medina 1999; Nas and Jaffe 2004; Medina 2005; Scheinberg and Anschütz 2006; Wilson, Velis and Cheeseman 2006; Chaturvedi 2007; Durand 2010; Fernández 2010; Gerdes and Gunsilius 2010; Scheinberg, Wilson and Rodic 2010; Scheinberg, Simpson, Gupt et al. 2010). However, wastepickers are only one part of a larger picture of non-governmental recovery agents and systems at work today in cities in emerging countries, where a diversity of informal and formal private recovery approaches now treat urban solid waste as a valuable resource. As a result, the so-called "modernisation" process brings clashes between the actors: some experts talk of "*competition*" (Scheinberg, Simpson, Gupt et al. 2010: 8), others refer to "*interceptions*" (Besen 2006: 6) or "*disputes*" (Demajorovic, Besen and Rathsam, 2005) and others invoke "*contested*" waste (Fahmi and Sutton, 2010).

We argue that appropriation conflicts over MSW arise, pitting a variety of actors against each other: municipal authorities, private operators, wastepickers, residents' associations, industrial companies, etc. In this chapter, we ask two questions: *what are the factors underlying these urban solid waste appropriation conflicts*? And *what do they tell us about emerging forms of post-networked urban services*?

To explore these questions, we need to broaden the SWM perspective beyond the municipal actors alone. We will use a systemic approach (as defined in UN-Habitat 2010) that allows us to account for the fact that SWM is not only concerned with neutralising a nuisance, but also, increasingly, with recovering a valuable resource. To put it into economic terms, the municipal SWM service is no longer engaged only with dealing with a public good (or 'public evil'), namely 'garbage', but with managing private goods (resources) *as well*. Furthermore, we will consider three major functions of SWM: undifferentiated collection, selective recovery, and disposal.

This approach was implemented within an in-depth empirical study conducted in India and Brazil, in the context of the author's PhD research (Cavé 2013). "Ordinary cities" (Robinson 2006) were preferred to the biggest megacities, so as to be able to capture a picture of the city in all its diversity and complexity:

> *...whereas global and world city approaches focus on small elements of cities that are connected into specific kinds of economic networks, and developmentalist approaches tend to emphasise the poorest, least well provisioned parts of the city, the ordinary city approach brings the city 'as a whole' back into view.*
>
> (Robinson 2006: 10)

After a thorough selection process based on several criteria (per capita GNP, diversity of actors, evolving MSWM situation, political stability, lively industrial activity, accessibility for field work, etc.), two cities were chosen (cf. also Table 8.1):

- Vitória, situated on the south-eastern coast of Brazil, in the state of Espírito Santo. The city lies at the heart of a large urban conurbation composed of three other cities. It is Latin America's biggest natural harbour and the main gateway for the worldwide export of ores from Minas Gerais;
- Coimbatore, located in Tamil Nadu state in South India. The third largest city in India's most industrial state has a long history of textile and mechanical industry, and was once called "The Manchester of India".

Empirical material was mainly gathered from over 130 qualitative semi-structured interviews. All groups of actors dealing with solid waste, not only municipal service agents, were sampled.

Although estimates of solid waste quantities often lack transparency and reliability, it can nonetheless be inferred that the inhabitants of Vitória produce substantially more waste than their Indian counterparts. This difference may be related to per capita GDP, which is seven times higher in Brazil than in India.

The argument will be unfolded in two steps.

First, the total solid waste streams in both cities will be described, with a focus on conflicts about the appropriation of waste, and theories from the field of institutional economics will be deployed to unpick the socio-territorial dynamics of this essential urban service. This analysis will consider the entire "deposit" (by analogy with deposits of natural resources such as gas or coal) of urban solid waste, apprehended through the concept of Common Pool Resources (CPR) developed by Ostrom (1990). Indeed, in the academic literature, there have been attempts by a few researchers to analyse solid waste as commons (Bose and Blore 1993; Ruet 2002; Chaturvedi and Gidwani 2010; Lane 2011),but these analyses exclude part of the sector and focus exclusively on recyclable waste. We contend that it is analytically more fruitful to treat the entire deposit of urban solid waste as CPR, and that it is also normatively more relevant to manage solid waste through the allocation of usage rights, as the CPR approach suggests.

Second, we will discuss how changes in this "soft" urban service in emerging countries resonate with contemporary debates on post-networked urbanism. Because of its inherent characteristics, the SWM sector enables us to re-explore the relationship between the delivery of essential services and a centralised and monopolistic infrastructure. We will argue that resource economics provides a

Table 8.1 Comparing Vitória and Coimbatore conurbations

	Population	*Waste generation*	
		Tons per day	*Kg per person per year*
Vitória	1,484,800	1,111	271
Coimbatore	1,482,000	882	217

(Source: Cavé 2013: 54)

more heuristic perspective on this subject than network economics and suggests that a usage-rights-based semi-decentralised regime of solid waste management could prove both socially more progressive and functionally more efficient.

Ultimately, this chapter aims to make a contribution to a political economy of the gradual shift from a network economy based on a 'linear' metabolism and the presumption of unlimited resources, to a more circular metabolism acknowledging resource scarcity and growing global demand for raw materials.

Conflicts over Waste Flows in Two Ordinary Cities

In our exploration of contemporary changes in SWM, we focus on conflicts that reveal the tensions at work. Conflicts over solid waste arise in particular when public authorities try to incorporate recovery objectives into a municipal service previously structured around two main activities – undifferentiated collection (upstream) and dumping (downstream) – in contexts where market economy waste recovery mechanisms are already in place, working on source-segregated waste (upstream) together with recovery and recycling circuits (downstream). As a recent report puts it, the point is that "*while the commodity value of materials is taken for granted, the service aspect of recycling is relatively new everywhere*" (Scheinberg, Wilson and Rodic 2010: 2).

Waste management in Vitória: institutionalised informality

In Vitória's urban area, a variety of solid waste management mechanisms are in place.

First, the municipal government has contracted out the delivery of the MSWM service to a private Brazilian company, Vital. Vital operates collection throughout the city and transports the SW to a private sanitary landfill, operated by another private firm named Marca. The private operator is paid by the Municipality on the basis of the weight of the waste collected. This municipal service is partially financed through a tax paid by households.

Second, the *catadores* (Brazilian wastepickers) pick solid waste items (i.e. recyclable waste: paper, cardboard, plastics and metals) out of domestic waste and sell them to local merchants. These merchants also obtain items directly from local people who bring them to their warehouses to sell. The merchants then sell their own stock to a few large brokers for eventual nationwide industrial recycling.

Third, some *catadores* have formed co-operative self-help groups. The Vitória municipal authorities recognise these structures and give them the solid waste that they have collected separately throughout the city for free. The self-help groups also get materials from their own network of suppliers. They then sell the solid waste to specialised brokers, at a better price than independent *catadores* get from smaller dealers.

Figure 8.1 gives an overview of SW flows in Vitória.

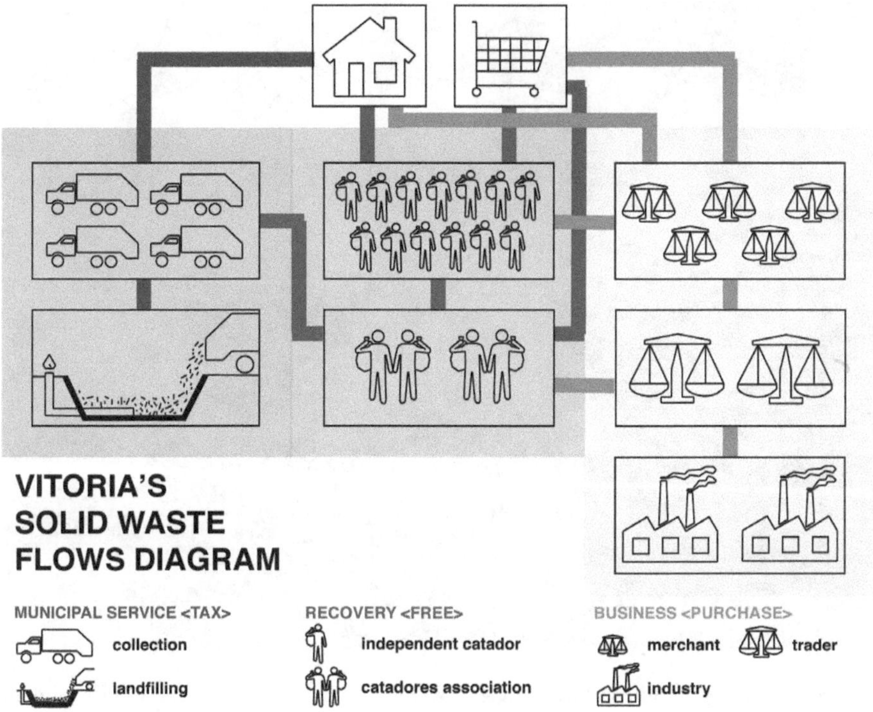

Figure 8.1 Vitória's solid waste flows: systemic diagram
Source: Cavé 2013: 125.

Each of these solid waste management processes is characterised by a specific form of access to waste and interaction with citydwellers:

- One is based on a tax: Vitória's residents have to pay for household waste to be collected and taken away by the municipal solid waste management service.
- Another is a real business sector, where materials are bought and sold. In this field, residents can actually sell their solid waste items to local traders and obtain revenue from them (Figure 8.2).
- In between is a domain characterised by free access to waste, occupied by *catadores*, either organised or independent. In this field, informal actors pick solid waste items out of household bins without any economic transaction with households.

These different mechanisms coexist throughout the conurbation and maintain mutual relations; none is completely autonomous from the others. The most coveted segment of the MSW deposit is solid items: metals, papers, cardboards and plastics.

Figure 8.2 A paper and cardboard trader's warehouse in the outskirts of Vitória. Unlike merchants, traders need a large space to store materials and cannot settle downtown. In front of the building an empty wooden cart can be seen, which belongs to a wastepicker

Source: Cavé 2011.

Competition in the access to and appropriation of waste by various agents from these three different fields has provoked a series of clashes that hamper the municipal SWM service.

Recovery agents vs. municipal service operators: The actors in the recycling chain have everything to gain by extending the exploitation of urban household waste. They perceive solid waste as a resource and acquire it free of charge (gift) or following a commercial transaction (exchange). Nonetheless, under the contracting-out process, the municipal authority pays a private service provider for the collection, transport and storage of all municipal waste. For that reason, these private operators, who are paid in proportion to the tonnage they handle, see all sources of solid waste as a resource and have every interest in maintaining at a high level the flow of waste destined for landfill. This has led to clashes, for example in the wealthy residential district of Jardim Camburí, between *catadores* and municipal service personnel, when the latter found that the *catadores* are "creaming off" the most lucrative items just ahead of the municipal garbage collection.

Municipal selective collection service vs. individual catadores: Second, once the Municipality proposes to establish a selective collection service, the informal solid waste recovery role performed by the *catadores* ceases to be ignored and begins to be seen as a problem. This phenomenon was highlighted in Vitória by the former director of the municipal department, in a study relating to aluminium cans (Baptista 2001). The author shows how, when the resale price of used cans doubled between February and September 1999, rising from 700 BRL/T to 1500 BRL/T, the quantity collected by the municipal service fell by more than 50%, from eight tons per month to only three tons. In other words, when aluminium prices rose, the *catadores* realised that collecting cans from waste was becoming lucrative, with the result that these items were intercepted in massive quantities. So the upstream skimming performed by the *catadores* prevents the municipal service from acquiring solid waste at precisely the time when the resale value is high. It makes absolutely no economic sense to run a system intended specifically to collect solid household waste, if the most lucrative contents of that waste are diverted. As a result, therefore, the municipal authorities began to combat wastepicking.

The catadores' *associations vs. the brokers*: Finally, many administrative and commercial institutions were in the habit of disposing of their packaging waste by giving it to the *catadores*' associations. It was a practical way to resolve a logistical problem while also appearing charitable. However, since 2006, commercial organisations have been obliged to draw up a waste management plan. This notably includes the requirement to pay in order to ensure that their refuse ends up in the right place. To offset this new cost, the increasing tendency among supermarkets, shopping centres and industries is to sell their solid waste to brokers.

> *The firm Aparas Vitória*[2] *buys* [...] *packaging waste from supermarkets. For example, the Carone supermarket used to give its solid waste to the* catadores, *but now they have stopped, because they sell it!*
>
> (Ana – Caixa Econômica Federal[3] – 07/04/10)

> *During the last holidays, we received nothing for a day and a half... Why? Because the Vitória Shopping* [mall] *stopped giving away* [its packaging waste] *and began to sell it. That represented 5 tons per month for us!*
>
> (Almeirinda – municipal social worker supporting the Ascamare *catadores* association – 10/08/10)

> *In the condomínios*[4] *they separate solid and liquid waste. And a lot of condomínios are selling their waste! So, will they go on giving it to the catadores?* [...] *The catadores are going to have to adapt...*
>
> (Célia – SEBRAE[5] – 10/08/10).

Even certain public institutions are shifting from donation to commercial exchange by selling to brokers, despite being legally obliged to give their solid waste to the *catadores* associations.

Waste management in Coimbatore: a sought-after business

In Coimbatore, the implementation of a new municipal SWM scheme in 2008 also revealed the presence of several coexisting waste management mechanisms.

First, the municipal authorities provide a public solid waste collection service: public agents from the Coimbatore Municipal Corporation (CMC) collect household waste from street bins and bring them to four transfer stations. The treatment and landfill service has been contracted out to a private Indian consortium (UPL Environmental Engineers Limited) in a twenty-year public-private partnership (PPP). Besides transporting solid waste to a site in the outskirts, UPL's task consists of composting organic waste and burying residual waste in a landfill (Figure 8.3). The private operator gets its revenues from gate fees, compost sales and solid waste resale.

Second, there is also a highly structured informal recovery sector in the city. This can be divided into four different types of practitioner: i) the wastepickers, now banned from the old dump sites, who pick up solid waste in the streets for nothing and resell it; ii) the merchants (sedentary or itinerant), who buy solid items previously sorted by the inhabitants themselves (metal items, old newspapers, plastic bottles etc.) and then sell them to brokers (see Figure 8.4); iii) the brokers, who are located in the outskirts and who aggregate materials, sort them, crush them, compact them and store them for selling to industry; and iv) the local recyclers or national industries, which use the recovered materials as inputs in production processes.

Figure 8.3 Collective selection in Coimbatore. A municipal employee operates the selective collection in one of Coimbatore's nine model wards. The narrowness of streets makes it difficult for trucks to circulate. An old lady prevents a sheep from eating the organic waste that the agent tries to aggregate

Source: Cavé 2010.

Figure 8.4 An old newspaper merchant's shop in the streets of Coimbatore. The materials gathered are sorted and packed, ready to be sold to a trader

Source: Cavé 2011.

Third, the Indian Tobacco Company (ITC), a huge Indian industrial group, has started to run an innovative scheme called "*Wealth Out of Waste*" (WOW). Its original aim was to extract used paper and cardboard to supply its own paper plants. ITC has joined forces with two partners: the waste management firm Ramky and the federation of residents' associations RAAC. The former is responsible for collection and sorting operations; the latter helps ITC to gain the trust of local people and to roll out its scheme without waiting for approval from the municipal authorities. As a result of this alliance, the system has been extended to all solid waste, including the recovery of plastics and metals. In practical terms, Ramky agents distribute big bags to households and then make frequent door-to-door passes to buy the solid items, pre-sorted by residents or domestic servants. The waste paper collected goes to the ITC factory, whereas Ramky sells the other solid waste to big brokers.

Figure 8.5 provides an overview of SW flows in Coimbatore.

The same three forms of access to waste and interaction with urban residents identified in Vitória are found here: tax, free access and market-oriented business. The free access area is smaller than in Vitória, given that wastepickers are not

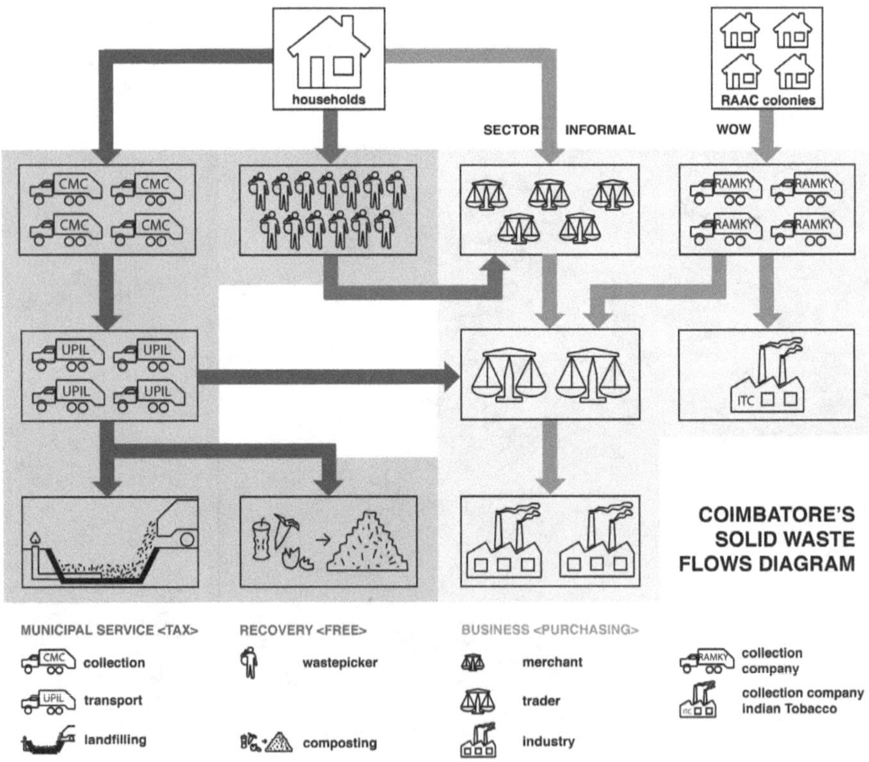

Figure 8.5 Coimbatore's solid waste flows: systemic diagram
Source: Cavé 2013: 160.

organised into collectives. Furthermore, the business segment is reinforced by the WOW scheme, which is designed to provide resource-recovery through formal door-to-door purchasing.

It is apparent that the actors in all three sectors are keen to get their hands on all solid household waste. As a result, there are inevitably clashes and eviction strategies. Specifically, three types of conflict can be identified.

The redesigned municipal scheme vs. the wastepickers: While the dumping site was still operating, UPL officials allowed some thirty wastepickers to access it. Since the landfill service began operations, downstream access is no longer free: the dumping sites have been closed and the wastepickers are barred from the new transfer platforms. In addition, access to the upstream waste is harder, since door-to-door collection has been instituted. The collection agents are thus moving onto the territory of the informal collectors. This practice is particularly damaging to the wastepickers, because the merchants pay households for their waste.

> *Ragpickers*[6] *are not allowed* [in the district], *because they would ruin our work. With the help of the authorities, we stop them coming.*
>
> (Mr S. – CMC – 18/12/09)

The WOW programme vs. the spontaneous recyclers: ITC's management presents WOW as a corporate social responsibility (CSR) scheme. According to them, the aim of this approach is to prevent recyclable waste being buried in landfills. Not everyone agrees with this claim:

> *It is not social responsibility! It's a question of raw materials: they want paper! They are killing the ragpickers.*
>
> (Mr P. – UPL – 21/12/09)

Behind the CSR rhetoric lies the wish to get hold of potential raw materials, which constitutes a threat to the livelihoods of wastepickers. In other words, WOW, which is implemented via the RAAC in concert with the leaders of wealthy neighbourhoods, ultimately gives ITC a territorial monopoly over household solid waste in these gated communities. This damages not just the informal waste collectors but also UPL (see below), which explains why its managers – at least in what they say – take the side of the wastepickers.

Municipal service vs. the WOW programme: In the wealthy residential district of Race Course Road there were clashes in 2009 between municipal collection staff and WOW operatives, who were on their rounds at the same time to buy the same recyclable material.

> *They* [ITC] *have nothing to worry about: a certain source of supply at a price they have set themselves, without being affected by fluctuations* [...]. *They are becoming a competitor!*
>
> (Mr P. – UPL – 21/12/09)

> *We said to CMC: "let us help you with the work."* [But] *They treat us like competitors, not like an NGO that wants to help!*
>
> (Mr R. – RAAC – 17/12/09)

It would seem that RAAC members have got the wrong end of the stick. In the PPP agreed with UPL, waste is no longer seen as a problem to be handled. Quite the contrary – it is about separating out the potentially valuable parts of the waste stream in order to generate additional revenues. In fact, the contract between CMC and UPL explicitly gives UPL an "exclusive right" over the flow of household waste handed over to it by CMC (CMC, 2008, Clause 2.2). That is why CMC was very quick to oppose the development of the WOW programme in the municipality, as evidenced by the positions expressed in several interviews:

> *The Municipal Commissionner* [...] *says: "this is my waste, not yours". I reply: "it's neither your problem nor mine: it's the city's problem."*
>
> (Mr R. – RAAC – 24.01.11)

> *It's nothing more than advertising, it's commercial...* [...] *No one is doing it for charity, you know... it has a cost... they get something out of it!* [...] *The Municipal Commissioner will go to court, if it turns out that this reduces our revenues.*
>
> (S. K. – CMC – 14/12/09)

> *CMC has precedence! The waste belongs to CMC. We won't give up a single item!* [bursts out laughing]. *Because there is the contract with UPL.*
>
> (Ravi K. – CMC – 18/12/09)

Given CMC's opposition to the WOW programme, in 2010 the latter was run exclusively in the city's "colonies",[7] as well as in the conurbation's periurban communities.

Appropriation conflicts at the core of (changes in) solid waste management

The empirical observation of urban SWM, at a time when public authorities are implementing separate municipal collection services, supports four major conclusions:

i in Vitória as in Coimbatore, three main solid waste management mechanisms coexist, above and beyond the municipal collection and disposal service;
ii the objective of all of them is to appropriate the residents' valuable solid waste (metals, paper and cardboard, plastics);
iii this kind of convergence generates rivalries and causes appropriation conflicts since, at city scale, the same object (an empty plastic bottle, for instance) can be collected by the municipal service (therefore representing a cost for the waste generator), sold to a merchant (therefore representing revenue for the

waste generator) or left to recovery agents (neither cost nor revenue for the waste generator).

iv however, regardless of whether they sometimes compete, there is always entanglement between the different actors and mechanisms.

When public authorities introduce separate collections, appropriation conflicts arise, making solid waste management a disputed sphere. In seeking to collect recyclable waste from city residents, municipal services encroach on the activities of other actors. This service, often outsourced to private operators, deprives informal recovery agents (or wastepickers) of resources. These conflicts reveal a whole chain of previously neglected stakeholders: informal merchants who buy solid waste from inhabitants, brokers who collect, sort and store materials, and sell them back to industry, and also large industrial groups which see waste as a reservoir of substitute raw materials and, ostensibly on public interest grounds, are becoming involved in their recovery. In other words, appropriation conflicts arise because of a converging desire for recyclable waste, which leads to eviction effects for the weakest or the least flexible actors.

We argue that such appropriation conflicts arise because of the dual character of a waste product:

i first, the particular status of waste: by definition something thrown away, waste is an object that no longer belongs to anyone. Waste is what has been abandoned, i.e. *res derelicta*, things over which their former owners have renounced their property rights.[8] This is why appropriation conflicts happen: it is a situation of resource overuse in a sector where property rights are not clearly defined;

ii second, however, the nature of waste is by no means intrinsic. As S. Lupton puts it: "*the status of solid waste fluctuates: a commodity becomes waste if it is abandoned, and becomes a private commodity again if it is re-appropriated*" (Lupton, 2011). Legally, an object always has a status, yet that status is constantly requalified, as it largely depends on existing recovery mechanisms. A specific plastic package may be seen as a useless residue until there is a shortage, at which point it is re-qualified as a commodity. As M. O'Brien puts it: "*this is why what is waste today will not be waste tomorrow and why what was, common-sensically, waste yesterday is now incorporated as an economic 'sector'*" (O'Brien 1999: 278).

A Political Economy Analysis of Urban Solid Waste Management

Precisely because of this fluctuating status, we suggest it is time to stop treating waste as an object and SWM issues in terms of the garbage/resource dichotomy, which ultimately implies a public/private good dialectic. Instead we shuld start considering urban solid waste holistically as a "deposit" (analogous to a natural gas or coal deposit). What, then, would be the (economic) characteristics of the urban solid waste deposit?

Considering solid waste as a common good...

The diversion of solid waste by non-governmental actors represents an obstacle to the implementation of separate collection under the municipal service. The potential economic revenues from the trading of solid waste items are substantial, as an extensive comparative study has clearly shown (Scheinberg, Simpson and Gupt 2010). The amounts at stake – together with the ownership claims to which they give rise – favour the proliferation of competing mechanisms. In Coimbatore, according to our calculations (confirmed by UPL representatives), the sale of solid items represents up to 15% of UPL's total revenues. Such an amount may explain why municipal authorities prohibit any upstream diversion. This economic perspective also explains why the municipal authorities opposed ITC's "*Wealth Out of Waste*" scheme, despite the fact that this initiative would apparently lighten their collection burden. Solid waste is considered as a source of profit rather than as a burden.

Door-to-door collection of source-separated materials is costly, as it entails the duplication of rounds, vehicles and bins. It is therefore desirable that flows should be as 'clean' (i.e. well segregated) as possible and that no "skimming" should occur prior to collection. Nonetheless, the centralised municipal service is always vulnerable to interceptions by independent operators who are much more flexible, both spatially (door-to-door; in certain neighbourhoods only) and temporally (before collection trucks arrive). In Vitória as in Coimbatore, non-governmental actors operate at neighbourhood scale, relying on local hubs (i.e. merchants and brokers). In Coimbatore, UPL is in charge of the end of the SWM chain: from transfer facilities to landfill. The prior steps are undertaken by municipal employees. However, many non-governmental actors can operate on streets and doorsteps to capture part of the solid waste stream and particularly the most lucrative part: solid items. Thus, whereas under its contract UPL should receive 100% of Coimbatore's waste, part of the SW flow is *de facto* intercepted by other more flexible actors: wastepickers, door-to-door junk buyers, local merchants and even municipal collection employees. In Coimbatore, ITC has introduced a mechanism that cuts the ground from under their rivals' feet: WOW staff distribute a large jute sack free to households, so that they can separate their solid waste items before they even put out their garbage.

Our empirical analysis of spatial and economic dynamics demonstrates that urban solid waste has the characteristics of a common good:

- From an economic point of view, the diversion of recoverable solid waste constitutes a coveted objective for several actors. To this extent, waste is a rival good.
- From a spatial perspective, the most lucrative items are gradually extracted from the SW deposit ('cherry picked') through several stages of upstream interception. The fact that this urban service is provided on the streets (and not through an underground network, like wastewater for instance) makes it possible and relatively easy for non-governmental actors to capture part of the flow. In other words, it is in practice impossible to exclude anyone from its appropriation.

Neither strictly a private nor a public good, urban solid waste is always simultaneously both. It always contains a mix of useless garbage and valuable items. In economic terms, the totality of solid waste within a city is therefore a common good.

... And the Urban Solid Waste Deposit as a Common Pool Resource

Our further contention is that the deposit of urban solid waste, in Coimbatore as in Vitória, should be regarded more precisely as a common pool resource.

Waste as stock and flows

As stated in the introduction, previous attempts to conceptualise solid waste as commons have focused exclusively on the "cream" of the deposit, the most sought-after component, solid items (Bose and Blore 1993; Ruet 2002; Chaturvedi and Gidwani 2010; Lane 2011). We argue, however, that a city's entire solid waste deposit should be taken into consideration. For that purpose, we suggest that a distinction should be made between waste flows and stocks: "flows" refer to the resource units that circulate (because of their value), whereas "stock" refers to the units that are not extracted from the resource (because they are regarded as valueless).

One portion of the urban solid waste deposit is quickly recovered or purchased and never actually ends up in a landfill. This part that is sufficiently valuable not to be discarded can be equated with flows. The rest of the deposit is permanently discarded and abandoned, of interest only to the municipal authorities. This fraction of the deposit, of zero or negative value, corresponds to a stock. There is no spontaneous incentive for anyone to get involved in its circulation. When municipal authorities outsource the waste collection and management service to private firms, they make this stock attractive by paying them according to the amount they collect. The deposit-stock then becomes a flow.

Mixed together, flows and stock constitute the solid waste deposit. This contiguity generates cross-contamination between the individual properties of the constitutive elements of the deposit, generating trash. For instance, a sheet of white paper brought into contact with a rotten tomato will soon become wet and stained, unusable for all intents and purposes. One could have been composted and the other recycled, but through mixing, the tomato and the paper become merely garbage. It is therefore primarily the act of mixing that turns materials that might have been valuable flow into valueless stock.

Recognising this flow-stock structure in the SW deposit enables us to penetrate further into the internal dynamics of this common good.

Addressing appropriation conflicts

In the wake of major empirical research on environmental resources that are managed outside both the State and market spheres (e.g. fishing grounds, groundwater, pasture), E. and V. Ostrom forged the concept of "common pool resources" (CPR) (Ostrom 1990), in an attempt to go beyond the "flat" notion of common goods. CPR are defined as a "*...resource system that is sufficiently large as to make it costly (but not impossible) to exclude potential beneficiaries from obtaining benefits from its use*" (Ostrom 1990: 44).

The concept of CPR is particularly useful in the analysis of the problems of resource overuse. Its main analytical input lies in the distinction between the resource as stock and the resource as flow. According to E. Ostrom, any resource system is composed of these two interdependent elements, flows and stocks; the resource flow corresponds to the unit withdrawal from the resource stock.

Such a distinction converges with our characterisation of the SW deposit: we have observed both eviction effects (within the service modernisation process) and resource overuse problems (of which appropriation conflicts are the symptom). We may therefore be able to use the CPR approach to tackle appropriation conflicts. According to E. Ostrom, "*appropriators* [of common pool resources] *are faced with some problems that are similar to those of appropriating private goods* [that is, rivalry] *and other problems that are similar to those of providing public goods* [that is, non-excludability]" (Ostrom 1990: 47).

According to Elinor Ostrom, such problems are of two main kinds. The first is rent dissipation. This refers, for instance, to the (economic) risk involved when separate door-to-door collection is implemented and the deposit is at the same time significantly creamed-off upstream. This problem can, in principle, be solved through the "*way of attributing a fixed, time-independent quantity of resource units* [to the various appropriators, so as] *to reduce uncertainty and conflict over the assignment of rights*" (Ostrom 1990: 64). The second problem consists in the attribution of spatial or temporal access to the resource, as reflected in the interception dynamics observed in Vitória and Coimbatore. According to E. Ostrom, these problems arise "*because spatial and temporal distributions of common resource units frequently are heterogeneous and uncertain*" (Ostrom 1990: 65).

Usage rights as a way to regulate appropriations

Assignment of usage rights as a way to regulate CPR issues

Given that the problems are due both to a degree of "*uncertainty*" and "*heterogeneity*" in the distribution of resource units, the double problem of rent dissipation and access to the resource may be addressed by mobilising usage rights. Usage rights have recently been theorised within the framework of Institutional Resource Regimes (IRR), which adds the understanding of public policies to Ostrom's analysis of the regulation of property rights (Varone, Nahrath and Gerber 2008: 4–5). In their effort to conceptualise institutional resource regimes, Varone, Nahrath and Gerber provide a definition of usage rights as an analytical and operational tool. Usage rights are realised in the privileged access to a flow of resource units. Unlike property rights, usage rights determine "*who might have what use of which quantity of the resource, in the form of which goods and services derived from it*" (Varone, Nahrath and Gerber 2008: 7). Usage rights may derive from public policies that assign such rights to beneficiaries who are not formally owners. They thus refer to resource unit management and withdrawal rules that do not grant absolute freedom in the use of the resource.

Emanating from a combination of public and private law, usage rights seem an appropriate prism through which to tackle CPR. Taking into account the multi-segmented nature of the service and the economic value of part of the deposit, which make interceptions unavoidable, the assignment of targeted usage rights could offer an innovative way to resolve solid waste appropriation conflicts. This assignment would be officially guaranteed: "*a usage right is considered as such if some institutions protect its holder from other users potentially interested in the same flow of benefits*" (Varone, Nahrath and Gerber 2008: 5).

Treating the solid waste deposit as CPR points to a unifying principle: the proper disposal of refuse (the stock) in controlled landfills. The implementation of usage rights may allow this principle to be put into practice. Non-governmental solid waste interceptions could be accepted – and even stimulated – on condition that the refuse from every collection is strictly channelled towards adequate landfill facilities.

A semi-decentralised SWM service through the assignment of usage rights?

Would it then be possible to reconcile the solid waste recovery chain with a territorial public service? 'Waste' in the strict sense of the term (garbage) is the core of public SWM service. However, this should not prompt the authorities to exclude other players from its use; rather it should entail the obligation on those players to process all waste (gas, liquid and solid) appropriately. The notion of usage rights, applied spatially through the allocation of collection zones to the different actors, makes technical integration possible: waste collectors would be authorised to take part of the deposit, provided that the residues from their recycling processes are channelled to outlets approved by the municipal authorities. This would constitute genuine "integrated solid waste management".

Given their responsibilities, instead of claiming a monopoly over the entire solid waste deposit, urban public authorities could assign usage rights to different actors, public and private, formal and informal. This would give rise to a semi-decentralised service. In such a system, all diversion schemes would be implemented in a centralised way, so as to include not only the wastepickers, but also the informal brokers as local recycling agents. Furthermore, nonrecyclable waste, as well as all the residues arising from the different recycling schemes, would be channelled towards a centralised landfill infrastructure.

Apart from the functional integration of the actors in the recovery chain, this kind of approach would also guarantee integrated pricing for users: the revenues from the resale of recyclable waste could be used to offset the tax for disposing of garbage. This new form of integration would be a way of implementing a "pay-as-you-throw" charging policy in the cities of the global South.

SWM and the debate on post-network urbanism

Networks challenged: a resource perspective

The field of research into service delivery in urban areas through network infrastructure has been grounded in Northern countries, with a primary emphasis on water and energy provision (Tarr and Dupuy 1988; Coing, Lara (de), and Montaño 1989; Offner, 1993; Barraqué 1995; Lorrain, 1995; Coutard, 1999), i.e. basic urban supplies flowing through "physical" networks of pipes or cables.

Today, this perception of the urban fabric as being structured by unitary networks that satisfy basic needs is facing strong challenges. On the one hand, these kinds of supply-based, linear, centralised infrastructures are questioned on environmental grounds, at a time when they face a number of difficulties, notably financing infrastructural renewal and adapting to falling consumption in shrinking cities (Petitet & Schneier-Madanes 2005; Coutard & Rutherford, 2009; Barles 2010; Coutard, 2010; Féré and Scherrer 2010). On the other hand, network-based urbanism has never become universal in the cities of the global South, where this kind of normative perspective has, in practice, generated socio-spatial segregation rather than urban cohesion (Navez-Bouchanine 2002; Bakker, 2003; Swyngedouw, 2004; Jaglin, 2005; Petitet and Schneier-Madanes 2005; Giraud et al., 2006; Botton, 2007; McFarlane, 2008).

The SWM sector can shed interesting light on these issues. First, it offers methods of public service delivery that generate social inclusion, not by providing the same service to every urban resident, but rather by incorporating informal recovery agents into the municipal service. From the user perspective, it could even be argued that the separate collection service should primarily focus on those urban areas that produce the largest quantities of the most lucrative waste, whether commercial (shopping centres, offices) or upper-class residential (Indian *colonies*, Brazilian *condominios*) zones. Second, the municipal SWM service would appear to be a "soft" service shaped by environmental engineering (Barraqué 1993; Melosi 2000) and to be moving away from the "sanitary engineering" model. Urbanites are no longer just recipients of a service, but also producers of a valuable resource, the management of which should include recovery loops rather than a single, linear evacuation route.

Some may argue that the diversity of management and recovery mechanisms observed in both cities is nothing but an intermediate step in urban and industrial development, and that the public service mechanism will ultimately take over. In a global context of increasing raw material shortages and rising industrial demand, such a linear historical outlook seems somewhat vulnerable. At the local level, appropriation conflicts raise questions about the viability of a centralised and monopolistic SWM system. We thereby argue that the SWM issue should not be understood through the prism of classical network economics (toll goods), but rather through the lens of resource economics (common goods, common pool resources).

Unlike water or electricity, waste does not much resemble a flow with uniform characteristics, which lends itself to solutions based on circulation through channels and on processing by techniques that exploit economies of scale. In addition, it is the act of mixing different materials and substances that generates trash. Consequently, the objective is not to maximise the quantities of waste (in order to generate economies of scale) but, on the contrary, to try to minimise the quantities of mixed garbage and to maximise the exploitation of segregated materials by getting as close to their source as possible (where they are still 'clean').

However, at the very least such materials have to be left outside houses for collection by the municipal service, and this leaves time for outside agents to intercept them. The fact that every household can sell its solid waste, instead of giving it away or disposing of it through payment of a tax, maintains the interception risk at a high level. When money can be made out of selling solid waste, recovery diversions become almost impossible to prevent. This is why the network approach does not work for SWM, as with any other service that deals with a situation where citydwellers not only receive a public service but also generate flows that have to be recovered and possess economic value.

Trying to treat waste as a flow is a denial of the portion of stock contained in the deposit. Most research on urban services pays little attention to the production/processing centres to which the pipelines are connected; they consider only the flows. However, the study of a network is in reality the study of a system, comprising on the one hand, a centralised production or processing centre linked to extra-urban sources (stock), and on the other hand a set of interlinked urban distribution channels (flows), which form an integrating territorial mesh. The notion of stock adds something to the analysis of networks.

Toward* semi-decentralised *solid waste management systems?

In the case of waste, the inclusion of both stock (for minimisation) and flows (for maximisation) within the analytical framework of a CPR suggests the idea of a *semi-decentralised* SWM system. In such a system, non-governmental recycling initiatives would not be eradicated in favour of a monopolistic and centralised service focused on landfill solutions. Recovery agents would be incorporated as local players able to efficiently capture as much waste as possible at source (subject to the existence of "*pay-as-you-throw*" type incentives). However, they would be included provided that they would channel their own waste residues to the centralised treatment facilities in order to cope with environmental and sanitary externalities. Hence, stock management would be centralised, and flow management decentralised.

As we have seen, in Coimbatore as in Vitoria, the different agents and mechanisms are always entangled. In other words, they do form a network, but a network that is neither physically rigid nor politically regulated. We argue that its regulation, through the assignment of practicable usage rights, could generate a new form of network – not a physical network, but a network of different politically and economically regulated agents around the city. Treating the solid waste deposit as a common pool resource, composed of both flows and stocks with a fluctuating boundary between the two, would be a way to maintain public service delivery while openly addressing appropriation conflicts and environmental issues.

This kind of regulation appears all the more necessary today in a context of global resource shortages, in which the waste recovery sector is becoming a strategic target for the manufacturing industry. In other words, a genuine worldwide market for recovered raw materials is emerging. The swing of the needle from the 'garbage' side to the 'resource' side of the dial has been spotted by the big, global industrial players, who are beginning to implement "urban mining" practices that intensify local appropriation conflicts (see Cavé 2014), increasing the need for fair arrangements and effective regulation.

Notes

1 Wastepickers are destitute people who live from picking over rubbish. The things they find are used either for personal consumption (food, clothes, furniture) or for resale (materials).
2 One major solid waste trader in Vitória
3 Brazilian public bank
4 Gated communities
5 Non-profit support structure for young Brazilian girls.
6 The more pejorative term "*ragpickers*" is commonly used in India to refer to "*wastepickers*".
7 Gated communities that can be managed, up to a certain degree, by their residents' associations.
8 In both cities, municipal authorities have the "responsibility" of dealing with waste; they do not own the waste.

References

Bakker, K. (2003) "Archipelagos and networks: urbanization and water privatization in the South." *The Geographical Journal*, **169**(4): 328–341.

Baptista, F. (2001) *Caracterização fisica e comercial do lixo urbano de Vitória-ES em função da classe social da população geradora*. Dissertação de Mestrado, Programa de Engenharia Ambiental da Universidade Federal do Espírito Santo (UFES/CT), Vitória.

Barles, S. (2010) Ecologies urbaine, industrielle et territorial. In O. Coutard and J.-P. Levy (Eds.), *Ecologies Urbaines*. Paris: Economica-Anthropos, 61–83.

Barraqué, B. (1993) Qu'est-ce que le génie de l'environnement? In B. Barraqué (Ed.), *La Ville et le Génie de l'Environnement.* Paris: Presses de l'ENPC, Paris, 13-32.

Barraqué, B. (1995) *Les politiques de l'eau en Europe.* Paris: La Découverte.

Bertolini, G. (1992) "Les déchets: rebuts ou ressources?" *Economie et statistique* **258**: 129–134.

Besen, G. R. (2006) *Programas municipais de coleta seletiva em parceria com organizações de catadores na Região Metropolitana de São Paulo: desafios e perspectivas*. São Paulo: USPEditor.

Bose, A. and I. Blore (1993) "Public waste and private property. An enquiry into the economics of solid waste in Calcutta." *Public Administration and Development* **13**(1): 1–15.

Botton, S. (2007) *La multinationale et le bidonville: privatisations et pauvreté à Buenos Aires*. Paris: Karthala.

Cavé, J. (2008) *La construction d'une politique publique non conventionnelle dans un pays « sous perfusion »: le cas des petits opérateurs privés de la distribution d'eau à Maputo.*

Mémoire de Master 2 Recherche, Ecole des Mines, Paris X Nanterre, ESCP-EAP, Polytechnique, ENA, ESSEC, September, Paris.

Cavé, J. (2013) *La gestion disputée d'un mal public impur: Economie politique des ordures.* Thèse de Doctorat en Aménagement de l'espace et Urbanisme, Université Paris Est, Ecole des Ponts Paris-Tech, Champs sur Marne, 458p.

Cavé, J. (2014) "Who owns urban waste? Appropriation conflicts in emerging countries." *Waste Management and Research,* **32**(9): 813–821.

Chaturvedi, B. (2007) *Wasting Our Local Resources: The Need for Inclusive Waste Management Policy in India.* New Delhi: Chintan Environmental Research and Action Group, 64p.

Chaturvedi, B. and V. Gidwani (2010) "The Right to Waste: Informal Sector Recyclers and Struggles for Social Justice in Post-Reform Urban India." *India's new economic policy: a critical analysis* **15**: 125–153.

Coing, H. and I. Montaño (1985) *Villes et Déchets dans le Tiers-Monde: Technique et Société – Tunis et Caracas, la gestion du service.* ENPC, Noisy-le-Grand.

Coing, H., P. de Lara and I. Montaño (1989) *Privatisation et régulation des services urbains: une approche comparative.* Noisy-le-Grand : LATTS-ENPC/MELTE-Plan Urbain.

Coutard, O. (Ed.) (1999) *The Governance of Large Technical Systems.* London and New York: Routledge.

Coutard, O., and J. Rutherford (2009) "Les réseaux transformés par leurs marges: développement et ambivalence des techniques 'décentralisées'." *FLUX,* Aux marges des réseaux, **76/77**: 6–13.

Coutard, O. (2010) Services urbains: la fin des grands réseaux? In O. Coutard and J.-P. Levy (Eds.), *Ecologies Urbaines.* Paris: Economica-Anthropos 102–129.

Debout, L. (2012) *Gouvernements urbains en régime autoritaire: Le cas de la gestion des déchets ménagers en Egypte.* Thèse de Doctoraite en Géographie, Amenagement, Urbanisme (PhD thesis), Université Lumière, Lyon.

Demajorovic, J., G. R. Besen and A. A. Rathsam (2005) *Os desafios da gestão compartilhada de residuos solidos face a logica do mercado,* in: P. Jacobi and L. Ferrei Ra (Eds.). Diálogos em ambiente e sociedade no Brasil. São Paulo: Annpas, Annablume.

Durand, M. (2010) *Gestion des déchets et inégalités environnementales et écologiques à Lima : entre durabilité et vulnérabilité.* Thèse de Doctorat en Géographie et Aménagement de l'espace, Université de Rennes 2, Rennes.

Fahmi, W. and K. Sutton (2010) "Cairo's contested garbage: sustainable solid waste management and the zabaleen's right to the city." *Sustainability* **2**:1765–1783.

Féré, C. and F. Scherrer (2010) L'eau urbaine après le réseau? L'injonction contradictoire d'une gestion durable des services urbains de l'eau. In G. Schneider-Madanes (Ed.): *L'eau mondialisée. La gouvernance en question.* Paris: La Découverte.

Fernández, L. (2010) *Dynamiques du recyclage spontané: Regards croisés sur les villes de Montevideo, et Paris au XIXe.* Mémoire de Master 2 recherche, Laboratoire des métiers de l'histoire de l'architecture, École Nationale Supérieure d'Architecture de Grenoble, September, Grenoble.

Furedy, C. (1992) "Garbage: exploring non-conventional options in Asian cities.". *Environment and Urbanization* **4**(2): 42–61.

Gerdes, P. and E. Gunsilius (2010) *The waste experts: enabling conditions for informal sector integration in solid waste management: lessons learned from Brazil, Egypt and India.* Eschborn, Germany: GTZ (German Technical Cooperation).

Giraud, P.-N., J. Ruet, B. Lefevre and A. Maria (2006) Le financement des services essentials dans les villes pauvres. *Revue d'économie financière* **85**: 285–308.
Jaglin, S. (2005) *Services d'eau en Afrique sub-saharienne: La fragmentation urbaine en question*. Paris: CNRS Editions.
Lane, R. (2011) "The Waste Commons in an Emerging Resource Recovery Waste Regime: Contesting Property and Value in Melbourne's Hard Rubbish Collections." *Geographical Research* **49**(4): 395–407.
Lorrain, D. (Ed.) (1995) *Gestions Urbaines de l'Eau*. Paris: Economica.
Lupton, S. (2011) *Economie des déchets: une approche institutionnaliste*. Bruxelles: de Boeck, Ouvertures Economiques.
McFarlane, C. (2008) Governing the Contaminated City: Infrastructure and Sanitation in Colonial and Post-Colonial Bombay. *International Journal of Urban and Regional Research* **32**(2): 415–435.
Medina, M. (1999) *Globalization, Development, and Municipal Solid Waste Management in Third World Cities*. El Colegio de la Frontera Norte.
Medina, M. (2005) "Waste Picker Cooperatives in Developing Countries." Conference on Membership-Based Organizations of the Poor, Ahmedabad, India: WIEGO, Cornell, SEWA, 22p.
Melosi, M. V. (2000) *The Sanitary City: Urban Infrastructure in America from Colonial Times to the Present*. Baltimore: Johns Hopkins University Press.
Nas, P. J. M. and R. Jaffe (2004) "Informal Waste Management: Shifting the focus from problem to potential." *Environment, Development and Sustainability*, **6**: 337–353.
Navez-Bouchanine, F. (2002) *La fragmentation en question: des villes entre fragmentation spatiale et fragmentation sociale*. Paris: L'Harmattan.
O'Brien, M. (1999) "Rubbish values: Reflections on the Political Economy of Waste." *Science as Culture*, **8**(3): 269–295.
Offner, J.-M. (1993) "Le développement des réseaux techniques: un modèle générique." *FLUX*, **9**(13): 11–18.
Ostrom, E. (1990) *Governing the commons. The evolution of institutions for collective action*. Cambridge: Cambridge University Press.
Petitet, S. and G. Schneier-Madanes (2005) "Le "modèle du réseau" face aux enjeux du développement durable." *Économie and Humanisme*, **373**: 19–24.
Robinson, J. (2006) *Ordinary cities: between modernity and development*: London, New York: Routledge.
Ruet, J. (2002) *Water supply and sanitation as 'urban commons' in Indian metropolis: how redefining the State/Municipalities relationships should combine global and local de facto 'commoners'*. Paper presented at the Ninth Conference of the International Association for Study of Common Property, Victoria Falls, Zimbabwe, June 17–21.
Scheinberg, A. and J. Anschütz (2006) "Slim pickins: Supporting waste pickers in the ecological modernisation of urban waste management systems." *International Journal of Technology Management and Sustainable Development* **5**(3): 257–270.
Scheinberg A., M. Simpson, Y. Gupt et al. (2010) *Economic aspects of the informal sector in solid waste management*. Eschborn, Germany: GTZ (German Technical Cooperation).
Scheinberg, A., D. C. Wilson and L. Rodic (Eds.) (2010) *Solid Waste Management in the World's Cities*. London and Washington DC: Earthscan for UN-HABITAT.
Swyngedouw, E. (2004) *Social Power and the urbanization of water*. Oxford: Oxford University Press.
Tarr, J. and G. Dupuy (1988) *Technology and the Rise of the Networked City in Europe and America*. Philadelphia: Temple University Press.

The World Bank (2012) *What a Waste! A Global Review of Solid Waste Management.* Washington, DC: Knowledge Papers 15, Urban Development Series.

Varone, F., S. Nahrath and J.-D. Gerber (2008) "Régimes institutionnels de ressources et théorie de la régulation." *Revue de la Régulation* **2**: 26pp.

Wilson, D. C., C. Velis and C. Cheeseman (2006) "Role of informal sector recycling in waste management in developing countries." *Habitat International* **30**: 797–808.

9 Is the network challenged by the pragmatic turn in African cities?

Urban transition and hybrid delivery configurations

Sylvy Jaglin

What is a post-network city, and does this notion apply to the cities of sub-Saharan Africa? If the term means urbanised areas in which large sections of the built environment function, in the long term, without networks, then Africa's cities could be so described. This empirical observation has given rise to numerous studies, and prompted reflection on ways to remedy the situation (Banerjee et al. 2008; Banerjee and Morella 2011; Foster and Briceño-Garmendia 2010; Eberhard et al. 2011). The vast majority of existing studies analyse the situation as a failure of the networks, often seen as a consequence of backwardness: the network model remains both an ideal and a goal of public action, in which the focus is on overcoming the obstructing factors (decentralising, privatising, improving governance…). These studies are based on the assumption of urban convergence (Cohen, 1996), in which Africa's urban spaces are considered as representing different stages of the pre-network city. However, a recent branch of these studies has emphasised the fact that a variety of technological advances, combined with the growing decentralisation of urban management could – by bypassing the obstacles to the extension of the integrated network – help to improve access to essential services (Botton and Blanc 2014; Foster and Briceño-Garmendia 2010; Lighting Africa 2013), thereby raising the possibility of *off-network* cities.

However, this model of the relations between cities and networks raises numerous questions, particularly regarding the dynamics of change in urban services, their integration into Africa's particular forms of urban transition and their feedback effects on urban change itself. This chapter explores these questions by focusing on the reasons for and the nature of the recent interest expressed in academic and policy circles in the so-called informal service actors, who operate outside the perimeter of the conventional network operators (Ahlers et al. 2013; Botton and Blanc 2014; Katusiimeh et al. 2013; Naulet et al. 2014; Van Dijk et al. 2014). I try first to explain the timing of this new interest in an old phenomenon and why that interest has taken so long to emerge, and second to explore what it reveals about the interdependent transformations of infrastructures and African cities.

Let us recall three key realities: across the continent, the privatisation of services to the benefit of big companies has failed to extend levels of coverage; informal service activities not only continue but are expanding; and a "wide range of contextualised solutions", especially in the water sector (Lorrain and Poupeau 2014: 11) are meeting urban demand. Forty years after academic research and development institutions became aware of the informal dynamics of small-scale commercial production (Hugon 2014), twenty-five years after they became aware of their role in transport (Godard and Teurnier 1992; Pradeilles et al. 1991), urban services for water, sanitation, electricity and – a little in advance of the others – waste are now approached with the tools used for the analysis of the informal economy. This time lag is linked in particular with the natural monopoly characteristics of network services and with the legal monopoly framework in which they are organised: by making informal modes of service production illegal, this legal organisation has hindered the recognition of activities perceived as residual and destined to disappear with time. Recognition and attempts to formalise such activities thus reflect a profound change in perceptions. Indeed, including informal services implies acknowledging the plurality of demand and of supply options, and therefore accepting the exclusionary rather than universal nature of the network, leaving space for nonconventional operators in urban governance and giving up the (official) hegemonic power of the public authorities as sole rule setters. Recently, these changes have begun to find expression in policies – for example, in the redefinition of statistics on the populations served, in the greater diversity of actors involved in reforms and projects, and in the design of regulatory frameworks that include "second choice solutions [which] seemed to offer a happy compromise" (Foster and Briceño-Garmendia 2010: 98).

As yet, however, this movement would seem to represent more a catch-up reaction than a strategy to remodel the framework of action. I propose to call this movement a 'pragmatic turn' (PT) and to examine its causes, nature and consequences for Africa's urban socio-technical regimes.[1] What does it tell us about ongoing infrastructural and urban change? In African contexts, governmental capacity to regulate is notoriously weak and therefore its capacity to influence, exploit or govern the informal sector is equally so. However, it is not zero. The aim is therefore to analyse the new forms of interaction between the state and informal actors involved in service delivery, and the shift in the power balance that ensues.[2] By contrast with the profusion of academic research on privatisations and the extension of new public management into the sphere of urban services, the PT has led to relatively low academic interest and theoretical effort so far, with an emphasis on three main areas: the analysis of infrastructural change through an extension of STS theories to the conditions and characteristics of developing cities (Furlong 2014); the conceptualisation of modes of service governance in contexts that are heterogeneous and normatively plural (Baron and Bonnassieux 2014); and the critical study of the role and nature of the "informal", especially in water services (Ahlers et al. 2013; Ahlers et al. 2014; Botton and Blanc 2014). These studies offer a conceptualisation of 'hybrid' socio-technical systems, which bear different names – "delivery configurations" (Olivier de Sardan 2010), "modernised mixtures" (Letema et al. 2014), "co-production" (Ahlers et al. 2014) – but say little about the way their operation is shaped by local conditions of urbanisation.

Following the perspective adopted by a few other works (Ahlers et al. 2013; Jaglin and Zérah 2010; Lorrain and Poupeau 2014), the approach developed in this chapter abandons the focus on infrastructure (large integrated networks, mini-networks, autonomous systems). It re-territorialises the question of services by re-embedding, first, the network in the global supply of services and, second, the diverse delivery configurations in their social, material and spatial urban environment. In other words, the aim is to treat delivery configurations as urban assemblages made up of actors, technical objects, institutions, economic interests, social practices and representations, but also of material spaces, all interlinked and making it possible for a service to be provided in a given urban context. In this approach, the analysis of the "resistance of the milieu" (Dupuy 2011) to the deployment of the network gives way to a study of the relative position of the different components, of their individual working and the interactions between them, and of the overall functioning of the assemblages. The latter are not cut loose from the African urban context, but rather incorporate the features of that context: the socio-economic heterogeneity, which is increasing differences between luxury and necessary goods and services (Hugon 2014); the plurality of standards and rules (Chauveau et al. 2001), which is unconducive to homogeneous frameworks for collective action; and the dynamics of socio-technical hybridisation, which is potentially more favourable to permanently heterogeneous assemblages than to processes of substitution between a current pre-network city and a (dreamed-of) post-network city.

In this respect, this article puts forward two hypotheses. The first is that the acknowledgement of an enduring hybridity of delivery configurations and its current incorporation into urban design thinking is a response to new urban socio-spatial dynamics: on the one hand, extensive and rapid expansion in the form of a heterogeneous urban-rural continuum rules out any hope of networks catching up in the medium term; on the other hand, the emergence of "middle classes", though an imprecise and disputed category (Darbon and Toulabor 2014), has been associated since the 2000s with transformations in the demand for and consumption of urban services. The second hypothesis is that the formalisation of hybrid delivery configurations is an experimental method of regulating urban diversity in response to the socio-political tensions associated with this ongoing urban transition. The concept of regulation here refers to two overlapping dimensions of action: the sectoral regulation of services, i.e. the set of technical, economic, legal and organisational factors that govern their operation; and political regulation, understood as the set of mechanisms employed by public authorities to stabilise antagonisms and to ensure the reproduction of a social system (Jaglin 2007). The mode of regulation itself remains unstable and contingent, subject to multiple social and political tensions likely to affect its dynamics and its aims. For this reason, several repertoires of arrangement coexist, selectively picked by urban authorities, often in relation with outside players, donor agencies and NGOs. Nonetheless, with regard to the PT, they are invariably a response to two imperatives: reconciling service policies with macro-political frameworks drawn from the neoliberal arsenal, often in partial and truncated ways; and maintaining room for manoeuvre and resources that will give the dominant socio-economic classes a lifestyle and living conditions that match their expectations. Accordingly, formalising hybrid delivery configurations seems to be, in African cities, one of the manifestations of the neo-liberalisation of urban policies (Morange and Fol 2014) with new groups of winners and losers.

Drawing on recent research in Africanist urban geography and sociology on the one hand, and social studies of technology focusing on cities of the North on the other, the paper critically analyses the conditions of service provision in African cities by adapting the lenses of the literature on sociotechnical transitions to urban conditions in the South. It does not rely on a comprehensive case study, but combines my own research on urban services in several African cities with contributions from other recent empirical works. The argument is organised in three stages. The first part of the chapter defines hybrid delivery configurations by analysing the way in which they reflect the socio-spatial conditions of Africa's urban transition, and explains how they are changing under the PT. The second part explores the nature of the socio-technical changes triggered by the formalisation of hybrid delivery configurations, their capacity to transform urban infrastructures and the relevance of the analytical frameworks developed in the field of STS for the study of changes in large technical systems when it comes to African urban contexts. The third part explores the politics of formalising delivery configurations and shows that the PT is ultimately part of an attempt at the political regulation of urban heterogeneity, primarily through the tools of commodification.

From informality to the regul(aris)ation of hybrid delivery configurations

Improving the conditions of access to essential services in African cities remains a challenge, which the policies and projects implemented in recent decades have tackled with very limited success. The operators of conventional networks have not succeeded in providing services suited to the needs and resources of most city dwellers. Fewer than 40% of African urban households have running water at home, a septic tank or improved latrines, and in almost half of African countries scarcely 50% of city dwellers have access to electricity, whereas with demographic growth in urban areas, provision rates for all urban services (particularly drinking water) dropped between 1990 and 2005 (Foster and Briceño-Garmendia 2010: 126–127).

Having analysed elsewhere why the "modern infrastructural ideal" has proved unsuited to urban expansion in Africa (Jaglin 2005, 2014), I will focus here on the way in which hybrid delivery configurations correspond to the socio-spatial contexts of Africa's urban transition. Although diverse, sub-Saharan African cities have certain shared features: among the highest sustained demographic growth in the world (3.7% per year from 2000 to 2010), fragile and poorly equipped public institutions, the influence of colonial legacies and processes of institutional and cognitive transfer emblematic of "projected societies" (Darbon 2008), socio-economic inequalities and poverty that generate high levels of social frustration, an urban economy and production that are largely – sometimes primarily – informal, and rapid and diffuse spatial expansion that raises challenging problems of infrastructure and governance.

The range of essential services in these cities is today marked by three structural factors. First, in all sectors, demand is growing fast, even more so since urbanisation began to be accompanied by a consolidation of the middle classes, a significant proportion of whom live in the new suburbs. Next, access to resources and services remains rationed: this means that there is still latent unmet demand, which is both a source of economic opportunities, as evidenced by the interest and attraction of these new markets in a very dynamic demographic context (Damon 2014), and a factor of risk for urban authorities and, through them, for states. Finally, the political question of democratisation of access to urban resources (and of the right to the city) remains profoundly marked by a moral economy in which clientelism and patronage are disputed but relatively effective ways of sharing resources within societies with stark social disparities.

From networks to hybrid delivery configurations

In the mid twentieth century, the dominant configuration was a juxtaposition of a conventional network, in respect of which regulation was conceived, and a disparate collection of informal services. At best, the latter were tolerated and at worst they were opposed, with periods of alternation between the two in most cities. This configuration was first dominated by public actors and state-centred regulation, and then, from the 1990s, by reorganisations orchestrated within the framework of neoliberal reforms that favoured commodification, partnership contracts with private companies and the introduction of various forms of competition. The slow and circumscribed spread of these public-private partnerships has brought only very partial responses to urban demand, which remains partly dependent on public operators in various stages of modernisation, and, above all, on informal operators. The latter provide private commercial services, whether individual or collective, all of them illegal with respect to the official operators' exclusive contracts. Exploiting the diversity of demand and spending power, as well as of living conditions in highly heterogeneous urban societies, they offset the deficiencies of the conventional service and – depending on the type of urban area – target well-off customers dissatisfied with the official service or poor customers excluded because of lack of resources, physical distance from major infrastructure or illegal land occupancy (Jaglin 2014).

There are multiple examples in every sphere: water markets have been extensively described (see issue 7/1 of *Water Alternatives* 2014; Baron and Bonnassieux 2013; Botton and Blanc 2014; Jaglin 2005, 2013; Naulet et al. 2014), as have informal transport services (Godard 2002; Godard and Teurnier 1992; Lombard 2006; Pradeilles et al. 1991). Waste collection and recycling are also the focus of numerous studies (Ginisty 2014; Katusiimeh et al. 2013; Ngambi 2015;

Sory 2013), while the integration of private sanitation solutions into value chains that ensure long-term funding is attracting growing attention (Trémolet and Evans 2010; Van Dijk et al. 2014). Less well-known, alternative energy services are a burgeoning field (Clément 2009; Franks and Prasad 2014; Gaunt et al. 2012; Lawaetz and Smyser 2010; Lighting Africa 2013; Smyser 2009), whereas the inventiveness of telecommunications and associated services is now well demonstrated (Cheneau-Loquay 2012). These arrangements are not infinite in number and variety and not all are found everywhere or with the same intensity. However, in all cities the provision of services relies on a combination of several of them, depending on the state and performance of the network, the dynamism of the small-scale private operators, the forms of urbanisation and the spending power of households. The technical and operational pluralism of delivery configurations is therefore the result of contingent co-construction processes, through which service functions are extended to end-users excluded from the conventional networked service. These are the hybrid configurations that the policies associated with what I call the PT seek to formalise, or even to encourage.

The dominant pattern of reorganisation of the waste service within the framework of "global schemes" planned and managed by municipalities thus reflects different mechanisms of cooperation between informal refuse collectors (at the households level), who sometimes also have recycling activities, a formal public service of collection (from public containers or transfer stations) and transport to final semi-controlled landfill centres, the management of which is increasingly outsourced to private subcontractors. Studies emphasise the fact that one of the challenges of this form of arrangement is to coordinate – across the whole chain – heterogeneous actors whose activity has in common neither practices and representations, nor technical knowledge and regulatory mechanisms, nor results (Katusiimeh et al. 2013; Ngambi 2015; Pierrat 2014; Sory 2013). In the sphere of water, where urban markets are highly diverse, different experiments are underway to find appropriate arrangements. Some choose contracting out to private operators, acting as wholesalers, i.e. simultaneously "major customers" of the conventional operator and "retailers" for the general population (Botton and Blanc 2014: 112); others seek to regulate small-scale independent providers (Ahlers et al. 2013; Botton and Blanc 2014; Valfrey-Visser et al. 2006) or mini-networks whose management is assigned to user associations (Naulet and Biteete 2014). In sanitation, research is still in its infancy and is looking to identify segments of the activity where public action could be effective (Trémolet and Evans 2010). Studying sanitation management in two informal districts in Dar es-Salaam and Kampala, Van Dijk et al. identify three functional segments – latrine construction, maintenance and emptying – and suggest that the main obstacles to the development of a satisfactory service could be removed by a pragmatic policy recognising existing private solutions, enhancing government support in terms of technical development and training for small-scale local contractors and designing appropriate financing mechanisms (Van Dijk et al. 2014).

In all these examples, small-scale informal private operators are seen as players in service markets: "Increasingly, policy perspectives are moving away from seeing hybrid forms of service delivery governance as dysfunctional, and instead seeing them pragmatically as 'arrangements that work'" (Ahlers et al. 2014: 10). The aim of formalisation policies is to regulate their activities and coordinate them with those of the conventional operator, in particular through new forms of institutional contracting out or micro-partnership arrangements. By making urban service markets more secure, and even if this entails institutional and socio-technical bricolage (Criqui 2015), these policies open up the markets to new providers and reassure middle-class customers, who are more demanding about service quality and reliability.

New socio-spatial dynamics of service demand: emergence of the middle classes and urban sprawl

Hybrid delivery configurations have generally been associated with service to poor city dwellers, but they need to be seen more broadly in relation to the socio-economic and spatial dynamics of demand which are also influenced by a recent dual trend: the rise of the urban middle classes, and African-style urban sprawl.

Vague and open to dispute, the notion of the middle class is at the heart of the debate in African cities, where the emergence of new, rising social formations has been empirically observed (Darbon and Toulabor 2014). This category encompasses millions of urbanites with daily earnings typically ranging between US$ 2 and US$ 20 per person, who share two characteristics: having escaped from poverty through a capacity to generate a surplus; but facing the persistence of a form of vulnerability (Darbon 2014). More precisely, Africa's middle classes encompass three disparate subcategories: the global middle classes (US$ 10 to US$ 20 per day), with substantial revenues and significant assets; the stabilised middle classes (US$ 4 to US$ 10 per day), who have escaped from poverty but lack sufficient assets to be sure of maintaining their social position from one generation to the next; and the floating class (US$ 2 to US$ 4 per day), who have disposable income but remain at risk of downward social mobility (Darbon 2014).

These populations have spending power, although in varying proportions, and aspire to consume, particularly commercial urban services which consolidate their new social position. One of the problems is how this demand can be satisfied. Indeed, while the notion of the mathematical "averaging" (Damon 2014) of revenues has the advantage of highlighting a continent-wide dynamic, it in no way indicates uniformity of behaviours and values, since statistical aggregation masks very different national dynamics and local situations. More a code name than a concept (Darbon 2014), the notion of a middle class is nevertheless a way of drawing attention to a dual connection between urban social change and service consumption practices. First, these middle classes "invent themselves" in terms of their own self-representations and their social advancement, in which desires and norms of consumption are important and, in particular, place value on individualised access. Next, the prices of these services and the way they are sold (in units and in the form of split purchases) directly contribute to the fulfilment of these middle-class aspirations: by providing services that are more "affordable" (although not less expensive), these forms of commercialisation contribute to the improvement in living standards which, generally speaking, owes much to the goods and services produced by the informal economy, as well as to the large-scale arrival on the markets and in day-to-day life of low-cost technologies (and goods) from China and India (Darbon 2014). The PT proceeds from a political acknowledgement of the fact that hybrid delivery configurations contribute to the 'invention' of the middle classes, by trivialising access to services, and to their 'self-realisation', by selling the services in a form that makes them affordable.

These new urban markets are to a large extent located at the suburban interfaces, where land and real estate development activity today is intense. Sub-Saharan urban Africa is not exempt from the propensity of the city to dissolve into the urban, identified several decades ago in industrialised countries, where it is associated with post-Fordist economic changes. In a very different economic context, African urbanisation also produces distended, discontinuous and heterogeneous built spaces, well described for example in Guinean West Africa (Dorier-Apprill and Domingo 2004; Héraud-Arouna 2005). Densely populated, this huge cross-border regional area is characterised by a very high level of commercial and migratory exchanges, intense land speculation generating transformations that are clearly visible in the landscape, physical and economic urbanisation (new activities targeting new markets) driven by a spectacular increase in mobilities, and sociocultural changes revealed in the diversification of practices, ways of life and representations. Stretching along the coast from Abidjan to Lagos, this "African-style urban sprawl" is caused less by demographic growth than by a redistribution of residential function in a region well provided with roads and tracks covered by a particularly efficient network of informal transport operators, open trucks and *zémidjan* (moto-taxis).

This steadily spreading urban expansion, variously described as a "multi-polarised conurbation" (Dorier-Apprill and Domingo 2004), African-style *desakota* (Calas 2007; 2013) or suburbanisation (Buire, 2014), is accompanied by declining densities, which further raise per capita infrastructure costs. According to the calculation by Foster and Briceño-Garmendia, in big cities in Africa, those with populations of more than 3 million and an average density of 5000 people per square kilometre, water and sanitation represent the highest share of infrastructure (54%), followed by roads (28%), electricity (17%) and ICT (1%) (Foster and Briceño-Garmendia 2010: 129). When population density is low, particularly on the edges of cities, the cost of building infrastructure becomes prohibitive and experts agree on the need to think of alternative, cheaper technologies, such as solar panels, boreholes or in situ sanitation. However, the problem is not so much to do with employing these technologies as with embedding them in appropriate and durable socio-technical systems, especially as these areas are usually located outside the scope of coverage of the conventional operator, or indeed of an urban public authority. For this reason, delivery configurations in these areas take different operational forms depending on the resources available to real estate developers and occupants, ranging from rich, super-equipped, autonomous enclaves to vast areas characterised by informal solutions (Allen 2014).

The PT is thus a response to urban transition, in its socio-economic (demand for services among emerging middle classes), spatial (periurbanisation, urban sprawl, encroachment into farmland) and political dimensions. The last of these is crucial: much more than the very poor city dwellers locked in day-to-day survival strategies, and the well-off or rich households whose spending power and mobility allow them to find satisfaction on globalised markets, it is the middle-class sections of society which today are the main social force demanding public action. Their requirements in terms of quality of life and domestic comfort are not fully satisfied by the available commercial types of supply, while their lifestyles and aspirations prompt them to reject the mutual support and DIY solutions from which they are trying to escape. Since the conventional operators are unable to respond to these urban demands, which cannot be met simply through service differentiation (Jaglin 2008), formalising hybrid delivery configurations is another attempt to answer urban heterogeneity.

Before returning to this political dimension in the third section, I will now explore the link between the PT and a possible transition in African urban socio-technical regimes.

From formalisation to infrastructural transition in African cities?

The essential services studied here are produced by means of socio-technical systems that combine technical artefacts, sociopolitical organisations and expertise, but also practices and representations. Some of them depend on a capital intensive material infrastructure network, others on networks of actors and scattered technologies that require little capital. Taken together in relation to delivery configurations, however, these socio-technical assemblages share no infrastructure, in the sense employed by Star and Ruhleder (1996),[3] and do not have the capacity to form strong communities of practice. My view here is that the purpose of the experiments in formalising hybrid delivery configurations is to bring such an infrastructure and community of practices into existence.

In that case, can they be analysed as together constituting a major technological transformation in the way in which the essential functions of urban metabolism are performed in African cities? In other words, can the formalisation of hybrid delivery configurations be considered as both the driver and the outcome of an infrastructural transition, and if so, a transition to what? In order to explore these questions, I draw on a handful of works at the intersection between urban studies and STS, which examine the urban dimensions of infrastructural transitions (Bulkeley et al. 2014; Monstadt 2009; Monstadt and Wolff 2015; Rutherford and Coutard 2014). Employing the concept of socio-technical regimes, defined as "relatively stable configurations of institutions, techniques and artefacts, as well as rules, practices and networks that determine the 'normal' development and use of technologies" (Smith et al. 2005: 1493), they seek to understand "the ways in which the dynamics of infrastructure systems are configured by and through urban places" (Bulkeley et al. 2014: 1473). Examining the conditions and modes of change, the resistances and political games they give rise to, they analyse when and how the destabilisation of an urban socio-technical regime triggers a transition.

Apart from the fact that the theoretical frameworks of these studies, which are primarily "derived from circumstances in the North", cannot be immediately transposed to cities in the South (Furlong 2014), their perspective differs somewhat from my own, which is not so much to question the conditions of change and how it is managed than to examine the nature of the transformations under the PT: does it drive a transition from one urban socio-technical regime to another?

Let us begin with three observations inspired by the analysis of socio-technical transitions (Furlong 2014; Geels 2002). First, the processes observed relate less to a movement of transformation than to attempts to formalise and coordinate the disparate components that make up the particular assemblages which, in each city, constitute a delivery configuration: the change comes not from the destabilisation of an established system, but from an attempt to consolidate what is already there. Next, the changes examined arise less from processes of innovation as such than from a series of incremental adaptations that facilitate the – relative and unequal – functional inclusion of actors and practices long partitioned in sealed socio-technical worlds.[4] As Furlong points out, these modest changes "involve integration as opposed to innovation" (Furlong 2014: 145). Finally, while many initiatives initially take the form of a "niche experiment", they in no way claim to be in "opposition to dominant values and practices" (Bulkeley et al. 2014: 1474). On the contrary, numerous studies show that, in Africa (Kjellèn and McGranahan 2006; Ginisty 2014) and elsewhere in the South (Criqui, 2015), the conventional – integrated – network model is less challenged by these initiatives than consolidated by a socio-technical bricolage, which seeks to address its deficiencies, and by regulatory adjustments which help to stabilise hybrid delivery configurations. In order to reconcile the conceptual frameworks of socio-technical studies with the conditions that prevail in cities of the South, Furlong observes that a stable socio-technical system can remain enduringly hybrid, and that the transition does not need to culminate in a process of substitution, but rather may bring with it diversification and heterogeneity (Furlong, 2014). However, is this process still a transition?

Empirical observation shows that the changes target certain components of hybrid delivery configurations without disrupting them, and that their occurrence is discontinuous and scattered, not affecting the conventional network's subsystem. Thus, the spread of low-cost Chinese technologies and products on African markets has transformed certain segments of supply on the water markets: in Maputo in the 1990s, it facilitated the emergence of small-scale private operators who developed mini-networks; in Douala, by reducing costs, it caused a proliferation of privately financed boreholes and a reorganisation of the urban water markets. The example of electric batteries, photovoltaic panels, low-cost electricity meters, polyethylene water tanks and pipes, but also entry-range and second-hand mobile phones, testifies to the successful introduction of low-cost technologies that ease the pressure on the network rather than disrupt its regime. Incremental change, through the adoption of compensatory practices – e.g. the gradual proliferation of individual water storage tanks, generators (long-standing in Nigeria) or solar water heaters (recent in South Africa) – works in the same direction: by mitigating the recurrent dysfunctions of the networks, it makes their failings acceptable.

Considered at the level of delivery configurations, change above all highlights growing technical inequalities between the various channels of supply and the complexity of their interactions. On the one hand, network infrastructures have high inertia and lack flexibility in the face of urban growth and sprawl; on the other hand, they retain an astonishing power of attraction, recapturing territories when investment is available and pushing out informal activities. This explains why the components of the same hybrid delivery configuration often show differentiated dynamics of evolution, particularly in the presence of outside actors who promote different models and target distinct parts of the assemblage. Maputo provides an example (Ginisty 2014). There, the water service underwent an in-depth reorganisation, the most visible aspect of which affected the network operator, which was split into a public asset-holding company (FIPAG) and a private company (AdeM) which was granted an operating licence for the service in the *Cidade*. However, in addition to this central socio-technical nucleus, there were two further water supply subsystems. First, there were some 300 hydrants, "confiscated" more than managed by committees, which users rejected in favour of buying water from individuals (Ginisty 2014: 196); second, there was a market of around 450 small-scale private providers delivering underground water through standpipes and household connections to 'spaghetti' networks. The delivery configuration therefore comprised at least three components, and its formalisation since the mid-2000s has entailed the organisation and professionalisation of small-scale private operators, the coordination of all the services, but also the scripting of methods for the negotiated eviction of the small-scale operators wherever the network reaches (Blanc et al. 2009; Botton and Blanc 2014). By choosing the small-scale operators rather than the hydrants, residents confirmed their acceptance of the commodification of water and more generally of commercial urban practices. However, the households surveyed by Ginisty also stressed their preference for the conventional model of the network, the epitome of "successful" city dweller status, for which they make or have made significant financial efforts (Ginisty 2014).

Is this a case where the transition "is less about innovation and more about integration into the physical as well as the Latourian network" (Furlong 2014: 144). or where the very idea of transition should simply be abandoned? My analysis suggests that informal services are rarely an alternative solution, completely emancipated from the network; it also suggests that emerging local initiatives try to orchestrate partial and reversible changes in delivery configurations in cities that are themselves undergoing rapid metamorphoses – not to begin a move away from the network as a socio-technical regime, but, on the contrary, to maintain its existence, even if confined to a few exclusive urban showcases. What is at stake is much less the transformation and substitution of socio-technical regimes, than the way in which they coexist and co-evolve. Rather than transition, the central question relates to the conditions of emergence of an "effective" coordination between heterogeneous worlds. Indeed, as hybrid delivery configurations bring together material infrastructures, immaterial frameworks (cognitive, procedural, normative, symbolic) and lastingly separate communities of agents, it is crucial to understand whether, and how, some form of coordination can produce a valid and functional arrangement between these heterogeneous worlds (Star and Ruhleder 1996). That is the full challenge of the formalisation attempts under the PT.

The Pragmatic Turn: commercial regulations in heterogeneous urban environments

In order to address urban heterogeneity current initiatives rely on commodification and extension of the service markets, but they are rarely equipped to identify and correct the environmental and social externalities (on the difficulties of sectoral regulation, see Jaglin 2014). Initially intended to bring about a short-term political readjustment in terms of access to resources and amenities, geared to the socio-economic metamorphoses of the urban transition, these initiatives are indifferent to urban environmental "sustainability" and their regressive social effects receive no more consideration.

It is undoubtedly a difficult exercise since, while bringing integration and better living conditions to certain urban areas, the formalisation of hybrid delivery configurations also raises issues of social inequalities and environmental consequences in circumstances that are not conducive to the expression of objections, for reasons that relate to political legacy (for example, in Mozambique; Ginisty 2014), or to the characteristics of the emerging middle classes. Apart from their membership of essentially horizontal social networks, these classes are characterised, according to Darbon (2014), by a strong individualism linked to the demanding pursuit of upward social mobility and to a conformism marked by limited interventions in the public sphere. Sometimes associated with a "post-political" urban condition (Choplin 2014), this attitude is nevertheless in no way an expression of apathy; rather, it reveals the concentration of energies on individual advancement through the maintenance of activities and revenues that are always under threat. In this respect, Darbon notes that the middle classes are exposed to a fourfold pressure: economic (against insecurity); psychological (for self-discipline); social (the breaking of old social bonds); and normative (to adopt the rules and standards of the wealthy classes). However, political pressure is absent (Darbon 2014). Their lack of involvement in and control of the public sphere, together with their relative aversion to collective action, explains the rarity of collective mobilisation by the African middle classes, but this low level of political engagement should not be interpreted as a lack of interest in the state. Indeed, so-called 'middle-class' groups are concealed behind state power, which enables them through action or inaction to pursue their efforts for advancement (Darbon 2014). In other words, the potential anti-establishment political force of the emerging middle classes, slated to rise with the urban transition and as a result of electoral factors that reinforce the influence of urban social constituencies, is today neutralised and in a sense diverted into a massive and astonishingly creative engagement with all sorts of urban activities, in particular informal ones. Against this backdrop, formalising hybrid delivery configurations appears as a way for African state authorities to negotiate the urban transition: a partial regularisation of the informal sector facilitates the provision of essential services, in exchange for a civil order that also benefits the urban elites and dominant classes.

Regulating urban services or governing the informal?

The fact that the formalisation of hybrid delivery configurations constitutes an appropriate response to urban demands does not mean that it is free of problems, as noted by Ahlers et al. – "embracing the informal in the form of hybrid arrangements is no shortcut to progress" (Ahlers et al. 2014: 10). Indeed, as a political regulation process, the PT is at present sorely lacking in technical and economic expertise and tools to tackle externalities (environmental, sanitary, social). In the absence of economies of scale, informal services are often costly for floating class households, for whom the level of effort required to keep up payments may ultimately prove unsustainable; without appropriate methods of control and technical regulation, the quality and continuity of services are both uneven and variable; markets are unable to maintain societal and territorial solidarity, raising the spectre of urban fragmentation; and finally, informal private actors take no account of environmental externalities, which leads to the risk of overexploitation of the commons to the detriment of wider ecological balances. These failings are not resolved by a formalisation limited to the granting of a form of recognition, but nor are they a structurally inevitable result of the hybrid nature of the configurations or the presence of primarily informal actors within them. In reality, very different results are observed depending on how and by whom the arrangements are politically managed. Two examples provide an illustration.

In Ouagadougou, twenty years after the Sankarist political compromise of "shared urban management" (Jaglin 1995), under which land ownership was regularised in exchange for the (economic and human) engagement of city dwellers whose weariness was beginning to be apparent in the early 1990s, the introduction at the beginning of the 2010s of a policy to contract out the water supply service to small-scale operators in five peripheral areas of the city constituted a paradigm shift. It is now the commercial service, in a modified but classic example of outsourcing operations to small-scale private operators under contract with the ONEA (national water and sanitation office), which is responsible for meeting the needs of more than 86,000 suburbanites (Botton and Blanc 2014). The operating profits recorded by at least four of the five operators, none of them with previous experience in the water sector, are no doubt evidence of their management skill, but also of the social change underway in these districts: consumption remains modest but is on the rise, with households acquiring not just individual connections but also showers and toilets. This configuration, in which the service delivered by small-scale private operators is 'regular' and 'solicited', reflects a deliberate policy by the main operator to extend its activity to the edges of the conventional network, without losing control over the public water service: by relying on small-scale operators under contract, it marginalises other access alternatives (in particular street selling of water from hydrants), and maintains relative homogeneity in service principles (water quality, pricing), while the technical specifications and shortness of the contract (five years) maintain the prospect of future reintegration into the conventional network.

In Maputo, small-scale independent informal producers operate in an apparently competitive market (Blanc et al. 2009) and their activity, often cited as evidence of the informal sector's capacity to replace the failing public service, is the subject of an original experiment in formalisation (Botton and Blanc 2014). Ahlers et al. (2013) nevertheless criticise "the myth of 'healthy' competition" which underpins this policy and the presuppositions associated with it. The sharp inequalities in size and economic power existing between the operators contradict the idea of an "open" market, and, as in the conventional water sector, "the combination of sunk costs and economies of density leads to monopoly-like characteristics" (ibid.: 181). Furthermore, market deficiencies are exacerbated by the regulations (on prices, conditions of entry to the market and of spatial expansion of the mini-networks) enforced by the two operator associations, which impose cartel practices that serve their interests as producers, in particular the biggest ones, rather than those of users. In other words, "[w]hat is hailed as 'free entry' to the market, or 'healthy' competition, is in reality a power play determined by economically unequal actors, locationally determined social relations, and historic uneven development, resulting in uneven service provision" (Ahlers et al. 2013: 181).

The cartelisation of water services (Jaglin and Bousquet 2012), deals between informal waste collectors (Sory 2013), or the corporatism of the informal taxi syndicates (Pradeilles et al. 1991; Wilkinson 2010) all attest to the limitations of imperfect competition on informal markets. These limitations affect not only the regulation of the conditions under which users access the service, but also working relations within production units, competition between unequal producers, and to an even greater degree the extent of negative externalities (overexploitation of groundwater, uncontrolled evacuation of waste into urban interstices, street congestion…). A recent study on informal waste collectors in Kampala emphasises the contrast between their collection service, much appreciated by households, and their complete environmental inefficiency (Katusiimeh et al. 2013). Another study, on the sale of water from private boreholes in Nairobi, also shows the limitations of the service in terms of social equity and resource protection (Chakava et al. 2014). For example, the poor spend 6% to 30% of their monthly income to buy water that does not meet the WHO's drinkability criteria at a price ten times higher than that paid by wealthy households for the first block of the mains water tariff. Moreover, surveys show that the overexploitation of underground water, which is poorly and ineffectively managed and monitored, is leading to a rapid lowering of the water table in certain districts. The authors nevertheless conclude that this solution is probably the "least bad" of the strategies of access to water for the many poor households in the Kenyan capital city who receive little or no pipe water provision, and the only one that ensures a continuous supply of sufficient quantities, in particular during the recurrent droughts.

The formalisation of delivery configurations thus follows very different outlines: whereas in Burkina Faso the state initiates, controls and regulates, in Mozambique the government hands over the "policing" of the water market to a "shadow regulator" which cares more about the interests of the big producers than of the public (Ahlers et al. 2013), while in Nairobi the sectoral authorities turn a blind eye. How should these differences be interpreted? The questions raised by hybrid delivery configurations and their formalisation are neither about the private status of the operators nor, intrinsically, about the informal nature of their activities. They point to the character of public regulation, the political intentions that underlie it, and the capacity for action of the authorities responsible for its implementation. The selectiveness and unequal aspirations of the formalisation policies apparent in African cities reflect the way political powers perceive their opportunities in relation to the tensions generated by the urban transition and to the robustness of their political alliances: as a result of new modes of managing the relations between formal and informal within hybrid delivery configurations, the PT is one expression among others of the way in which the authorities seek to manage and regulate urban change.

Toward a political economy of hybrid delivery configurations

The production of cities where rapid demographic and spatial growth is taking place, which are marked by sharp socio-economic inequalities, against a background of persistent destitution in numerous urban areas still devoid of basic services and infrastructures, raises significant challenges in terms of regulation.

Part of contemporary African urban realities, often considered as evidence of the return to growth and of improvement in economic conditions, is expressed through urban policies that reflect a revival of "modernist" visions and projects. These take the form of a proliferation of strategic plans, leading Watson to wonder whether they represent the dreams of the continent's inhabitants or embody the return of the worst nightmares of planning (Watson 2014). For Myers, they represent above all "a return to faith in elitist, modernist visions for metropolitan development wildly at odds with what grassroots organizations say, think, or plan for in these cities' majority areas" (Myers 2014: 16). After first documenting these disparities through a rapid overview of projects in a handful of African capitals, he goes on to dissect them in detail in his analysis of the *Nairobi Metro 2030* plan, whose central message – "Nairobi is open for business" – is unmistakable, consonant as it is with the precepts of entrepreneurial urban planning and the vocabulary of neoliberal globalisation (Myers 2014: 10). While the terms environment, sustainability and inclusion are also present, "this appears to be a form of green-washing the document, since minimal space is devoted to actually planning for waste management, water resources, or genuine means of inclusivity" (ibid.), while the informal neighbourhoods and their catastrophic lack of services are virtually absent from the content of the Plan. Allen emphasises a similar tension when examining "the peri-urbanization of differential sustainability": through the production of eco-enclaves, like the vast luxury Airport Hills estate in Accra, the peri-urban interfaces are today simultaneously one of the primary locations of and a mine of resources for an urban transition which is both desirable and exclusionary; they are part of a twofold current involving the spread of an eco-environmental culture that reflects the aspirations of the urban middle classes and the dispossession/displacement of populations marginalised by the consumption of space and resources by these emerging classes (Allen 2014). She suggests that "[f]rom this perspective, the peri-urban interface can be apprehended as a laboratory where interventions to regulate the urban metabolization of nature might selectively safeguard certain environmental resources or services while consolidating unjust patterns in the distribution of positive and negative environmental externalities" (Allen 2014: 524).

Although different in its forms, this discriminatory urban planning, with its production of socio-spatial inequalities, is not new. However, its context is novel, because large neglected or coveted city areas are now heavily populated by city dwellers, escapees from poverty ready to pay for improved services, courted by local political leaders seeking social constituencies in local electoral contests. A laissez-faire approach, abandoning the urbanisation of these new middle classes territories to informal actors alone, is therefore no longer always an economic and political option. By drawing on the now well-mastered repertoire of neoliberal urban policies – formalisation of property rules and rights, extension of the market regulations, promotion of entrepreneurial principles and competition – the formal acknowledgement of hybrid delivery configurations creates the conditions for a transfer of resources, risks and opportunities between groups of city dwellers, consumers and producers, while reasserting state control over certain informal activities in return for legal and political protection. In so doing, it not only satisfies the existing small-scale operators, but makes their activities more attractive for new entrepreneurs, who may potentially be closer to the authorities as is illustrated by the emblematic case of Douala (Nantchop, in press).

The urban transition constitutes a scale shift in the dynamics of urbanisation and brings profound social changes, as well as sharp competition for space and resources. A crucial challenge for urban policies and the primary objective of local governments is to regulate these contradictory processes and to devise solutions that can resolve them in a context dominated, since the 1980s, by so-called neoliberal policies that consist, like the land policies inspired by Hernando De Soto, in regularising the informal. The PT is one of the manifestations of this political programme for integration *by* the market.

Conclusion

Let us return to the original question: are African cities post-network cities? I have shown that the network model exists within hybrid delivery configurations, and is not undermined by their formalisation: re-embedded in the material and social environment of cities, made more functional by low-cost extensions or off-network devices, the model seems both to be consolidated within limited perimeters and subject to long-term competition outside these perimeters. In these circumstances, the formalisation of delivery configurations of essential services should be understood as a process of regularisation similar to those affecting numerous other aspects of the urban economy: land markets, informal trading, etc. It is encouraged or tacitly accepted by states, but is located in the overlap between the strategies of many other actors: local urban entrepreneurs, multinational utility firms, upwardly mobile households… It is facilitated by adaptations (technical, commercial, institutional) and financially assisted by development institutions and NGOs that also provide support, experience and expertise which capitalise and channel the projects. Finally, it is founded first on principles of combination and hybridisation between heterogeneous socio-technical systems, and second on a "normalisation' of commercial practices that span the previously separate categories of public service and informal markets, because the challenge for governments is both to promote integration via the market and to control the conditions of that integration.

So while it plays an active part in the political regulation of the urban transition, the PT does not for the moment seem to be contributing to a profound transformation in urban socio-technical regimes in Africa. Enacting 'differential production' of services within the framework of formalised hybrid delivery configurations could nevertheless constitute the basis for a third generation of reform in network services (after the modernisation-decentralisation and then the privatisation of public utilities). The real innovation here lies less in the way this reform combines public and private actors, commercial and public service principles and a degree of user involvement in governance, than in the fact that it imports the principle of heterogeneity into the operational sphere, by incorporating informal activities and the quest for a grammar of coordination between the distinct socio-technical subsystems of a given delivery configuration.

Hence, while the network may not be 'soluble' in the PT, its hegemony may be. This is not a minor change, and its benefits, in terms of extended service provision and support for the emergence/consolidation of the new urban middle classes, should not be underestimated. In general terms, however, whether the formalisation of hybrid delivery configurations contributes to the creation of a city that is more equitable in terms of access to public services and more sustainable in terms of environmental protection is a question that remains open. The recognition of informal actors in fact says nothing about the identity of the beneficiaries of these policies, about their links with urban and national authorities, about the profits and revenues they can hope to obtain. Likewise, the expansion of informal urban service provision, the conditions of its coexistence with other forms of provision (public and private, community-based) or the likelihood of its driving out other modes of access to resources – for example, free resources – can vary greatly. Ultimately, the regularisation of informal operators is less a goal than a transactional tool for the liberalisation of the markets of essential services and for their extension to emerging middle-class populations previously disqualified by the exclusionary paradigm pursued by the conventional operators: neither its robustness as a method of regulating the heterogeneity of urban societies, nor its capacity to ensure the trickle-down of social progress beyond the big gainers from the urban transition, are certain. They will need to be tested and assessed over time.

Notes

1 This notion comes from recent studies which, at the interface between social studies of technology and research on cities in the North, have suggested and discussed the idea of a distinct 'urban socio-technical regime' (Bulkeley et al. 2014; Hodson and Marvin 2010; Monstadt and Wolff 2015).

2 Although African conditions are very different from those studied by Schoon and Altrock in China, we drew on their concept of "conceded informality" and the way in which it can be used to analyse "a flexible management of diverse informal practices depending on their relevance, usefulness and potential threat towards state authority" (Schoon and Altrock 2014: 216). Through five strategies, described as "modes of governing 'the informal'", the authors show how the state reconciles two necessities: finding practical and effective solutions to rapid urban and economic changes; and managing the transition from a dominant state-party to a state-party that is seeking legitimacy and forms of cooperation with non-state actors in governing the cities.

3 According to Star and Ruhleder: "Infrastructure is a fundamentally relational concept. It becomes infrastructure in relation to organized practices" (1996: 113), and "infrastructure occurs when the tension between local and global is resolved. That is, an infrastructure occurs when local practices are afforded by a larger-scale technology, which can then be used in a natural, ready-to-hand fashion" (1996: 114). It is then precisely about cooperative activities through the frontiers of heterogeneous worlds.

4 At least within the official frameworks of sectoral policies and the official exercise of professional practices. By contrast, a number of studies have emphasised that "Sunday plumbers" and electricity "poachers" share their time between formal work in service companies and informal, on-request bricolage, thus establishing a bridge between these two worlds which long precedes its incorporation into policy.

References

Allen, A. (2014) Peri-urbanization and the political ecology of differential sustainability., In S. Parnell and S. Oldfield (Eds.), *A Routledge Handbook on Cities of the Global South*. London: Routledge, 522–538.

Ahlers, R., F. Cleaver, M. Rusca and K. Schwartz (2014) "Informal Space in the Urban Waterscape: Disaggregation and Co-Production of Water Services." *Water Alternatives* **7**(1): 1–14.

Ahlers, R., K. Schwartz and V. Perez Guida (2013) "The myth of 'healthy' competition in the water sector: the case of small scale water providers." *Habitat International* **38**: 175–182.

Banerjee, S. and E. Morella (Eds.) (2011) *Africa's Water and Sanitation Infrastructure: Access, Affordability, and Alternatives*. Washington: The World Bank.

Banerjee, S. et al. (2008) *Ebbing Water, Surging Deficits: Urban Water Supply in Sub-Saharan Africa*. Washington: The World Bank (Background Paper 12).

Baron, C. and A. Bonnassieux (2013) "Gouvernance hybride, participation et accès à l'eau potable Le cas des associations d'usagers de l'eau (AUE) au Burkina Faso." *Annales de géographie* **5**(693): 525–548.

Blanc, A., J. Cavé and E. Chaponnière (2009) *Les petits opérateurs privés de la distribution d'eau à Maputo: d'un problème à une solution?* Paris: AFD (coll. Regards croisés, working paper 85).

Botton, S. and A. Blanc (2014) "Un service public marchand de proximité: L'action des petits opérateurs privés pour la desserte des quartiers périurbains en Afrique." *Actes de la recherche en sciences sociales* **203**(3): 106–113

Buire, C. (2014) "Suburbanisms in Africa? Spatial Growth and Social Transformation in New Urban Peripheries: Introduction to the Cluster." *African Studies* **73**: 241–244.

Bulkeley, H., V. Castán Broto and A. Maassen (2014) "Low-carbon Transitions and the reconfiguration of urban infrastructure." *Urban Studies* **51**(7): 1471–1486.

Calas, B. (2007) "Dynamiques métropolitaines d'Afrique orientale." *Les Cahiers d'Outre-Mer* [online], **237**(January–March), DOI: 10.4000/com.624.

Calas, B. (2013) "Mondialisation, Clusterisation et recyclage colonial." *EchoGéo* [online], **26**, DOI: 10.4000/echogeo.13586.

Chakava, Y., R. Franceys and A. Parker (2014) "Private boreholes for Nairobi's urban poor: the stop-gap or the solution?" *Habitat International* **43**: 108–116.

Chauveau, J.-P., M. Le Pape and J.-P. Olivier de Saran (2001) "La pluralité des normes et leurs dynamiques en Afrique." In G. Winter (Coord.), *Inégalités et politiques publiques en Afrique*. Paris: Karthala-IRD, 145–162.

Cheneau-Loquay, A. (2012) "La Téléphonie Mobile dans les Villes Africaines. Une Adaptation Réussie au Contexte Local." *L'Espace Géographique* **41**(1): 82–93.

Choplin, A. (2014) "Les 'Sans fiche sans photo': Déplacements forcés et (non) mobilisation citoyenne à Nouakchott (Mauritanie)." *L'Espace Politique* [online], **22**(2014-1), http://espacepolitique.revues.org/2926; DOI: 10.4000/espacepolitique.2926.

Clément, P. (2009) *Erythrée: une approche fondée sur la mobilisation du "capital social": la coopérative villageoise*. Paris: Axenne/AFD.

Cohen, M. (1996) The Hypothesis of Urban Convergence: Are Cities in the North and South becoming more alike in an Age of Globalization? In M. Cohen, B. Ruble, J. Tulchin and A. Garland (Eds.), *Preparing for the Urban Future. Global Pressures and Local Forces*. Baltimore: The Johns Hopkins University Press, 25–38.

Criqui, L. (2015) "Infrastructure urbanism: Roadmaps for servicing unplanned urbanisation in emerging cities." *Habitat International* **47**(93): 93–102.

Damon, J. (2014) *Le marché des classes moyennes dans les pays émergents: quelle réalité? Quelles opportunités?* Paris: CCCI Ile-de-France (Collection "International" n° 25).

Darbon, D. (2008) "Réformer un inexistant désiré ou supprimer un inopportun incontournable? Le service public confronté à l'Etat et aux sociétés projetées en Afrique." *Téléscope* (ENAP/Québec) **Winter**: 98–112.

Darbon, D. (2014) "Nom de code 'Classes moyennes en Afrique'. Les enjeux politiques d'une labellisation de groupes invisibles et vulnérables. In D. Darbon and C. Toulabor (Eds.), *L'invention des classes moyennes africaines: Enjeux politiques d'une catégorie incertaine*. Paris: Karthala, 15–59.

Darbon, D. and C. Toulabor (Eds.) (2014) *L'invention des classes moyennes africaines: Enjeux politiques d'une catégorie incertaine*. Paris: Karthala.

Dorier-Apprill, E. and E. Domingo (2004) "Les nouvelles échelles de l'urbain en Afrique." *Vingtième Siècle. Revue d'histoire* **1**/81, 41–54, DOI: *10.3917/ving.081.0041*.

Dupuy, G. (2011) "Fracture et dépendance: l'enfer des réseaux?" *Flux* **83**(January–March): 6–23.

Eberhard, A. et al. (2011) *Africa's power infrastructure: investment, integration, efficiency*. Washington: The World Bank.

Foster, V. and C. Briceño-Garmendia (Eds.) (2010) *Infrastructures africaines: Une transformation impérative*. Washington: Banque mondiale et AFD.

Franks, L. and G. Prasad (2014) *Informal electricity re-selling: Entrepreneurship or exploitation?* Cape Town: University of Cape Town/Energy Research Centre.

Furlong, K. (2014) "STS beyond the 'modern infrastructure ideal': Extending theory by engaging with infrastructure challenges in the South." *Technology in Society* **38**: 139–147.

Gaunt, T. et al. (2012) *Informal Electrification in South Africa: Experience, Opportunities and Challenges*. Cape Town: Sustainable Energy Africa.

Geels, F. W. (2002) "Technological transitions as evolutionary reconfiguration processes: a multi-level perspective and a case study." *Research Policy* **31**(8–9): 1257–1274.

Ginisty, K. (2014) *Inégalités et (in)justices spatiales à Maputo. Pratiques des services urbains*. Paris: Université Paris Ouest Nanterre La Défense.

Godard, X. (Ed.) (2002) *Les transports et la ville en Afrique au sud du Sahara*. Paris: Karthala-INRETS.

Godard, X. and P. Teurnier (1992) *Les transports urbains en Afrique à l'heure de l'ajustement – Redéfinir le service public*. Paris: Karthala/INRETS.

Héraud-Arouna, M. (2005) *La réponse d'un pays entre terre et eau à l'essor des besoins alimentaires des villes. Le cas des pays du Bas-Ouémé (sud Bénin)*. Paris: Université Paris X-Nanterre (doctoral thesis).

Hodson, M. and S. Marvin (2010) *World cities and climate change: producing urban ecological security*. Maidenhead: Open University Press

Hugon, P. (2014) "L'informel' ou la petite production marchande revisitée quarante ans après." *Mondes en développement* **2**(166): 17–30.

Jaglin, S. (1995) *Gestion urbaine partagée à Ouagadougou: pouvoirs et périphéries (1983–1991)*. Paris: Karthala-ORSTOM.

Jaglin, S. (2005) *Services d'eau en Afrique subsaharienne: la fragmentation urbaine en question*. Paris: CNRS Éditions.

Jaglin, S. (2007) "Décentralisation et gouvernance de la diversité: Les services urbains en Afrique anglophone." In L. Fourchard (Ed.), *Gouverner les villes d'Afrique: Etat, gouvernement local et acteurs privés*. Paris: Karthala/CEAN, 21–34.

Jaglin, S. (2008) "Differentiating networked services in Cape Town: echoes of splintering urbanism?" *Geoforum* **39**(6): 1897–1906.

Jaglin, S. (2010) Gouvernance des réseaux et accès des pauvres à l'eau potable dans les villes d'Afrique subsaharienne. In J. Fisette and M. Raffinot (Eds.), *Gouvernance et appropriation locale du développement: au-delà des modèles importés*. Ottawa: Presses de l'Université d'Ottawa, 107–135.

Jaglin, S. (2013) "Les dispositifs hors réseau des villes en développement: quels enseignements?" In J.-C. Deutsch and I. Gautheron (Coords.), *Eau pour la ville, eaux des villes: Eugène Belgrand XIXe-XXIe siècle*. Paris: Presses des Ponts, 400–421.

Jaglin, S. (2014) Regulating Service Delivery in Southern Cities: Rethinking Urban Heterogeneity. In S. Parnell and S. Oldfield (Eds.), *A Routledge Handbook on Cities of the Global South*. London: Routledge, 434–447.

Jaglin, S. and A. Bousquet (2012)"Conflicts of influence and competing models: the boom in community-based privatization of water services in sub-Saharan Africa., In B. Barraqué (Ed.), *Urban water conflicts*. Leiden: Taylor and Francis/UNESCO-IHP, 169–193.

Jaglin, S. and M.-H. Zérah (2010 "Introduction: Eau des villes: repenser des services en mutation." *Revue Tiers Monde* **203**: 7–22. English version: "Urban Water: Rethinking Changing Services. Introduction", online at cairn.info.

Katusiimeh, M., K. Burger and A. Mol (2013) "Informal waste collection and its co-existence with the formal waste sector: the case of Kampala, Uganda." *Habitat International* **38**: 1–9.

Kjellén, M. and G. McGranahan (2006) *Informal Water Vendors and the Urban Poor*. London, IIED (HSDP Series, Theme: Water-3).

Lawaetz, S. and C. Smyser (2011) Challenges and Opportunities in Electricity Service Provision for Urban BOP Communities. In P. Márquez and C. Rufín (Eds.), *Utilities at the Base of the Pyramid*. Cheltenham, England, and Northampton, Massachusetts: Edward Elgar Publishing, 134–156.

Letema, S., B. van Vliet and J. B. van Lier (2014) "Sanitation policy and spatial planning in urban East Africa: Diverging sanitation spaces and actor arrangements in Kampala and Kisumu." *Cities* **36**, 1–9.

Lighting Africa (2013) *Rapport Lighting Africa sur les tendances du marché. Présentation du marché de l'éclairage hors réseau en Afrique*. Nairobi: Lighting Africa/IFC/Banque mondiale.

Lombard, J. (2006) "Enjeux privés dans le transport public d'Abidjan et de Dakar." *Géocarrefour* **81**(2): 167–174.

Lorrain D. and F. Poupeau (2014) "Ce que font les protagonistes de l'eau. Une approche combinatoire d'un système sociotechnique." *Actes de la recherche en sciences sociales* **203**(3): 4–15.

Monstadt, J. (2009) "Conceptualizing the political ecology of urban infrastructures: insights from technology and urban studies." *Environment and Planning A* **41**(8): 1924–1942.

Monstadt, J. and A. Wolff (2015) "Energy transition or incremental change? Green policy agendas and the adaptability of the urban energy regime in Los Angeles." *Energy Policy* **78**, 213–224.

Morange, M. and S. Fol (2014) "Ville, néolibéralisation et justice." *Justice spatiale/Spatial Justice* **6** http://www.jssj.org/article/neoliberalisation-ville-et-justice-spatiale/ (last accessed 14 September 2015).

Myers, G. (2014) "A World-Class City-Region? Envisioning the Nairobi of 2030." *American Behavioral Scientist*, published online 29 September 2014, DOI: 10.1177/0002764214550308.

Nantchop, V. (in press) "L'action publique urbaine à l'épreuve des réformes du service public d'eau à Douala (Cameroun)." *GéoCarrefour*.

Naulet, F. and L. Biteete (2014) *Cahier n°6 – Quelle régulation pour les réseaux autonomes en RDC ? Promouvoir l'extension des mini-réseaux et des bornes-fontaines en RDC*. Paris: Gret.

Naulet, F., C. Gilquin and S. Leyronas (2014) *Eau potable et assainissement dans les villes du Sud: la difficile intégration des quartiers défavorisés aux politiques urbaines*. Paris: GRET (Débats&Controverses, n°8).

Ngambi, J. (2015) *Déchets solides ménagers dans la ville de Yaoundé (Cameroun): de la gestion linéaire vers une économie circulaire*. Le Mans: Université du Maine (doctoral thesis).

Olivier de Sardan, J.-P. (2010) *Local Governance and Public Goods in Niger*. London: ODI (APPP working Paper No. 10).

Pierrat, A. (2014) *Les lieux de l'ordure de Dakar et d'Addis Abeba. Territoires urbains et valorisation non institutionnelle des déchets dans deux capitales africaines*. Paris: Université Paris I Panthéon-Sorbonne (doctoral thesis).

Pradeilles, J.-C., G. Garcia-Oriol and I. Tall (1991) *L'organisation corporative des taxis collectifs à Bamako et Lomé: richesse et perspectives*. Lyon: Codatu.

Rutherford, J. and O. Coutard (2014) "Urban Energy Transitions: Places, Processes and Politics of Socio-technical Change." *Urban Studies* **51**(7): 1353–1377.

Schoon, S. and U. Altrock (2014) "Conceded informality. Scopes of informal urban restructuring in the Pearl River Delta." *Habitat International* **43**: 214–220.

Smith, A., A. Stirling and F. Berkhout (2005) "The governance of sustainable socio-technical transitions." *Research Policy* **34**: 1491–1510.

Smyser, C. (2009) *Slum Electrification Programmes: An Overview Of Global Versus African Experience*. Background paper for the workshop Promoting Energy Access for the Urban Poor in Africa: Approaches and challenges in slum electrification, Nairobi, Kenya, October 26–27, 2009.

Sory, I. (2013) *"Ouaga la belle!" Gestion des déchets solides à Ouagadougou: enjeux politiques, jeux d'acteurs et inégalités environnementales*. Paris: Université Paris I Panthéon-Sorbonne.

Star, S. L. and K. Ruhleder (1996) "Steps Toward an Ecology of Infrastructure: Design and Access for Large Information Spaces." *Information Systems Research* **7**(1): 111–134.

Trémolet, S. and B. Evans (2010) *Output-Based Aid for Sustainable Sanitation*. Washington, DC: GPOBA/WSP.

Valfrey-Visser, B., D. Schaub-Jones, B. Collignon and E. Chaponnière (2006) *Access through innovation: Expanding water service delivery through independent network providers. Considerations for practitioners and policymakers*. London: BPD/AFD.

Van Dijk, M. P., S. Etajak, B. Mwalwega and J. Ssempebwa (2014) "Financing sanition and cost recovery in the slums of Dar es Salaam and Kampala." *Habitat International* **43**: 206–213.

Watson, V. (2014) "African urban fantaisies: past lessons and emerging realities." *Environment and Urbanization* **1**(26): 561–567.

Wilkinson, P. (2010) "Incorporating informal operations in public transport system transformation: the case of Cape Town, South Africa." *Brazilian Journal of Urban Management* **2**(1): 85–95.

10 Enabling urban energy

Governance of innovation in two UK cities

Janette Webb

Introduction

The renewable and low carbon energy initiatives described in the policy documents of many European city authorities convey a sense of increasing ambition to move away from urban infrastructures reliant on intensive use of energy from fossil fuels to a future where 'green' cities take responsibility for sustainable energy production and consumption. This impression is reinforced by European national and supra-national energy and climate policies which position cities as having significant responsibility for governing societal innovation, and by the transnational knowledge exchange networks of municipal authorities intended to support 'smart' sustainable cities,[1] where depoliticised 'route maps' and 'tool kits' are presented as offering a standardised framework for rational problem solving. However, these ambitions to reshape urban energy systems have to be understood in relation to twenty-first century globalising markets oriented to large-scale exploitation of fossil fuels, where transnational corporations control significant assets. In the last century, as urbanisation intensified and energy use accelerated, energy production in affluent parts of Europe generally became more geographically remote from cities, was less likely to be governed directly by municipal enterprises and was frequently invisible to inhabitants. In addition, European crises over public finances, and the decreasing capacities and powers of many local governments, raise questions about the extent to which municipal authorities can and will respond strategically to the need to transform energy production and consumption for future sustainability.

Projects concerned with transforming current energy infrastructures are situated in the resulting unequal power relations of energy market, institutional and governance structures, which are themselves embedded in local politics; the trajectories of such projects may variously reinforce or disrupt such structures (Rutherford and Coutard 2014). This chapter explores the contested legitimacy of urban projects to recreate localised energy supply in two cities in the UK, where local authorities have limited discretion over policies and finance, and have had no direct responsibility for energy since the early twentieth century. Municipal electricity and gas companies largely disappeared in the 1940s, with post-war nationalisation pursuing universal networked, standardised services on a regional, and then central, basis prior to privatisation in the 1990s. The chapter will explore why and how some twenty-first century city authorities are engaged in projects aiming to re-localise components of energy supply, and discuss the embedding of the same technology of combined heat and power and district heating in political processes which result in contrasting governance structures, business priorities and trajectories. Case studies of Birmingham in England and Aberdeen in Scotland provide insight into the ways in which localised energy actor networks build capacity to innovate in the context of the UK's privatised, centralised, socio-technical energy infrastructure.

Decisions to pursue local energy projects are powerfully constrained by incumbent interests controlling large-scale energy infrastructure. However, such urban political processes are neither separate from, nor completely conditioned by, national and supra-national political dynamics, but are both shaped by and contribute to reshaping these (MacLeod and Jones 2011). Any localised provisions successfully developed in these circumstances may remain small-scale, incremental experiments which are vulnerable to co-optation through the powerful logics of capital accumulation in a centralised model. Alternatively, such large-scale systems may be susceptible to challenge from the political momentum generated by localised energy developments, potentially contributing to a longer term systemic transition to sustainable energy (Rohracher and Späth, 2014). The chapter will therefore consider the extent to which such local projects can contribute to opening up the 'closed' energy infrastructures characterising the UK's privatised, liberalised system.

The following section considers the energy policy and governance frameworks and centralised energy markets in which UK city authorities are simultaneously constituted as particular kinds of actors, and seek to gain powers and action potentials. The third section notes the multiple, and potentially conflicting, rationales at work behind the resulting low carbon energy policies, and their potential for deployment by city authorities to differing political ends. The fourth section introduces the concepts of social innovation and social capital, as necessary for the assembly of financial, technological and business knowledge and capacity when codified governance techniques and established access to resources are absent, as is the case when urban energy users engage in changing large-scale energy infrastructures and privatised supply relationships. The fifth section develops the case studies of Aberdeen and Birmingham through comparison of the differing articulation of local energy strategies and purposes. Attention is paid to the discursive deployment link of strategies in relation to contested material developments using combined heat and power (CHP) and heat networks or district heating (DH). Both cities are unusual in the UK context in going beyond incremental energy saving, and project teams struggle to mobilise and enrol a viable actor network, and to assemble knowledge and finance to justify development. Locally mediated responses to UK neo-liberal political economy produce contrasting business models and trajectories. The sixth section reviews energy developments in each city and the social and political significance of their different governance and ownership structures. In the concluding section, the chapter evaluates to what extent, and in what forms, material innovations in energy infrastructures are being embedded in each city, and asks whether these are likely to be consequential for UK centralised energy infrastructures and low carbon transition, or not.

Neo-Liberal Political Economy and Low Carbon Urban Energy Projects in the UK

As European energy policy has moved towards market liberalisation, municipal control over urban energy infrastructures has generally declined, which suggests more constraint over localised innovation. Market concentration and consolidation, through mergers and acquisitions, has been the norm (Petz et al. 2012), with more than two thirds of the European market controlled by eight large corporations (Jamasb and Pollitt 2005: 26). Energy infrastructure, fuel supplies and services are thus subject to financial market logics, where the focus is on maximising short-term rates of return on investments, insulated from the particularities, and uncertainties, of place-based history, politics and culture (Torrance 2008).

Despite the pattern of increased concentration and centralisation, liberalised regulation has not resulted in European convergence on one particular market formation. In countries such as Sweden, Denmark and Germany, where a social-democratic consensus has supported integrated, and in some instances cross-subsidised, public services including energy, municipal authorities have had significant regulatory and operational control until very recently. Such variation persists, with municipal and regional businesses continuing to operate localised generation and/or supply, but with increasing numbers of public private partnerships, franchise arrangements and more fragmentation of local services (Helm 2010; Monstadt 2007; Rutherford 2008). Where municipalities have sold shares in energy companies to privatised utilities there is some evidence that these are likely to be managed according to commercial criteria,[2] rather than as a means of balancing social and environmental objectives with profit-making goals. The overall impact of these complex and shifting political dynamics on urban energy infrastructures and services is uncertain, but the pattern of ownership concentration indicates the difficulties of reconstructing locally-controlled provision, governed by local priorities.

In the UK the disappearance of localised and municipal energy supply long predates privatisation and liberalised regulation. Local companies were nationalised in the 1940s under planning for universal standards and efficiency improvements; regional rationalisation was followed by increasing centralisation. The dominance of neo-liberal political economic ideologies in the 1980s led to the privatisation of an already centralised infrastructure, commencing with the sale of shares in the British Gas Corporation; electricity privatisation followed in the 1990s. Subsequent mergers and takeovers have resulted in a supply infrastructure dominated by six large-scale corporations. The UK state re-engaged directly with energy policy only at the beginning of the twenty-first century, as increasing scientific evidence about the impact of fossil fuel use on climate change gained political traction. The UK

Climate Change Act (2008) included provision for a new government department of energy and climate change (UK DECC), with responsibilities including the design of incentives for low carbon electricity supply. The process was significantly shaped by incumbent business interests, indicating the 'designed in' weakness of public sector coordination capacity (Winskel et al. 2014). Policy instruments to incentivise decarbonisation focused on subsidies in the form of price support for large-scale renewable electricity generation. However, financial crisis, combined with increasing UK dependence on imported gas and rising energy prices, have destabilised what appeared to be political consensus over transition to low carbon energy. The relative priority to be given to economic growth, energy affordability, security and independence of supply, and climate change is the subject of conflicting political rationales, and uncertainty over the future of UK energy infrastructure is pervasive (UK DECC 2012).

In many ways these circumstances have intensified debate about the political-economic and carbon rationalities of established energy systems. One subordinate policy strand in the UK Electricity Market Reform (EMR) White Paper (2011), for example, notes the potential for decarbonisation of the energy system from distributed, or decentralised, infrastructure in the form of urban-scale combined heat and power and district heating networks:

> We recognise that integrated, local-level distributed energy systems could be an important step towards a more coordinated approach that includes, for example, transport and waste.
>
> (UK Government Department of Energy and Climate Change (DECC), 2011: 104)

The marginal status of such policy is evident in its inclusion as a single text box in the lengthy White Paper, and it has been suggested this was included solely because of the requirement on UK governments to respond to recommendations by the Royal Commission on Environmental Pollution (RCEP)[3] whose report on energy and climate (RCEP, 2000) included evidence on the long-term social, environmental and economic benefits of decentralised energy, particularly CHP and DH, as components of infrastructure. Of course, locally-embedded generation of energy does not necessarily require a break with a private finance model, where control is remote from energy users. The White Paper text nevertheless signals such potential with reference to long-term financial benefits, improved local control through community involvement and investment, and a more resilient, affordable and secure energy supply (pp.104–5). Localised generation of heat, and in many cases electricity, is also typically promoted on the basis of its techno-economic system efficiencies: by reducing the load on centralised generation and transmission systems, it reduces the need for expensive stand-by, high carbon, generation plant and grid reinforcement (Kelly and Pollitt, 2010). In the context of UK centralised electricity and gas networks, however, such infrastructure would entail significant new investment, which would pose challenges to the interests of incumbents in continuing to extract financial returns from sunk investments in gas and electricity networks; hence the attributed social, environmental and economic value of urban retrofit of CHP/DH infrastructure in the UK is highly contested.

Positioning UK Cities as Developers of Low Carbon Energy

Political struggle over the future of energy systems and the conflict between simultaneous demands for market expansion through further exploitation of fossil fuels, climate protection, energy affordability and security confront governments with complex collective action problems. Municipal leadership, supposedly 'unencumbered with the "paralysis" afflicting national governments in responding to resource security and climate change issues' (Hodson and Marvin 2009: 196), has returned to the fore as part of the solution. The tentative steps towards re-localising energy infrastructure, particularly for heat supply, aired in the EMR White Paper (2011) have been further developed in subsequent UK policy, with city authorities positioned as intermediaries: 'Local authorities are critical players in increasing the deployment of heat networks as they can create a supportive environment … and support or sponsor specific projects' (UK DECC 2013: 50). Urban authorities are expected to work as intermediaries building strategic partnerships, coordinating levels of governance, joining up policy across functions, and acting as standard bearers for best practice in relation to transparency, inclusivity and accountability (Betsill and Bulkeley 2006; Guy et al. 2011). These are high expectations in at least two senses: first in the context of globalising energy markets with limited incentive to attend to regional and local interests, or environmental objectives; and second in the context of models of local governance informed by market-driven, short-term cost reduction criteria, where new projects mean competing with other cities for diminishing public finance, or using public assets as a bargaining tool to attract private investment.

In the conventions of neo-liberal political economy, a market model of local governance, using competitive contracting, outsourcing and public-private partnerships (PPPs) rather than direct service provision, is equated with efficiency and effectiveness (Osborne and Gaebler 1992). Governance through competitive contracting is expected to work as a vehicle for combining a democratic agenda with mobile expertise and capital, and modulating the tensions between them. In practice, evidence from case studies of a number of European cities suggests that the market model for public services is associated with increased fragmentation, reduced quality of provision, and less inclusivity and equity (Bulkeley and Kern 2006; Coutard and Rutherford 2010; Monstadt 2009; Rutherford 2008; Rutherford and Coutard 2014). Some research indicates that social and environmental protection goals are potentially commensurable with the use of private finance for low carbon infrastructure. For example, Hodson and Marvin (2009) conclude that the London Hydrogen Partnership (LHP) did mediate competing objectives of social equity and profitability. Local actors facilitated flows of knowledge, bridging conflicting expectations and managing conflict between a local progressive/inclusive politics and a business-led 'exogenous' model of the city as a showcase. Nevertheless, they conclude that such projects risk public co-optation as a showcase for mobile capital, without necessarily securing long-term local benefits. Monstadt (2007) also argues that energy privatisation has not precluded urban governance of low carbon innovation, but he concludes that city authorities such as Berlin struggle to identify and/or develop the necessary capacities: 'urban governance of energy systems in Berlin is characterised by … an institutional void: a lack of generally accepted rules, procedural norms and organisational capacities guiding policy-making to protect the public interest' (Monstadt 2007: 340).

The politics of low carbon energy are evidently subject to recurring controversies and reversals. Superficial consensus over energy plans and projects may disguise embedded conflict over their strategic significance and future, with the likelihood that dominant private sector energy interests prevail over a more locally-mediated and orchestrated low carbon system (Bulkeley et al. 2014). Coutard and Rutherford (2010) document the struggles of the Ile de France regional authority in strategic planning for sustainable energy and local economic resilience, in the face of established commitments to high-carbon economic growth, embedded in state governance and international markets. In the absence of cohesive national and supra-national political commitment to redirection of finance for low carbon infrastructure, much city governance of sustainable energy transitions may remain aspirational, with actions focused on incremental energy savings rather than more radical innovation (Bulkeley and Kern 2006; Hawkey et al. 2014). Even in the city of Stockholm, contest over resources for clean energy and the diversion of finance into high-carbon agendas remains potent, disrupting earlier commitments to urban sustainability (Rutherford 2014). The schisms apparent in urban energy politics suggest that there is considerable potential for different interest groups within cities to exploit the tensions to serve a variety of local political agendas, with uncertain consequences for material change.

Localised Innovation in Energy Infrastructures and the Role of Social Capital

Questions about whether, and how, city authorities with limited capacity, structural powers and resources in the neo-liberal UK can innovate in low carbon energy infrastructure hence remain. Rebuilding localised energy for a carbon-constrained era represents a fundamental break with the established centralising and privatising logics of high carbon energy, requiring societal, as well as technical, innovation. Much innovation theory focuses on macro-economic models oriented to top-down scientific and technological breakthroughs, but a strand of sociological theory examines innovation from the bottom-up perspective of situated practice. This focuses on the potential for socio-technical innovation in complex systems to be orchestrated by hybrid communities of practice comprising end users, experts, politicians and policy-makers in shifting alliances and coalitions (Steward 2012). Such 'complex actors' develop knowledge and capabilities through improvisation, or learning by doing, prompted by particular circumstances and opportunities. Infrastructure innovation from this perspective entails a seamless web of material artefacts, organisations, institutions and cultural values (Hughes 1983).

Understanding innovation in urban energy infrastructure, against the dominant norms set by established markets and regulatory incentives, requires insight into the formation of heterogeneous actor networks capable of material change at a meso-scale (Callon 1986; Latour 1986; Steward 2012: 333). This is unlikely to be a cohesive or linear process, given the differential power and capacity of different actors to impose their definition of the problem and its solutions, to define the agenda for change, and to mobilise resources. Given that no actor exercises overall control of complex socio-technical systems, such innovation processes are indeterminate, with potential for alliance-building among large energy users such as municipal authorities, civil society groups, engineering bodies and technical experts where shared constellations of interests can be crafted. In these circumstances established systems, and their functions, governance and control, may be challenged by new coalitions of interest (Summerton 1994).

Such coalitions are likely to rely on mobilising forms of social capital to identify and enlist necessary formal resources and expertise. The concept of social capital describes the potential of societies to learn, improvise and innovate through the indeterminate dynamics of social interaction (Portes 2000) represented in 'the networks … norms, values and understandings that facilitate co-operation within or among groups' (OECD 2001: 41). However, it is a contested concept, which is criticised for equating social interaction with a form of calculative exchange. Nevertheless, it can be used to identify the value and significance of non-monetarised sources of power and action capabilities (Tura and Harmaakorpi, 2005). It remains in circumstances marked by uncertainty over resources, capacity and the legitimacy of acting, when formal, codified technique and rules of governance are lacking (Bowles and Gintis 2002; Sennett 2012).

A distinction can also be drawn between *bridging* social capital, which connects different groups of actors with potentially competing or conflicting objectives through weak ties (Granovetter 1973), and *bonding* social capital, which sustains relationships within a group (Putnam 2000). Social capital of the bridging variety has been shown to be important for more radical or 'disruptive' innovation (Christensen 1997; Ehrnberg and Jacobsson 1997; Lundvall et al. 2002; Maskell 2004; Tura and Harmaakorpi 2005) and is regarded as a contributor to regional innovation capacity (Cooke et al. 2000). It is a form of social capital concerned with developing resources and options beyond those already known, and hence calls into being improvised solutions.

Assembling Innovation in Urban Energy: Aberdeen and Birmingham

Case studies of Aberdeen and Birmingham, based on semi-structured interviews and documentary analysis,[4] offer insight into processes characterising such capacity-building for localised innovation in a landscape framed by neo-liberal political economy, privatised energy infrastructure and weak local government.

Under UK devolved government, introduced in 1998, local authority responsibilities are decided within each constituent country. Although structures have remained very similar, a higher degree of (perhaps necessary) interdependence in local-central government relations is evident in Scotland, with a greater density of institutional links between civil society and policy networks in comparison with England (McGarvey 2011). In contrast, English central-local government relations are marked by more direct use of market mechanisms and competitive performance management (Downe et al. 2010). These somewhat different contexts for action are not expected to translate straightforwardly into overall differences in capacity for municipal energy innovation, if only because no UK local authorities have well-established municipal energy provisions, and their powers and budgets remain constrained by the respective central governments of each country. The cities of Aberdeen, in Scotland, and Birmingham, in England, are therefore treated as exemplars of how local authorities, operating in multi-level governance frameworks coloured by the different 'processes and power dynamics through which governing is orchestrated' (Bulkeley and Kern 2006: 2242), struggle to create capacity to innovate within a centrally controlled system of local government.

Aberdeen and Birmingham share histories of industrialisation, relative decline and post-industrial reconstruction. Aberdeen is a small city (population 217,000) in the northeast of Scotland, with a history in merchant shipping, food processing and fishing industries, now much reduced in significance. The exploitation of North Sea oil and gas since the 1970s means that Aberdeen harbour has been redeveloped as an offshore oil industry supply and service location, and the city has become known as the 'oil capital of Europe'. The oil and gas industries have brought investment in urban infrastructure, jobs and low unemployment. However, the benefits have not been evenly distributed, with 15% of households in relative poverty, and the highest income inequality in Scotland. The city of Birmingham, in the West Midlands of England, has a population of just over one million, and a history as a major centre of the industrial revolution, technological innovation and manufacturing, again much reduced in significance and scale. Main economic activities are now public, retail, leisure and professional services, and there are high levels of inequality, unemployment and inner urban poverty.

Although energy infrastructure and services were outside core council responsibilities, and there was little confidence in local energy governance capacity, each city authority has recently invested in local combined heat and power and district heating infrastructure, but with different primary objectives and resulting governance models. Project development was marked by internal dissent and dynamics of opposition, uncertainty, anxiety and risk aversion. Establishing the legitimacy of localised energy infrastructure entailed intensive and contested intra-council political negotiation focused on financial accountability and risk assessment, short-term cost criteria, legal requirements, and EU procurement and state aid rules. The marginality of energy management in UK city councils means that initiatives derive from service teams with different priorities and hence different orientations to energy as a means to address their primary objectives. Aberdeen developments were initiated by the city's housing team, and prioritised social objectives for affordable warmth through retrofitting 1970s electrically-heated multi-storey housing. In 2003 the city council established

Aberdeen Heat and Power (AH&P) as an independent non-profit company, limited by guarantee, with a volunteer Board of Directors. The purpose of the company is to work for the benefit of citizens through the provision of affordable heating. Under a fifty-year framework agreement with the council, and governed by a Teckal exemption[5] from EU market procurement rules, AH&P has developed three energy centres, supplying heating and hot water to around 1700 households in twenty-two of the city's fifty-nine multi-storey housing blocks, as well as twelve public buildings and community facilities. Some of the co-generated electricity is sold via a private wire to the school; the remainder is sold into the public network via a consolidator. Domestic users receive unmetered heat and hot water at a fixed tariff (paid with rent) reflective of costs. In Birmingham, localised energy projects, led by urban design engineers, were more conventionally framed by economic regeneration and competiveness strategies. In 2006, seeking to minimise exposure to perceived financial risk, the city council signed a twenty-five-year contract with a private company, Utilicom Ltd, to build, own and operate gas-fired combined heat, cooling and power networks for supply to the city authority and other large heat and power users in commercial and public services. This resulted in establishment of Birmingham District Energy Company Ltd (BDEC), a wholly owned special purpose (SPV) subsidiary of the parent company. Utilicom was subsequently acquired by GDF-Suez and restructured as a district energy utility, Cofely. BDEC directors are employees of Cofely, although there is also a partnership board, with large subscribers entitled to a profit share in the form of an energy rebate.

How can the emergence of localised energy innovation in each city be explained? Three key processes proved central to mobilisation of political support and assembly of a particularised actor network which rebuilt capacity to act on the material of energy infrastructure: first, the foundations for challenge to centralised energy markets and terms of provision were laid by urban politics stemming from anti-poverty campaigns; second, differently located local actors mobilised social capital through intersecting, multi-scalar socio-technical networks; and finally, the emerging actor network was able to build a locally-workable social, environmental and economic calculus of the value of urban CHP and DH infrastructure, geared to the politics of competition in Birmingham and the politics of welfare in Aberdeen. In each case, this formula proved sufficiently powerful to contest short-term least-cost decision criteria, and to justify investment. Each of these processes is examined in more detail below.

Urban Political Processes and Anti-Poverty Movements

Initial impetus for municipal investment in CHP and DH infrastructure in the two cities can be traced back to anti-poverty campaigns, and the associated inter-meshing of formal and informal political processes. In Birmingham, Tenants and Residents Association fuel poverty campaigners brought court cases against the council in the 1980s, forcing the housing authority to make unplanned improvements in the energy performance of its stock. City engineers, with

Table 10.1 Organisation of localised energy in UK cities of Aberdeen and Birmingham

	Aberdeen Heat and Power Ltd	*Birmingham District Energy Company Ltd*
City council-led objective	Affordable warmth for social housing	Urban regeneration in city centre commercial and public services district
Organisation structure	Independent company, limited by guarantee, under local ownership and control	SPV wholly-owned and operated by Cofely District Energy Ltd., a subsidiary of GDF-Suez
Business model	Non-profit ESCo; any surplus reinvested or used to secure low-cost heat tariffs	For profit ESCo with private shareholders
Governance structure	Volunteer board of directors including councillors, community and business organisations and former council officers	Directors from parent company, plus partnership board with profit share, in form of energy rebate, for founder subscribers
Heat tariffs	Cost-based	Market-based, indexed against consumer gas prices
Main customers	Tenants in public housing	Leisure and entertainment, retail, commercial and public services
Other customers	Sport, education, leisure health and public facilities	Tenants in public housing
Finance	UK and Scottish government grants, city housing regeneration capital, prudential borrowing, bank loan and overdraft	UK government grants, parent company internal finance plus borrowing
Risk mitigation	Loans guaranteed by city council; council long-term contract for purchase of energy	Loans guaranteed by parent company; council, NHS, University and commercial enterprises enter long-term energy purchase contracts

Source: author's research

technical knowledge about CHP/DH as an energy- and cost-saving measure, argued for direct investment by the council as a long-term solution for affordable heating in the city's social housing. Short-term, least-cost decision criteria prevailed, however, resulting in a decision to refurbish electric heating and improve insulation in the least energy-efficient multi-storey housing. Engineers nevertheless used the opportunity to gain support for a council-operated pilot CHP/DH scheme serving a community leisure centre and three multi-storey housing blocks. The project worked as a means to supply affordable heating to tenants, and generated favourable publicity and political capital for the council. Anti-poverty political action through the local courts hence created scope for experimentation in localised energy infrastructure.

Anti-poverty campaigns were also instrumental in innovation in Aberdeen, but this time long-standing Scottish cross-party political support for action to reduce poverty, led by Labour and Liberal Democrat coalitions, was given renewed momentum by legislation introduced by a UK Conservative government. During the 1990s, fuel poverty campaigns across the UK built momentum for home energy conservation legislation, at the point when the politics of climate protection also gained prominence. A by-election in England created the opportunity: the UK Home Energy Conservation Act (HECA) (1995) 'was established due to Diana Maddock, who was elected Lib Dem MP for Christchurch at a by-election in 1993. There was a Conservative Government plan to introduce VAT on fuel at 17.5%. Her Christchurch constituency had a high percentage of pensioners' (community energy consultant and member of AH&P Board). Hence she had direct cause to pursue legislation to protect householders from rising energy bills. The Act incorporated home energy-saving and climate protection targets, requiring local authorities to identify cost-effective measures for a reduction of 30% in home energy consumption and carbon emissions in the decade between 1997 and 2007. The requirements were patchily enforced, but in Aberdeen, where projects to reduce fuel poverty had been in place since at least the 1980s, the legislation was used to justify resources for a new Home Energy Coordinator post. An experienced anti-poverty campaigner, rather than an officer with a technical background in housing services, was appointed. Her political commitment created momentum for more radical measures, notably investment in localised CHP/DH, as a solution to fuel poverty.

In Birmingham, despite historical experience of addressing fuel poverty through CHP/DH, further impetus came not from the HECA legislation but from regeneration and competitiveness strategy, in the context of a ruling local Conservative-Liberal Democrat coalition wedded to cost reduction through market-contracting in local government. However, local Liberal Democrat manifesto commitments to *sustainable* development were a lever for city engineers, committed to the energy- and carbon-saving value of CHP/DH technologies, to secure support for urban energy investment as a component of economic competitiveness, and the Liberal Democrat deputy council leader was enrolled as the project's political champion.

Urban political processes were hence moulded through constantly evolving interaction between the different UK levels and scales of government 'where the positions and practices of local, regional and national actors are always interdependent and mutually constituted' (Coutard and Rutherford 2010: 722–3). Such processes gave differential shape to energy projects in each city: longstanding UK anti-poverty social movements, and localised campaigns against fuel poverty, were given form and traction by different kinds of actors, through different channels, and encountered contrasting local political coalitions with material consequences for subsequent urban energy infrastructure and supply relations.

Assembling an Actor Network and Social Capital

The long-term absence of capacity for energy governance in UK cities means that the institutional void identified by Monstadt (2007) in Berlin is far greater in the UK, where city authorities lack codified techniques, committee procedures, governance frameworks and organisational structures for energy policy and provision. Contesting, and breaking away from, established relations between passive energy users, large scale corporate suppliers and government regulators in a centralised system, materially embedded in the urban infrastructure, requires local actors to construct new capacities in socio-technical knowledge and resources in the complex political-economic and technical sphere of energy systems. Projects place demands on capacity to coordinate knowledge and resources across specialisms including planning, environment, roads, community services, housing, finance and economic development.

In both cities, identifying and discriminating between sources of trustworthy, usable knowledge, suited to local circumstances and objectives, and accessing all available resources, entailed intensive negotiation and alliance building. In the absence of codified procedures, lead officers drew on informal council networks to identify peers in other specialist teams who could be drawn in, if such initiatives could be shown to align with their purposes. In some cases this was a matter of common values and goals; in others it required an identification of some perceived instrumental benefit. In each case, the formation of an informal coalition of politicians and officers was critical to establishing and legitimating the public value of urban heat and power infrastructure. Members of this core group had to be willing to work beyond specialist remits and core competences, and to carry the risks and burdens of improvising sufficient technical, legal and financial knowledge of energy technologies and markets to re-localise elements of provision.

Where this activity mobilised 'bonding' social capital, it was matched by calling on forms of 'bridging' social capital through engagement in loosely-configured community energy and energy technology and market networks. In Aberdeen, UK HECA legislation provided the resources for an energy officer network, enabling improvised learning, with further connections to public agencies and technical expertise. Participation in these networks created pathways to formal legal and financial expertise and energy market knowledge. For example, the Aberdeen officer encountered the vice-chair of the committee set up to distribute HECA support funds. His biography included anti-poverty work as well as commercial expertise, and their complementarity and experience contributed to the articulation and justification of local investment in, and control over, energy assets as a means to tackling poverty through affordable heating provision for low income households. The Aberdeen engineering design consultant, who was also supportive of a non-profit business model, was identified through a public sector list of pre-approved companies. In addition, in Scotland, the density of institutional links between civil society and policy networks was evident in contributing to the emerging actor network of local officers and politicians and non-local energy systems expertise: the Scottish Community Energy Network funded by government coordinated local applications for grant funding when the UK government introduced a short-lived Community Energy Programme (CEP), with a £50 million budget in 2002. The significance of the CEP is discussed further in the next section.

In Birmingham the mobilisation of social capital and the constitution of a viable actor network shared similar qualities to that in Aberdeen, but different political priorities led to a socio-technical assemblage of knowledge and capacity focused on commercial goals. Notable impetus came from the technical knowledge of engineers, and their commitment to CHP/DH as a long-term energy and cost saving 'building block' in city economic strategy. The earlier success of the pilot CHP/DH infrastructure for social housing and leisure facilities was insufficient to overcome neo-liberal political aversion to public borrowing for direct investment, and (as in all UK councils) commitment to short-term least cost criteria for capital projects predominated in decision-making. However, regeneration strategy, combined with the stimulus of UK CEP funding for low carbon energy, created a critical opportunity for engineers to argue for investment in urban energy.

In the context of contentious investment decisions, Birmingham council engineers, whose reputational credibility was critically dependent on scrutiny and evaluation of technical recommendations, deployed forms of bridging social capital to build a sense of certainty over the technical-economic valuation of CHP/DH proposals. Distinctions were drawn between the value of 'useful knowledge' and its negatively-tagged counterpart 'theoretical knowledge'. The former, was regarded as deriving value from the practitioner knowledge of 'dirty hands merchants' (city engineer), while over-reliance on 'theory' was regarded as risking unnecessarily high costs. 'Theoretical knowledge' was characterised as the product of public and private consultants, perceived as potentially over-engineering systems through technical metrics which factored-in costs of protection against claims on professional indemnity. The mobilisation of social capital focused on the discovery of sources of practice-based 'useful knowledge' as a means of identifying effective commercial and financial expertise. These expertise networks were in turn seen as instrumental in progressive development of council capacity in energy contracting, across planning, legal, finance, procurement, and engineering specialisms. In this case, the more conservative politics of the council, the aversion to public borrowing, and the priority given to strategy for economic competitiveness, resulted in an emphasis on interaction with cities where long-term private contracts to purchase heat, cooling and power from commercial operators of CHP/DH were preferred. Engineering scepticism towards the 'theoretical' knowledge of public bodies similarly geared the search for expertise more to private sector networks. A critical turning point was interaction with politicians and officers from another city with an established private sector district energy company. Birmingham politicians and officers were less averse to a business model which located the main financial risk of project failure with a private sector owner, even though this required revenue guarantees from the council and other users, in the form of a a twenty-five-year contract with a monopoly for-profit private provider. Council agreement to proceed to technical evaluation of potential schemes followed.

In each case, despite limited capacities and resources for urban energy infrastructures, scope for material innovation was created by the mobilisation of social capital and the associated local actor networks whose constitutive parts reflected contrasting political agendas.

Constructing an Alternative Formulation of Social, Environmental and Economic Value

> It was my Saturday job for a long time.
>
> (City of Aberdeen Home Energy Coordinator)

In both cities project development was prone to recurring crises, with complex and painstaking multi-party negotiation centring on the translation of improvised knowledge and resources into material change in energy infrastructure. Moving forward relied on the construction of an alternative, locally-contextualised calculus of social, environmental and economic value, which had to be sufficiently powerful to overcome the force of normalised and standardised short-term least cost decision criteria.

In Aberdeen, greater political cohesion around anti-poverty and affordable warmth action under the Aberdeen Labour-Liberal Democrat coalition did not simply equate with support for investment in CHP/DH infrastructure. The key to disrupting cautious norms of low cost incremental action was council adoption of an Affordable Warmth Strategy. 'Cost in use' of heating to residents, rather than 'least short-term cost' to the housing budget, was established as the primary criterion for decisions, and a technical options appraisal identified CHP/DH as the best means of meeting the goal. Despite the progressive justification built by the local actor network, the decision was heavily contested, with the council legal team advising against municipal investment in energy infrastructure on the grounds of financial risk. At this juncture, political leadership was critical; the deputy leader, a Labour councillor and incomer who had moved north for the oil industry, chaired the council committee which considered the full case for the creation of an independent non-profit energy services company, with loans underwritten by the council: 'At the founding meeting he said that the advice of the council solicitors was "not to proceed, but we are not obliged to take their advice". So he put their report down and said "it is noted". So he had the political courage' (community energy consultant and member of AH&P Board).

In Birmingham, the focus of the Conservative-Liberal Democrat coalition on cost-based market contracting for services meant that the key to justifying localised energy was establishing legitimacy for a new evaluation metric using 'whole-life cost' (WLC) appraisal of energy services to contest the conventional cost-benefit equation. Despite the initial scepticism of council finance specialists towards the validity of WLC formulae, UK government private finance initiatives were supportive of such techniques, coincidentally leading to a requirement for officer training. The newly-trained finance officer became a conduit for local negotiation over inclusion of the cost to the council of replacements and failures in stand-alone gas boilers for each building, in a whole-life calculation of long-term energy costs. The unplanned breakdown of such equipment was subsequently reframed as a risk which could be formally costed and 'sold' to a private partner, as one component of a long-term energy supply contract. Forecast gas and electricity price increases were also a central factor in cost comparisons between costs of supply from centralised vs. localised infrastructure. When the economic calculus of future energy costs was assembled in WLC format, with component costs of network connection, future heat, cooling and power supplies, and equipment maintenance embedded, this proved sufficient to frame a legitimate case for the long-term economic benefit of localised CHP/DH, despite higher short-term costs.

In both cities, an element of public funding proved critical in enabling a fragile consensus to be made materially consequential. Following the 1997 UK election of a Labour government, commitment to mitigate climate change resulted in the creation of the short-lived Community Energy Programme (CEP), which ran from 2002 to 2007 with £50 million in grant finance from the UK Treasury for community heating infrastructure. In Aberdeen the Affordable Warmth strategy and pre-existing energy saving options appraisal meant that the city was able to proceed immediately. Housing regeneration capital was committed, in combination with three successful CEP grant applications, plus funding from energy utilities obliged by UK government to invest in energy saving. A Co-Operative bank loan to AH&P, guaranteed by the council, provided the balance of finance for the first phase, which increased council confidence sufficiently for subsequent investments to draw on prudential borrowing. While every phase of development proved extremely demanding for politicians and officers unaccustomed to managing strategic planning for heat networks, the third energy centre and network resulted in the first crisis. The terms of CEP grants were bounded by complex technocratic application and spending criteria, with specified carbon-saving targets and tight timetables, governed by political cycles and financial years. For its third CEP grant, Aberdeen had specified carbon-saving targets through connection of a fixed number of multi-storey housing blocks to new DH infrastructure. Poor building fabric condition resulted in some of the planned blocks being withdrawn, threatening the loss of the finance for the whole project. Fixed CEP timescales meant that a revised calculus for meeting the carbon target had to be devised quickly. The engineering consultant proposed, and the Council agreed to, the connection of a nearby municipal leisure complex with a significant heat demand, which contributed to CHP efficiencies. However, the resulting pipework had a higher cost without additional benefit to housing tenants, and created a cash flow crisis for AH&P. The relatively inexperienced Board agreed to the plan, but initially lacked understanding of its full financial significance, resulting in considerable anxiety, some distrust of municipal accountants and threatened resignations. A solution was found through recourse to external accounting expertise. The cash flow position was restructured through an overdraft facility with the existing lender, which the council agreed to underwrite, thus minimising its total cost.

In Birmingham, the emerging practitioner network, legitimation created by the WLC formula for energy services and the promise of a capital contribution from the CEP had proved sufficient to gain agreement for the procurement of CHP/DH infrastructure for the city centre regeneration area. Bidders and council engaged in a lengthy negotiation over financial and technical credibility, which resulted in the selection of Utilicom Ltd. as the preferred contractor. A further extended period of negotiation ensued, while splits between public sector partners and within council specialisms over the risks of proceeding, and doubts raised by city-centre businesses over the merits of a long-term monopoly heat supply contract with a private provider, were debated. Lack of council experience in governing energy services, and scepticism of other building owners expected to connect to the heat and cooling network, interacted with conflict over the interpretation of complex CEP rules and EU regulations for procurement and state aid, resulting in extreme caution and the withdrawal of one public sector partner. Agreement to proceed, and the associated CEP funding application, was pushed beyond the final deadline and the lead city engineer relied on informal social capital to create extra time. Local legal teams and UK government civil servants stayed on to fax documents and discuss terms with 'phones open to London with fifteen minute calls to update' (city engineer). The CEP application was faxed at midnight to the UK government, 'all the way to the wire' (city engineer), and with continuing scepticism and opposition by sections of the council.

Having proceeded this far, local heat, cooling and power infrastructure was introduced in the city centre regeneration district in 2007, followed by a second phase in the city east end at Aston University in 2009, and a third at the nearby NHS hospital in 2010, each with connections to nearby civil and commercial heat loads and with the potential for interconnection of networks.

Localising Urban Energy Infrastructures in the UK

The political, economic and practice-based demands of reconfiguring centralised, privatised, high carbon energy infrastructures and supply chain relationships, embodied physically and socially in UK cities, are demonstrated by CHP and DH developments in Aberdeen and Birmingham. Municipal innovation required enterprising constitution of the means to act, through a cost-benefit calculus customised to particular urban political configurations. Social, local economic and climate protection benefits of municipal energy investment were difficult to establish as legitimate factors, given the prevailing short-term least cost decision calculus embedded in the UK political-economic regime. Locally-committed actors bridged the gap between decontextualised financial markets and local interests, in an attempt to establish the validity of different 'orders of worth' (Boltanski and Thevenot 2007).

These two cases suggest the uncertain future of climate and energy policies. In each city, the 'interpretive flexibility' (Pinch and Bijker 1984) of the technology was exploited to suit different interests. The projects were not the expression of homogeneous political commitments to climate protection, either within or across urban authorities. The 'tropes' of climate change and carbon accounting were instead moulded to local politics, and were subject to contention and trade-offs in the evolving partisan politics which characterise cities. In Aberdeen a pocket of local control over poverty and emissions reduction in social housing had to be offset by the city's push for growth through increasingly hazardous exploitation of off-shore oil and gas, where ownership and control is almost entirely outside Scotland and is locked into the dynamics of global finance and capital accumulation. In Birmingham the uncertainties of a twenty-five-year market-mediated contract with a district energy subsidiary of a transnational corporation, with returns accruing to shareholders, intersects with local projects to govern carbon and energy. These are in turn presented to external investors as a reinvigorated model for growth, emphasising the urban cost efficiencies and securities of scaling up city centre energy networks.

In each case the malleability of the same material CHP/DH technologies to different governance and economic structures is evident, with contrasting primary interests and values being advanced. Within their own terms, both projects have proved effective, and there is a degree of convergence in each council around aspirations to a sustainable development 'governance fix' via an extension of low carbon energy systems. In Birmingham, urban energy planning has continued as a component of economic strategy. To some extent this has reversed its framing from financial risk to financial opportunity for municipal revenues, local employment and protection from volatile global energy markets. The city's Climate Change Strategy and Action Plan aspires to energy self-sufficiency, with targets to cut carbon emissions by 60% by 2026. Plans include extended heat networks, building on city centre schemes. Council estate biomass (now re-imagined as resource rather than waste) would be used as an energy source, and the industrial infrastructure of the canal network deployed for low-impact transport, alongside investment in energy from waste. Plans include connecting 20% of social housing to DH by 2020. However, the contract between the council and the private utility fails to guarantee social benefits for poorer households. The Cofely-controlled heat network infrastructure means there is no obligation to supply social housing, and the profit-maximising model means avoiding the capital costs, low returns and debt risk associated with poorer households. Connection of a small number of multi-storey housing blocks to the network has so far relied on further government grants. Some of this work has been managed directly by the council, and proposals under the UK government 'Green Deal' are being used for further municipal investment, again with a private partner, Carillion. Future use of the heat network for distributing heat from multiple sources is also fraught with uncertainty. Creating a framework for technical, legal and commercial standards and a degree of transparency over the distribution of costs and benefits requires multi-party negotiations between the Council, Cofely, property developers, and energy-from-waste contractors, as well as the privatised electricity distribution network operator, with high transaction costs and no clarity over shared interests.

In Aberdeen, CHP and DH infrastructure have also contributed to increasing ambitions to integrate low carbon energy development across council services as a component of economic competitiveness, and energy services have been integrated into Enterprise, Planning and Infrastructure. The council is using its success in carbon reduction through CHP/DH investment (variously assessed as between 30% and 40%) to claim leadership in climate policy, despite its role in the extension of the road networks and oil and gas, with their high carbon growth implications. The council *Carbon Management Plan* (2010–15) sets a carefully-bounded target for carbon neutrality by 2020 for its buildings and public housing. However, local control over the future direction of decentralised heat networks does create a basis for an alternative locally-accountable energy system. There is a long-term aim to create a heat main, referred to informally as 'the ring of fire', around the city, connecting multiple heat sources and users, including NHS, university and commercial sites. The informal coordination between Scottish central and local government facilitated funding for a network extension to the city centre in 2013, opening up the connection of private sector housing and commercial load. Plans for commercial contracting have resulted in the establishment of a wholly-owned profit-making subsidiary of AH&P, District Energy Aberdeen Limited (DEAL), with any surplus returned to AH&P for system extension and local welfare. Current AH&P projects are connecting a further eleven multi-storey housing blocks to district heating, making low-cost provision available in thirty-three of the city's fifty-nine multi-storey blocks. Heating costs have been reduced by 45%, taking low income households out of fuel poverty, and demonstrating the potential of local democracy to deliver social change.

In both cities the formation of customised, place-specific socio-technical networks, meshing urban politics and locally-embedded knowledge with non-local expertise, was critical to material change. With different initial objectives, and mobilising different actor networks, the private SPV in one city and non-profit company in another represent contrasting embryonic forms for wider systemic transformation in urban energy systems, but the different forms seem likely to be consequential for the future distribution of costs and benefits.

Conclusions

This chapter has examined the material expression of political processes and power relations in UK urban energy projects, in the context of European sustainable energy policies. The cases offer insight into the socio-technical qualities of cities as meeting points between locally-committed, under-resourced actors and global trade in finance, legal and business expertise. What do they suggest about the embedding of alternatives to centralised energy in cities, and their likely consequences for UK centralised energy infrastructures and low carbon transition? Clearly a very large gap remains between state expectations for city leadership in sustainable energy, the matching ambitions of many city authorities, and their capacities to realise such plans. In the UK, which has led the way in neo-liberal experimentation in local authority services (Le Gales 2002) and privatised energy infrastructure, city authorities struggle to mobilise support, and build legitimacy and resources, for material change. Neo-liberal politics and weak local government, combined with centralised, privatised energy dominated by transnational corporations, provide very constrained circumstances.

Nevertheless, urban energy retrofit in Aberdeen and Birmingham show that the neo-liberal model is not impervious to municipal agency and innovation. Capacities for energy governance have been enhanced in both cities, resulting in more extensive ambitions. In Aberdeen, the perceived deficiencies in the status quo centred on the affordability of energy, prompting radical change not just in establishing quasi-independent infrastructure but also in decisions to take ownership in a context where expertise is dominated by large scale and transnational corporations. In Birmingham, the disruption of conventional techno-economic cost-benefit metrics for energy supply led to a new political settlement intended to secure the city's future economic competitiveness and autonomy, in a globalising and volatile energy market. However, governance via long-term energy supply contract with a private district energy utility aligns such localised provision with the neo-liberal frame of private finance and profit-maximisation.

These two projects offer different visions of the energy futures of affluent cities, one emphasising 'sustainable' growth through private capital accumulation from investment in 'green energy' and the other oriented to a politics of citizenship and local participation. Each city is far from homogeneous with respect to these different rationalities, and there is no certainty over the trajectory in either case. UK city authorities have a tenuous grasp on energy policy and politics, despite the centrality of energy to economic activity and social reproduction. It seems more likely that the democratic moment in Aberdeen, establishing materially- and socially-embedded affordable heat supply, will be more difficult to step back from, and should have wider significance for UK centralised energy infrastructures and low carbon transition than the ring-fenced private contractor model in Birmingham. Aberdeen CHP/DH projects re-localise competence and control, offering material demonstration of the viability of an alternative system of provision, structured by principles of reciprocity and shared responsibility, rather than principles of centralised markets and unequal capital accumulation. Such local control does not equate to cohesion over climate politics, but it does contribute to re-politicising the rationalities of 'big energy' and gives legitimacy to democratic action for mitigation of climate change.

Current regulatory complexity and political uncertainty mean that there is no settled means of replicating or scaling up such innovation, however, and it remains unclear whether UK energy policy aspirations to an increased contribution from cities will be made to work more cohesively or to what ends, and with what share of costs and benefits. Energy market reform and price support mechanisms remain geared to centralised electricity generation, owned by large corporations that are legally required to prioritise shareholder interests. Lack of UK familiarity with CHP/DH, high initial infrastructure costs relative to sunk investment in gas and electricity networks and long-term payback periods situate decentralised energy as economically marginal at best, in a calculus historically shaped by progressive centralisation, and latterly by privatisation and economic short-termism.

This works against collaboration between cities, and between local, devolved and UK governments. Municipal capacities could be more effectively enabled through a supportive policy framework, counterbalancing short-term financial market metrics with measures which recognise the social, economic and environmental public benefit of decentralised urban energy. This requires systematic government action to mobilise investment and to introduce policy measures of the kind used in other northern European countries; examples are directive use of planning powers for low carbon heat supply, resources to map heat demand and zone urban areas suitable for district heating, licensing to regulate provision where developers can demonstrate integrated social, economic and carbon benefits, and loan guarantees or other risk underwriting measures matched with requirements for heat network operators and suppliers to be non-profit organisations. The resources to use local powers for planning and regulation in land use, transport and waste infrastructures for low carbon energy investments would thus be enabled, and social capital would be given more constructive expression. However, none of this seems likely to happen in the UK without a shift in the country's neo-liberal political economic regime, and there urban politics are likely to prove critical.

Notes

1 See for example http://ec.europa.eu/eip/smartcities/index_en.htm, http://www.covenantofmayors.eu/, http://www.iclei-europe.org, http://www.energy-cities.eu

2 A comparison of Swedish district heating companies shows that privatisation is associated with higher prices for heating, relative to those companies which have remained in municipal ownership (http://www.nilsholgersson.nu/). In the Report *Innehållsförteckning – Jubileumsutgåva av Avgiftsrapport Nils Holgersson 1996–2005*, the second point in the box on p.35 states: 'There is a clear difference in behaviour in terms of pricing, depending on who is owner [of the DH system] and the directives which the company has to work from.' This is illustrated by the last graph on p.43 where the green line represents the (lower) average price in municipal systems and the red line the (higher) average price in privatised systems over a five-year period (2001–2005).

3 This point was made by a contributor to a workshop on district energy leadership organised as part of the research reported, and held under the Chatham House Rule. The RECP was disbanded in 2011 under public spending reductions.

4 The paper draws on interviews with three local government officers, three financial advisers, five representatives of urban energy services companies, on-going discussion with four representatives of Aberdeen Heat and Power, and interaction with a network of UK urban authorities, policy-makers in UK and Scottish governments, and district energy practitioners, centred on four district energy workshops convened by the researchers. Internal and public local government documents are also analysed. This data is part of comparative research on urban development of district heating, www.heatandthecity.org.uk.

5 The *Teckal* exemption provides that, in certain circumstances, the award of a contract by one public body to another separate legal person will not fall within the definition of 'public contract', with the result that EU law will not require the contract to be put out to tender. The *Teckal* exemption comprises both a 'control test' and a 'function test': (1) the local authority must exercise similar control over the contractor to that which it exercises over its own departments; and (2) the contractor must carry out the essential part of its activities with the controlling local authority or authorities.

Acknowledgements

I am grateful to the UK Research Councils' (ESRC and EPSRC) Energy Programme for funding for the *Heat and the City* project www.heatandthecity.org.uk and to research fellow David Hawkey and co-investigators Andy Kerr, Heather Lovell, David McCrone and Mark Winskel. Particular thanks to Jonathan Rutherford and David McCrone for comments on earlier drafts. Thank you also to Jonathan Rutherford for translation of reports from the Nils Holgersson website regarding the analysis of the effect of private vs public ownership on DH tariffs. Lastly, particular thanks to the interviewees from local authorities, finance and district energy companies for their time and generosity in providing information, expertise and copies of documents.

References

Betsill, M. and H. Bulkeley (2006) Cities and the Multilevel Governance of Global Climate Change. *Global Governance* **12**: 141–159.

Boltanski, L. and L. Thevenot (2007) *On Justification: Economies of Worth*. Princeton: Princeton University Press.

Bowles, S. and H. Gintis (2002) Social capital and community governance. *Economics Journal* **112**(Nov.): 419–436.

Bulkeley, H. and K. Kern (2006) Local Government and the Governing of Climate Change in Germany and the UK. *Urban Studies* **43**: 2237–2259.

Bulkeley, H., V. Castán Broto and A. Maassen (2014) Low-carbon transitions and the reconfiguration of urban infrastructure. *Urban Studies* **51**: 1471–1486.

Callon, M. (1986) Some elements of a sociology of translation: domestication of the scallops and the fishermen of St Brieuc Bay. In J. Law (Ed.), *Power, action and belief: a new sociology of knowledge?* London: Routledge, 196–223.

Christensen, C. (1997) *The Innovator's Dilemma*. Boston, MA: Harvard Business School Press.

Cooke, P., P. Boekholt and F. Todtling (2000) *The Governance of Innovation in Europe: Regional Perspectives on Global Competitiveness*. London: Pinter.

Coutard, O. and J. Rutherford (2010) Energy transition and city-region planning: understanding the spatial politics of systemic change. *Technology Analysis & Strategic Management* **22**(6): 711–727.

Downe, J., C. Grace, S. Martin and S. Nutley (2010) Theories of public service improvement: a comparative analysis of local performance assessment frameworks. *Public Management Review* **12**(5): 663–678.

European Commission (EC) (2011) *Energy Roadmap, 2050*. COM/2011/885 http://ec.europa.eu/energy/energy2020/roadmap/index_en.htm (accessed 12 June 2012).

Ehrnberg, E. and S. Jacobsson (1997) Technological discontinuities and incumbents' performance: an analytical framework. In C. Edquist (Ed.), *Systems of Innovation: Technologies, Institutions and Organisations*. London: Pinter.

Granovetter, M.S. (1973) The strength of weak ties. *American Journal of Sociology* **78**: 1360–1380.

Guy, S., S. Marvin, W. Medd and T. Moss (Eds.) (2011) *Shaping urban infrastructures: intermediaries and the governance of socio-technical networks*. London: Earthscan.

Hawkey, D., M. Tingey and J. Webb (2014) *Local Engagement in Energy Systems Development*. Final Report to UK Energy Technologies Institute.

Helm, D. (2010) Infrastructure and infrastructure finance: The role of the government and the private sector in the current world. *EIB Papers*, **15**: 8–27.

Hodson, M. and S. Marvin (2009) Urban Ecological Security: A New Urban Paradigm? *International Journal of Urban and Regional Research* **33**(1): 193–215.

Hughes, T. (1983) *Networks of power: Electrification in Western society 1880–1930*, Baltimore: Johns Hopkins University Press.

Jamasb, T. and M. Pollitt (2005) Electricity market reform in the European Union: review of progress toward liberalization and integration. *The Energy Journal* **26**(1): 11–41.

Kelly, S. and M. Pollitt (2010) An assessment of the present and future opportunities for combined heat and power with district heating in the UK. *Energy Policy* **38**: 6936–6945.

Latour, B. (1986). The Powers of Association. In J. Law (Ed.) *Power, Action and Belief: a New Sociology of Knowledge?* London: Routledge, 264–280.

Le Gales, P. (2002) *European Cities: Social Conflicts and Governance*. Oxford: Oxford University Press.

Lundvall, B.-A., B. Johnson, E.S. Andersen and B. Dalum (2002) National systems of production, innovation and competence building. *Research Policy* **31**(2): 213–231.

Maskell, P. (2004) Learning in the village economy of Denmark. In P. Cooke, M. Heidenreich and H.-J. Braczyk (Eds.), *Regional Innovation Systems: The Role of Governance in a Globalised World*. London: Routledge.

McFarlane, C. and Rutherford, J. (2008) Political infrastructures: governing and experiencing the fabric of the city. *International Journal of Urban and Regional Research* **32**(2): 363–74.

McGarvey, N. (2012) Expectations, assumptions and realities: Scottish local government post-devolution. *British Journal of Politics & International Relations* **14**(1): 153–174.

MacLeod, G. and Jones, M. (2011) Renewing urban politics. *Urban Studies* **48**(12): 2443–2472.

Monstadt, J. (2007) Urban governance and the transition of energy systems: institutional change and shifting energy and climate policies in Berlin. *International Journal of Urban and Regional Research* **31**(2): 326–43.

Monstadt, J. (2009) Conceptualizing the political ecology of urban infrastructures: insights from technology and urban studies. *Environment and Planning A* **41**: 1924–1942.

Organisation for Economic Cooperation and Development (OECD) (2001) *The Well-being of Nations: The Role of Human and Social Capital*. Paris: OECD.

Osborne, D. and T. Gaebler (1992) *Reinventing Government: How the Entrepreneurial Spirit Is Transforming the Public Sector*. Boston: Addison-Wesley.

Petz, M., D. Edgar, B. Temple and K.-D. Maier (2012) *Dynamics of Industrial Consolidation: M&A Activities in the EU27 Electricity and Gas Sector*. Venice, IAEE European Energy Conference.

Pinch, T. and W. Bijker (1984) The Social Construction of Facts and Artefacts: Or How the Sociology of Science and the Sociology of Technology Might Benefit Each Other. *Social Studies of Science* **14**: 399–441.

Portes, A. (2000) The two meanings of social capital. *Sociological Forum* **15**(1): 1–12.

Putnam, R. (2000) *Bowling Alone: The Collapse and Revival of American Community*. New York: Simon & Schuster.

Rohracher, H. and P. Späth (2014) The interplay of urban energy policy and socio-technical transitions: the eco-cities of Graz and Freiburg in retrospect. *Urban Studies* **51**: 1415–1431.

Royal Commission on Environmental Pollution (RCEP) (2000) *Energy – The Changing Climate.* Twenty-second Report to UK Parliament, Cm47. Norwich, UK: HMSO.

Rutherford, J. (2008) Unbundling Stockholm: the networks, planning and social welfare nexus beyond the unitary city. *Geoforum* **39**, 1871–1883.

Rutherford, J. (2014) The vicissitudes of energy and climate policy in Stockholm: politics, materiality and transition. *Urban Studies* **51**: 1449–1470.

Rutherford, J. and O. Coutard (2014) Urban energy transitions: places, processes and politics of socio-technical change. *Urban Studies* **51**: 1353–1377.

Sennett, R. (2012) *Together: the Rituals, Pleasures and Politics of Cooperation.* London: Allen Lane Penguin.

Steward, F. (2012) Transformative innovation policy to meet the challenge of climate change: sociotechnical networks aligned with consumption and end-use as new transition arenas for a low-carbon society or green economy, *Technology Analysis & Strategic Management*, **24**(4): 331–343.

Summerton, J. (1994) Introductory essay: The systems approach to technological change, in J. Summerton (Ed.) *Changing Large Technical Systems.* Boulder, Colorado: Westview Press.

Torrance, M. (2008) Forging glocal governance? Urban infrastructures as networked financial products. *International Journal of Urban and Regional Research* **32**(1): 1–21.

Tura, T. and V. Harmaakorpi (2005) Social capital in building regional innovative capability. *Regional Studies* **39**(8): 1111–1125.

UK DECCC (2011) *Electricity Market Reform White Paper* http://www.decc.gov.uk/en/content/cms/legislation/white_papers/emr_wp_2011/emr_wp_2011.aspx (accessed 12 June 2012).

UK DECC (2012) *The Future of Heating: A Strategic Framework for Low Carbon Heat in the UK.* http://www.decc.gov.uk/en/content/cms/meeting_energy/heat_strategy/heat_strategy.aspx (accessed 12 June 2012).

UK DECC (2013) *The Future of Heating: Meeting the Challenge.* https://www.gov.uk/government/publications/the-future-of-heating-meeting-the-challenge

Winskel, M., J. Radcliffe, J. Skea and X. Wang (2014) Remaking the UK energy technology innovation system: From the margins to the mainstream. *Energy Policy* **68**: 591–602.

11 Volumetric urbanism

Artificial "outsides" reassembled "inside"

Simon Marvin

Introduction

Volumetric urbanism is the extension of forms of ecological enclosure into new dimensions of urban life. There are three critical aspects to these changes: first, the rescaling of enclosure from the capsule or building to the domic enclosure of whole cities or at least selected urban districts, comprising complex landscapes of infrastructure, buildings and natural ecologies; second, the move to substitute existing denuded, hostile, or threatening atmospheres with new artificial and technologically mediated outsides that are interiorised in domic ecologies; and third, the shift from existing domic ecologies, which are usually designed for human or non-human occupants, to the development of hybrid domic structures that attempt to increase the productivity of human environments by extending the efficiency of labour processes and the temporal and spatial exploitation of otherwise hostile environments. This chapter explores each of these dynamics and argues that these constitute a recently emerging and highly socially, spatially and ecologically selective logic of urban development that attempts to re-create new securitised atmospheric qualities within domic and volumetric structures that are intended to transcend eco-securitised threats and constraints on urban reproduction.

Initially the chapter sets out three short vignettes of domic enclosure to be better able to understand the logic of volumetric urbanism. Each exemplar represents a different urban imaginary that seeks to develop volumetric enclosure and the artificial creation of an interiorised outside with particular atmospheric qualities.

Dome Over Houston – "Move the city indoors": In 2009 the Discovery Channel series Mega Engineering aired a new programme 'Dome over Houston'.[1] Houston is the largest city in Texas and vulnerable to flooding, serious hurricanes and extreme heat events. The engineers advising the programme makers decided the most viable technological solution was to "move the city indoors".[2] A dome would be constructed over the downtown of Houston's skyscrapers, covering a surface area of twenty-one million square feet. At a height of 1500 feet it would be the biggest roofed structure in the world. It was claimed that the dome could withstand 180-mile hurricane winds, even though it would be made up of panels constructed of lightweight film. Although an engineering imaginary, the proposal stimulated an intense debate about the technical feasibility, the potential costs and the relevance of the ecological threats to Houston. However, residents in areas outside the downtown and not under the dome asked whether they would get their own dome, or whether they would be left to face to the hostile environments unprotected.

Mall of the world – "Climate controlled city": The property developer Dubai Holding is proposing to build a 450-hectare complex comprising a shopping district, a theatre district, a theme park, over one hundred hotels and serviced apartments buildings and a health resort within a green and landscaped environment. In order to protect the development from stifling summer temperatures that often reach 50°C, the facilities will linked together by 7 kilometres of streets covered by retractable roofs that can be opened in the cooler winter months. Dubai holdings are seeking to develop "an integrated city … within pleasant environments" that would create a year-round tourist market by creating a temperature- and climate-controlled volumetric space, hailed as the world's "first climate controlled city".[3] In a city that already contains an a range of artificial environments – snow slopes, offshore islands etc. – the Mall of the World seeks to provide an integrative temperature controlled complex that can integrate these individual developments in an integrated climate controlled environment.

Bubbles – a "controlled environment": A firm of London architects, Orprojects, has developed the Bubbles project to create domes of clean air for urban residents dealing with poor air quality.[4] According to project director Rajat Sodhi, the motivation for the project was the recognition that air quality in major cities in India and China had "crossed unacceptable limits" and that consequently "you really can't step out and be outdoors. You just move from one air-conditioned space to another".[5] The Bubbles project utilises EFTE, a lightweight material that encloses the public spaces – in particular, green spaces and parks – linking air-conditioned buildings such as residences, shops and offices. The interior environment can be temperature and humidity controlled in order to develop specific microclimates for different types of ecologies. Although proposed specifically in the context of the Beijing air pollution problem, the developers see this as a model of architectural responses to climate change meant for any urban context, "because what it fundamentally does is create a controlled environment" (see Figure 11.1).

The point of this chapter is not to argue that these are all new tendencies, since they all have much longer historical antecedents. However, what is new is the way in which more programmatic and systematic attempts are being made to construct premium ecologies in an effort to transcend specific urban limits and the constraints of denuded ecologies. These represent an attempt to build secessionary artificial environments to try to ensure the continued economic and material reproduction of selected users. At the same time the attempt to bound selected environments would seem to suggest leaving behind denuded environments for those unable to perform enclosure. This chapter is concerned with linking together debates about volumetric security with arguments related to artificial atmospheres in order to develop an understanding of emerging configurations of rebundled urban environments. The chapter looks at the intersection of three sets of debates.

Figure 11.1 Bubbles – a "controlled environment"
(Copyright: Orproject 2015)

The first is an emerging form of "volumetric urbanism" – the enclosure of the urban spaces under what are characterised as domic tendencies. This is the way in which segments of cities – usually (but not exclusively) premium spaces and ecologies such as high-end retail, residential, office complexes, shopping complexes and parks or gardens – are enclosed in semi-spheroid domic enclosures. The second theme is the emergence of "artificial environments". Of course, all environments in buildings and cars are to some extent artificial or constructed environments – often with sophisticated forms of climate control. The paper chapter suggests that there has been an extension of artificial life support systems into all the essential resources of human life – for instance, food production, more intense forms of atmospheric control (temperature, humidity, air quality etc.) and also forms of natural ecology such as gardens, parks and specialist biospheres. The argument here is that the commons is increasingly being recreated within and through artificial environments. The third theme is that these artificial domic environments increasingly have hybrid tendencies. There is a subtle merging of the techniques, technologies and practices of producing biospheric domes – domes designed to protect selected non-human ecologies – taking threatened biomes and reproducing them in urban settings for occupation by humans and non-humans. There appears to be a subtle merging of the specialist techniques of biospherics life support for zoos, birds, plants and so on, and the engineering of artificial life support environments for human reproduction.

The argument is structured in four sections. The second section examines a shift from capsular to domic urbanism. This briefly reviews the literature on enclosed urbanism and charts a shift from a focus on forms of capsular enclosure, which were essentially about short-term, micro-enclosed vehicles and individual buildings, to an up-scaled more comprehensive form of domic enclosure. The new features of urban domes are attempts to enclose all or selected parts of cities under some form of artificial enclosure that fulfils many different functions and aspects of everyday urban life. The third section focuses on the environments created by these domic enclosures and identifies the ways in which these are primarily designed to create artificial environments – the aspects that are primarily involved in the re-creation of artificial outsides inside a protected dome. Sometimes they are also about bringing denuded outside environments "inside". The chapter illustrates the ways in which new techniques, technologies and practices are being developed to essentially recreate clean, safe and reproducible external environments within domic ecologies. The fourth section then charts the emergence of hybrid domic ecologies, which are designed to create protected and total environments for human and natural environments. The concluding section identifies some key areas for further research.

Volumetric Enclosure – from 'capsular' to 'domic' ecologies

The argument here focuses on the ways in which there has been a rescaling of the form of enclosure of urban life support systems from the capsular to the more widespread domic enclosures of whole cities, urban cores or at least specialist urban districts. There are three main contributions here that are important to the argument – cabin ecologies, capsular urbanism and splintering urbanism. Here we can identify an emerging logic – domic or volumetric urbanism that attempts to extend the scale of enclosure.

First, the history of ecological architecture has helped uncover the critical role of what Peder Anker terms "cabin ecologies" in shaping contemporary understanding of the techno-scientific basis of architectural and urbanists' interest in the development of green buildings and eco-cities (Anker 2005, 2010). Central to this history of green urbanism was the development of "cabin ecologies" – enclosed human life support systems in the form of specialist ecologies, developed for the military for use in aircraft, submarines, bunkers and so on, and then further accelerated in the space programmes of the US and former USSR to create protected environments for astronauts in the hostile environment of space (Hohler 2008). Cabin ecologies developed a new technoscience of human life support: the analysis and modelling of the resource requirements and outputs required to sustain human metabolism; the development of new technologies for human life support – providing heating/cooling, clean air, treatment of wastes, food systems, and energy production that could be implemented in a (quasi-) autonomous enclosed ecology; and the development of these systems at a range of different scales – the space suit (De Mochaux 2011), the space cabin and the enclosed environments of biospheres and regenerative environments (Nelson et al. 2009;

Pickett et al. 2010). This technoscientific approach exerted a profound influence on the ecological designers – urbanists and architects – who extensively utilised these approaches from the design of cabin ecologies in their proposals for green architecture, semi-autonomous buildings, self-reliant districts and ecocities for application on earth (Cohen 2011; Leslie 2006; Luke 1997). What this work helps us understand is the development of a new technoscience of human life support that created an artificial atmosphere to temporarily support the human occupancy of stressed and hostile external environments. These techniques for creating the autonomous cabin also influenced the early champions of green architecture and the interest in self-reliant and recirculatory infrastructure – decentralised energy, reuse of water, exploiting waste resources etc.

Second, the work of De Cauteur and his concept of "capsular urbanism" helped shift the focus from the building to other forms of enclosure that make up the urban landscape (De Cauteur 2004). Capsular urbanisms refer not just to static buildings but also to a much wider range of mobile capsules –automobiles, other motor vehicles, trains, aircraft – that link the enclosed ecologies of buildings through complex mobility networks. Cities themselves are viewed as capsular environments of "the airport, shopping mall, theme park and gated community" that are then linked by other capsules (2004: 49). Although not specifically focused on questions on the ecologies of urban life support – the work is more focused on the processes of securitisation associated with the capsular urbanism – what De Cauteur helps us think about is the way in which logistical chains of capsular enclosures link together everyday life within some sort of wider atmospheric commons. The interesting question here is what happens to the artificial atmosphere of the capsules when the atmospheric commons itself becomes denuded and degraded.

Third, the work on "splintering urbanism" illustrated the critical role of enlarged urban spaces in forms of extended enclosure. It examined how the fabric of many cities is "starting to fragment into giant cellular clusters of packaged landscapes made up of customised and carefully protected corporate, consumption, research, transit, exchange, domestic and even health care spaces" (Graham and Marvin 2001: 5). The contribution was to link the development of these extended forms of spatial enclosure with the development of infrastructure networks that were unbundled from integrated systems to provide specialist and customised energy, telecommunications, water and mobility services to enclaves. The issue here is whether there is an extension from a focus on networked resources to the wider qualities of the atmosphere within which fragmented urbanism is contextualised. Is it possible to identify similar tendencies in the development of premium ecologies that extend from the infrastructure to essential life support of the atmosphere itself?

What this existing literature helps us understand is the ways in which urban enclosure is associated with different scales (cabin, capsule, building); the way in which urban life is mediated (but not for all urban residents) through occupation of different types of static and mobile capsules; and the way in which selected and customised infrastructure ecologies (water, energy, wastes etc.) are provided for these premium enclaves. The interest now is what happens when the scale of enclosure shifts from the static or mobile capsule itself to involve the spaces between capsules in extended urban landscapes – a form of 'volumetric enclosure' that encompasses multiple urban forms in a kind of spatial and atmospheric enclosure (Klauser 2010; Sloterdijk 2009). There are two forms of extension here that are important for the argument. The first is the shift from two-dimensional to three-dimensional forms of secure enclosure – this is less about the control of a boundary and more about the control of volumetric space (Klauser 2010). The second, which is addressed in more detail in the next section, is to extend the focus of life support from network infrastructure to encompass the atmospheric qualities of volumetric enclosure (Sloterdijk 2009).

There are at least three emerging styles of urban dome being developed – and possibly others – that start to provide a way of thinking about this rescaling of eco-securitised forms of urban enclosure. Here we will give a brief overview of these emerging forms – being careful to distinguish between materially constructed domes and those that are either virtual (electronic) or imaginary, such as design proposals. This sets out the shift towards developing an understanding of the scale of enclosure – either the enlarged city or a whole urban area – but also paying attention to the specificities of each style.

The first style is what are termed ***Urban Eco-Domes***. These domes are comprised of a material surface that permanently or temporarily encloses an internal climate controlled atmosphere primarily for human reproduction and life support. The analogue here is thinking about the rescaling of building services – heating lighting, air conditioning etc. – from a building to a wider urban complex that incorporates multiple indoor and outdoor spaces. Within architectural history there have been multiple proposals for urban domes to protect cities from hostile external environments. In the US there was Buckminster Fuller's iconic "Dome over New York", proposed in the late 1960s to protect parts of Manhattan Island from heat and cold. During this period the US Department of Housing and Urban Development funded further feasibility work on forms of domic enclosure for the city, until funding came to an end in the economic crisis of the early 1970s. In the former USSR there have been a number of design proposal for domic environments over a former gold mine, and in Moscow to protect an elite development from extreme cold weather. More recently in China there has been a series of proposals and even experiments with implementing domic enclosures to deal with air pollution. In Beijing a private international school has constructed a series of domes over school playing areas to allow children to play and undertake sports activities "outside" in domic environments with artificially processed clean air.[6] More recently other international schools have been following the same trend, simulating a normal environment with hospital-grade air-filtration systems and pupils entering and leaving through an airlock. There are other examples of domes in China that reverse the logic of the clean air domes by seeking to enclose the denuded outside environment within a domic enclosure. In Hangzhou the city has constructed a 36 metre high dome that covers a 20,000m^2 zone over a former pesticide factory that was emitting noxious smells.[7] Local residents argue that the dome is not effective in preventing odours because it only occupies half the former factory site. This is an example of the construction of a dome that attempts to internalise a denuded exterior environment in a new interior. This is part of a wider set of responses that seek to construct enclosed ecologies for human life support, utilising domic enclosures either to protect humans inside new outside environments, or to bring hostile outside environments inside.

The second type of dome enclosures is ***Urban Security Domes***. These are virtual domic enclosures, composed of radar and control systems (and even missile batteries) able to monitor, classify, anticipate and in some cases intercept ecological and military threats to urban areas. The 'Iron Dome' is an Israeli missile defence system developed with the support of US funding, designed to intercept and destroy short-range missiles and other projectiles fired at cities and strategic sites in Israel (Shapir 2013). The distinctiveness of these systems is that they are designed to provide a form of protection for towns and cities located close to threats. Although the accuracy and reliability of these systems is contested, analysis has shown them to be much more effective than previous systems – to the extent that Israelis watch the battery responses rather than moving to shelters (Shapir 2013). A number of mayors took the Israeli military to court to try to ensure their cities were given priority over strategic sites that the military wanted protected first. There are potential sales being discussed with cities in Europe, India and Singapore for urban protection, and the same technologies have been adopted for civilian uses in monitoring smart girds and water networks using the algorithms that control the iron dome system.[8] There are other domic virtual systems using complex control technologies. For example, the COR in Rio de Janeiro uses complex radar systems and NASA weather forecasting systems to monitor weather conditions that pose a particular threat to Rio, especially low income settlements on steep ground at risk of flooding and landslides from extreme rain events that are monitored from a control room (Luque and Marvin, in press). There is a set of responses constructing virtual domes of control systems to monitor and respond to eco-political threats to urban areas – largely developed by the information and defence industries.

The third category of domic ecologies is ***Urban Biospheres***. These are porous or sealed ecologies that are highly heterogeneous and place specific, designed to support particular ecologies in an urban context. Boundaries can be virtual through designation (a park), porous through access control, and or material based on domes that provide climate control for specific ecologies. These are primarily developed by ecologists with particular local and species specialisms – urban ecologists, ecological services, green infrastructure etc., utilising technological and environmental services expertise to construct artificial ecologies. The examples here include UNESCO urban biospheres to protect important habitats within or near urban areas, "sites of excellence where new and optimal practices to manage nature and human activities are tested and demonstrated". There are other more enclosed biospheres in the form of aviaries, aquariums and environments for specialist plant life – deserts, polar environments etc. The constructed biospheres use life support systems to construct and recreate specialist ecologies in an urban setting. Perhaps the best example of these is the cooled

Figure 11.2 Urban Biospheres – Gardens by the Bay, Singapore
(Source: WilkinsonEyre)

conservatories at Singapore's Gardens by the Bay, created by architects WilkinsonEyre (Lim 2014; see Figure 11.2). These are huge glasshouses containing plants and fauna from around the world, a striking recreation of alien microclimates such as Mediterranean, desert and mountain rainforest. Urban biospheres represent attempts to select, enclose and protect a particular ecology within a domic environment with material or virtual protection. The material is of primary concern to building a specialist environment inside, allowing plants and animals to survive in an urban context where they would not otherwise be found.

In summary, then, across a range of different disciplines – architecture, urban design, infrastructure, ecology, environmental services, military and information technology – a coalition of interest in the potential of urban domes has emerged as a way of rethinking the relationships between cities and their life support systems. These developments have started to mix up and hybridise the understanding of outside and inside.

Artificial Environments – bringing "outside" inside

Urban environmental studies has fundamentally challenged the old dualism of nature and the city, instead focusing on the relational and mutually co-constructed dynamics of cities and their wider socialised ecological relationships (Luke 1997, Swyngedouw 1996). These domic responses start to further disrupt the interrelations by intensifying the hybridisation of the relations between what were considered to be the outside and inside environments of cities. Volumetric urbanism attempts to create new technically processed, artificial atmospheres that disrupt the conventional notion of inside and outside. In fact, it seeks to create pristine, managed, constructed, monitored and processed 'outsides' in new inside environments. It seeks to do this in the following three ways.

The first understands cities as constructed 'urban anthromes' with atmosphere changing qualities – affecting carbon emissions, air quality and heat islands (Alessa and Chapin 2008; Ellis and Ramankutty 2008). Until recently the biosphere has been described by ecologists as units called biomes related to particular vegetation types and climate patterns. Biomes have recently been replaced by a new twenty-one-unit classification of global ecology based on the interactions between human processes and reshaped ecologies. Urban anthromes are classed as heterogeneous environments that display new syndromes – heat island effects, air pollution and carbon island effects – that are often characterised as domic effects over urban areas (see Figure 11.3).

Urban anthromes are a clear recognition within the ecological community of the importance of socio-ecological processes in reshaping urban ecologies. Of particular interest for us are the domic qualities associated with these changes, which build on the long-standing recognition of the heat island effect and air pollution – that qualities of the atmosphere are reshaped through the social process of urbanism through the transformation of material resources for heating and mobility and their interaction with the built and natural ecology. Recent work on carbon has identified a similar domic profile to carbon emissions – the important point here being that urban carbon emissions also subtly influence local atmospheric qualities which are associated with other ecological changes in plant and animal life, as well as disrupting human health (Jacobson 2010). What these domic structures of reshaped atmospheric qualities tell us is that urbanisation subtly reshapes the atmospheric qualities of the urban environment and therefore the context in which other social and material processes of urban reproduction take effect. In this sense the atmosphere is already artificially altered by urbanisation – a reshaped ecology created outside within a domic topology over urban areas.

The second way in which volumetric urbanism attempts to create artificial atmospheres understands cities as (attempts to construct) secured ecologies through the application of monitoring and security systems to create secure/resilient spaces through the development of domic spaces of security against natural and political threats. This response involves the creation of domic virtual

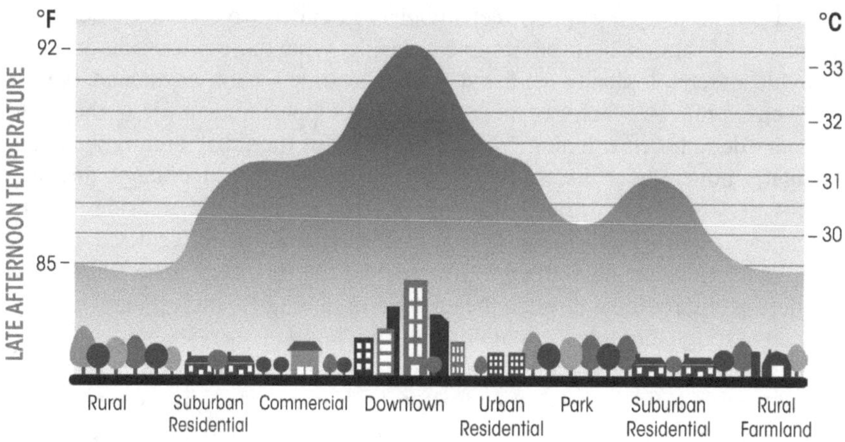

Figure 11.3 Urban Anthromes – domic syndromes
(Image courtesy of Clean Air Partnership (2010). Modified from US Environmental Protection Agency (2008))

structures – radar, monitoring systems – that span the urban context, attempting to monitor eco-security threats in the form of extreme weather, missiles and projectiles. This creates an artificial electronic domic capacity that claims to be able to monitor, forecast, anticipate, pre-empt and in some cases destroy urban threats. The clear intention here is to create a monitored domic space over an urban area with enhanced capabilities and capacities to see ecological and political threats, anticipate and even destroy threats in order to create a secure and protected atmosphere for daily life. In a sense this is about utilising technology to create an enhanced domic interior no longer at threat from exterior threats. Weather radar and even missile systems remove threats to urban circulation and enable the maintenance of the everyday through new capacities. The iron dome missile system is perceived as so effective that Israeli citizens have ignored the emergency alarms and rather than enter shelters, sit outside watching the interception of incoming missiles. Spectators have even been reported to be killed by projectiles. Palestinian citizens have begun to ask where their dome is.

The third dimension is the selective recreation of specialist ecologies for the protection of threatened species or eco-systems in cities. Of particular interest here is the specialist expertise in ecological life support needed to create artificial environments for selective planetary ecologies. These vary significantly. They can be about the creation of specially protected in-situ environments currently threatened by ecological change, such as human influence and encroachment, within established urban areas that are given special biosphere status to protect their ecological service functions. At the other end of the spectrum these are about the selection of a threatened ecology from another distant context, and its movement and recreation inside an urban context. This is a specialist ecological niche that is created inside an urban environment within an enclosed artificial ecology/atmosphere.

The final response is the creation of new forms of artificial environment that seek to re-create a new outdoors atmosphere inside a domic enclosure. Many of the eco-domes aspire to use life support technologies, extended from building services to construct whole atmospheric environments inside. What this represents is an attempt to provide a technologically mediated – clean, warmed, cool, moistened, wet, and dried – environment which can more effectively support human life in the context of a wider denuded threatening "natural" ecology of the urban anthrome. What is particularly novel about these shifts is the rather explicit and strategic attempt to create wider forms of artificial life support that recreate the 'outside' inside a protected space. It is clear that what is at stake here is the selective development of atmospheric responses that attempt to transcend the threats of a hostile – human influenced – environment through the creation of alternative atmospheric ecologies mediated through complex technological and building service systems. These ecologies are not always about creating a new outside inside; they may also be about keeping outside inside. The domic structure in Hangzhou, China covers a former heavily polluted industrial site that is leaking awful smells into the local city. These have been covered through a form of domic enclosure to keep the environmental bads contained "inside" a controlled and contained ecological enclosure. Thus there is continual mixing and movement between inside and outside.

In summary, the argument then is that increasingly technologies and ecological services are being rescaled both upward and downward from global and military monitoring to the metropolitan scale. There are increasing efforts to construct artificial environments that span extended urban areas in order to respond to the ecological and political threats of the anthropocene by creating new life support systems on an enhanced scale. No longer is this agenda only at the scale of capsules and cabins; it is also about new forms of volumetric urbanism, through which the strategic priority is to try to escape from and secure against ecological, military and political threats through artificially constructed environments.

Hybrid Domic Enclosures

A new tendency in domic responses is the emergence of new hybridised forms of volumetric environment. Rather than being solely designed for humans or non-human ecologies, there is an emerging tendency for the creation of synthetic environments that are designed for both human occupancy and selected non-human ecologies. Also, these environments are being caught up in the use of ecological technologies and techniques as well as using smart and virtual technologies for enhanced ecological control. There are three features of these hybrid environments that appear to be important in the logic of their development.

First, there is a sense of carefully selecting those aspects of non-human ecology that are the most productive ecologies for human priorities. There is an important analogue here for the type of hybrid environments being created in the work of Tim Luke on Biosphere 2 (Luke 1997; see Figure 11.4). This was an attempt to construct an enclosed environment that could support humans through the careful selection of productive ecologies that could produce clean air, food and water as well as processing waste. Peder Anker shows how experimentation with these enclosed productive ecologies came to shape conceptions of ecological urbanism – the origins of green buildings and self-sustaining or autonomous urban forms were the ecological and technologically rational principle of the spacecraft. In the case of volumetric urbanism, it is possible to see a reworking of enclosed ecologies around notions of productive ecology and transcended ecological constraints.

Second, hybrid enclosures seek to enhance productivity through a new work environment, extending the occupancy of urban time and space in a hostile environment. There are increasing examples of these productive enclosures – for instance, Amazon's new headquarters introduces eco-domes that are specifically designed to encompass humans and tree-based ecologies in more productive environments. The architects designing the scheme hope that a plant-rich environment will develop additional qualities that are not often found in a typical office setting, which are more natural and relaxing and which will allow employees to work more effectively. Trees and plant material will be selected for their "ability to co-exist in a microclimate that also suits people".[9] The Mall of the World in Dubai aspires to develop an artificial environment that enables year-round occupancy of what would otherwise be a hostile environment, through the creation of an integrated climate controlled city.

Figure 11.4 Biosphere 2
(Courtesy of Biosphere 2, University of Arizona)

Third is the notion of transcendence beyond normal ecological constraints through totally constructed artificial environments, utilising expertise from both human and non-human life support systems to ensure urban reproduction. Here, systems and techniques developed for the hostile environment of space, and sometimes for the specialist environments of the Arctic/Antarctic and deserts, are increasingly being trialled and transferred to the urban context. In the context of denuded and increasingly hostile atmospheres, new and intensified forms of volumetric enclosure are being developed to create new "outside" environments in interior enclosures.

In summary, these are new volumetric environments that create hybridised spaces for transcending ecological constraints. They are focused on the selective and partial ecologies needed for productive environments – a narrow and denuded form of ecological planning. Different forms of life support are rebundled to create artificial environments for the reproduction of humans and selected ecologies in premium spaces.

Conclusions

The argument, then, in summary form, is the following. There are important emerging tendencies in forms of enclosure in three critical respects. First, there is a rescaling from capsules, cabins and buildings to new accelerated and enhanced forms of volumetric or domic enclosure that seek to encompass atmospheres. Second, in the context of the denuded atmospheric commons there are new attempts to recreate artificially synthesised and managed outside atmospheres that are interiorised within new domic enclosures. Third, there are emerging hybrid forms that attempt to combine life support systems for both human and non-human ecologies within new productive environments that extend the efficiency of the labour process and the temporal and spatial utilisation of hostile environments.

There are three key areas for further research. The first is a more detailed empirical mapping of the emergence and practices of domic urbanism. While there have been a number of studies of particular forms of enclosure, there is little detailed comparative understanding of the purposes, processes and consequences of volumetric urbanism. The second is a more detailed analysis of the artificial 'premium environments of protection' and the spaces they leave behind – the 'denuded atmospheric commons/environments'. An understanding of the politics of production of these spaces, the tensions between them and the consequences for differential urban life support would be valuable. In particular, it will be necessary to engage in a critical assessment of the environmental and political costs of producing enclosures, and whether these transfer environmental bads and enhanced risks on to the 'global commons'. Do enclosures increase exposure and vulnerability for those unprotected from the commons?

The third key area is alternative forms of domic enclosure, based less on physical boundaries and the securitisation of environments and more on building mutual support and fair shares. Are there other types of domic support that are not based on tight boundaries? Consider, for example, another Chinese dome – the Bug Dome. This is a bamboo shelter created for the Shenzhen and Hong Kong bi-city Biennale of Urbanism/Architecture in 2009, China by a group of architects calling themselves WEAK! It was built using materials from the site and designed as a sheltered area for performance and discussion, but after the Biennale it was used as a gathering place for illegal workers from the Chinese countryside, many of whom were living in nearby shelters and abandoned buildings.

Notes

1 http://www.discovery.com/tv-shows/other-shows/videos/mega-engineering-saving-houston-with-a-dome/ (programme on YouTube).
2 http://www.greenpacks.org/2009/06/09/the-houston-dome-will-save-the-city-from-peril-may-help-environment/
3 http://www.dezeen.com/2014/07/10/mall-of-the-world-dubai-first-climate-controlled-city/
4 http://orproject.com/bubbles/
5 http://www.bbc.co.uk/news/blogs-china-blog-26984393
6 http://www.cbsnews.com/news/beijing-pollution-forces-students-to-play-under-dome/
7 http://in.reuters.com/article/2014/05/14/uk-china-environment-idINKBN0DU0UC20140514
8 http://www.metering.com/canada-to-use-israeli-missile-defence-tech-in-smart-grid-application/
9 http://www.theguardian.com/artanddesign/architecture-design-blog/2013/dec/20/amazon-futuristic-greenhouse-headquarters-seattle

References

Alessa, L. and F.S. Chapin III (2008) Anthropogenic biomes: a key contribution to earth-system science. *Trends in Ecology & Evolution* **23**: 529–531.

Anker, S. (2005) The close world of ecological architecture. *Journal of Architecture* **10**(5): 527–552.

Anker, P. (2010) *From Bauhaus to Ecohouse: A History of Ecological Design*. Baton Rouge: Louisiana State University Press.

Cohen, J.L. (2011) *Architecture in Uniform: Designing and Building for World War II*. New Haven: Yale University Press.

Colomina, B., A. Brennan, and J. Kim (Eds.) (2004) *Cold War Hot Houses: Inventing Postwar Culture from Cockpit to Playboy*. Princeton: Princeton Architectural Press.

Cosgrove, D. (1994) Contested Global Visions: One-World, Whole-Earth, and the Apollo Space Photographs. *Annals of the Association of American Geographers* **84**(2): 270–294.

De Cauteur, L. (2004) *The Capsular Civilisation, on the city in the age of fear*. Rotterdam: NAi Publishers.

De Mochaux, N. (2011) *Spacesuit Fashioning Apollo*. Cambridge, MA: MIT Press.

Driscoll, R.W. (1963) *Engineering Man for Space, May 15th, 1963 The NASA cyborg study (NASw-512)*. http://cyberneticsoo.com/wp-content/uploads/2012/01/cyborg-nasa-driscoll-1963.pdf.

Edwards, P.N. (1996) *The close world – computers and the politics of discourse in Cold War America*. Cambridge, MA: MIT Press.

Ellis, E.C. and N. Ramankutty (2008) Putting people in the map: anthropogenic biomes of the world. *Frontiers in Ecology and the Environment* **6**: 439–447.

Graham, S. and S. Marvin (2001) *Splintering Urbanism: Networked Infrastructures, Technological Mobilities and the Urban Condition*. London: Routledge.

Hodson, M. and S. Marvin (2009) Urban Ecological Security: A New Urban Paradigm? *International Journal of Urban and Regional Research* **33**(1): 193–215.

Höhler, S. (2008) Spaceship Earth – envisioning human habitats in the environmental age. *GHI Bulletin* **42**(Spring): 65–85.

Jacobson, M.Z. (2010) Enhancement of Local Air Pollution by Urban CO Domes. *Environ. Sci. Technol.* **44**(7): 2497–2502.

Klauser, F.R. (2010) Splintering spheres of security: Peter Sloterdijk and the contemporary fortress city. *Environment and Planning D: Society and Space* **28**(2): 326–340.

Leslie, T. (2006) Just What Is It That Makes Capsule Homes So Different, So Appealing? Domesticity and the Technological Sublime, 1945 to 1975. *Space and Culture* **9**(2): 180–194.

Lim, E.-B. (2014) Future Island. *Third Text*, **28**(4-5): 443–453.

Luke, T.W. (1997) *Ecocritique – contesting the politics of nature, economy, and culture*. Minneapolis: University of Minnesota Press.

Luque-Ayala, A. and S. Marvin (forthcoming) The maintenance of urban circulation: an operational logic of infrastructural control. *Environment and Planning D: Society and Space*.

Nelson, M., N.S. Pechurkin, J.P. Allen, L.A. Somova, and J.I. Gitelson (2009) Closed Ecological Systems, Space Life Support and Biospherics. In L.K. Wang et al. (Eds.), *Handbook of Environmental Engineering, Volume 10: Environmental Biotechnology*. NY: Humana Press.

Pickett, S., M. Cadenasso, J. Grove, C. Boone, P. Groffman, E. Irwin, S. Kaushal, V. Marshall, B. McGrath, and C. Nilon (2010) Urban ecological systems: Scientific foundations and a decade of progress. *Journal of Environmental Management* **92**: 331–362.

Shapir, Y.S. (2013) Lessons from the Iron Dome. *Military and Strategic Affairs* **5**(1): 81–94.

Sloterdijk, P. (2009) Foam City: About Urban Spatial Multitudes. In N. Turan (Ed.), *New Geographies*. Cambridge, MA: Harvard University Press.

Steinberg, P.E., E. Nyman, and M.J. Caraccioli (2011) Atlas Swam: Freedom, Capital, and Floating Sovereignties in the Seasteading Vision. *Antipode* **44**(4): 1–19.

Swyngedouw, E. (1996) The City as a Hybrid – On Nature, Society and Cyborg Urbanisation. *Capitalism, Nature, Socialism* **7**(2): 65–80.

12 Infrastructures and practices

Networks beyond the city

Elizabeth Shove

Introduction

Big infrastructures are complicated to design and operate and expensive to build and maintain. They have other features too, but these are the characteristics that dominate academic research and writing in science and technology studies, in urban and social theory and in various forms of historical and economic geography. Across these disciplines common projects include those showing how infrastructures are variously shaped by forms of urban and regional governance, by local and national politics and by the aspirations of city planners, engineers and entrepreneurs. In analysing these processes some commentators focus on the conflicts and tensions involved. Others highlight inequalities arising from the radically different forms of provision that characterise cities in different parts of the world. Meanwhile, for those interested in sustainability, urban infrastructures represent important sites of actual or potential renewal, and of transition towards more efficient or lower carbon regimes. Together, these and other such lines of enquiry constitute diverse but nonetheless skewed traditions in which matters of provision and supply take centre stage.

For various reasons there is much less academic debate about what infrastructures are *for*, or about the constitutive but also complicated relation between infrastructures and the patterns of consumption and practice they sustain.[1] In making the case for paying more concerted attention to infrastructures-in-use, I emphasise analysing the various ways in which they are implicated in the emergence, the disappearance and the cultural and spatial circulation of an array of social practices. In what follows I explore three themes that come to the fore when the agenda is turned around in this way. The first concerns the co-evolution of infrastructures and practices – a topic I consider with reference to a short stretch of road. The second has to do with the relation between infrastructures and the diffusion of social practices, here explored with reference to wiring and electric refrigeration. My third example, relating to cycling and driving, illustrates forms of obduracy and flexibility in what roads are and in how they are used. In combination, these three cases provide new insight into the spatial and temporal dynamics of infrastructures-in-use.

This method of approaching infrastructures as living systems, and as systems that are integral to the circulation and the emergence and dissolution of different aspects of daily life, has a number of intriguing consequences. In the context of this book, two are especially important. First, it is evident that distinctions between networked, post-networked or pre-networked cities fail to capture the multiple layering and historical co-existence of different possible conjunctions of infrastructures and practices. Second, when focusing on infrastructures-in-use, and hence on demand as well as supply, the city does *not* figure as an especially meaningful or self-evidently relevant point of reference. Instead, it becomes more important to see infrastructures as material and institutional arrangements through which complexes of social practice are constituted at various scales. This is not to deny their localised, materialised existence, but it is to highlight the large, spatially extended social phenomena of which infrastructures are also necessarily a part.

In bringing the chapter to a close I suggest that it is useful to distinguish between the geographies and politics of infrastructures-in-the-making and of infrastructures-in-use (Edgerton 1999). In particular, this distinction enables us to see trends and processes that are obscured by approaches in which infrastructural reconfiguration is conceptualised as a form of *urban* change. Before developing these ideas in any detail I begin by briefly reflecting on why infrastructures, many of which depend on complex forms of interconnection, have been analysed as distinctly urban phenomena.

Infrastructures as urban phenomena: priorities and preoccupations

In many ways it is odd to even question the urban focus of infrastructural research. On a number of counts it would be fair to say that infrastructures are *inherently* urban. In purely practical terms, the flow of materials into and out of densely populated areas calls for a level of organisation and management that does not pertain in other settings. This feature generates a host of interesting questions about how cities, as especially demanding sites of consumption, relate to surrounding hinterlands of resource production and waste management. In addressing these questions it is impossible to overlook the fact that infrastructures – whether of water, road, rail, power or waste – are implicated in the social and economic ordering of urban-rural relations.

A second, also obvious point is that cities are uniquely suited to infrastructural innovation, providing entrepreneurs with greater opportunities for profit than could ever be imagined or achieved in the countryside. As Harrison's work on the development of electricity in the USA demonstrates, issues of urban scale and density proved crucial for the flow of finance and for the economic calculations and rationalities involved in constructing profitable power stations and handling fluctuations in electrical supply and demand (Harrison 2013).

In short, cities need infrastructures, cities enable infrastructures, and in histories of infrastructural development, cities prove to be important actors in their own right (Schott 2010). In his beautifully crafted account of electrification in Western society, Hughes (1993 [1983]) compares the first forms of electrical provision, showing how these developed within Berlin, London and Chicago. While his work is in part about how knowledge circulated and travelled between experts and engineers, it is also about how networks were shaped by necessarily localised city-specific histories and politics. Though dealing with other systems and places, authors including Rosen (1986), Guy, Marvin et al. (2001) and Coutard, Hanley et al. (2005) contribute to this tradition, countering overly simple narratives of technological diffusion by showing that the historical development of infrastructures differs from one city to the next.

Today, as in the past, the urban is where the action lies. Every way we look, cities matter: they are the hubs around which road, rail and air transportation revolve; they are the nodes of which more extensive grids are made; they are the destinations to and through which cables are routed.

Despite all this I want to argue that as topics of academic analysis and enquiry, cities and infrastructures are not obvious, natural or necessary bedfellows. On reflection, the seemingly self-evident connections outlined above are better understood as outcomes of a distinctive and somewhat narrow approach. Brenner addresses much the same issue in a recent discussion of how "the urban" is mobilised in academic discourse. In an article entitled *Theses on Urbanization*, he contends that the "'unit-like' character of the urban" should be "understood as a structural product of social practices and political strategies, and no longer their presupposition" (Brenner 2013: 109). So what are the analytical and practical, political strategies that lie behind methods of conceptualising urban and infrastructure in the same frame?

Writing about networked and post-networked cities, about "splintering urbanism" (Graham and Marvin 2001) and cities as sites of infrastructural transition (Hodson and Marvin 2009) makes sense *if* we are in essence interested in methods of organising and governing the hardware of urban metabolic systems. It makes sense *if* we take city authorities and planners to be key actors and gatekeepers, capable of promoting or stemming innovation, and it makes sense *if* we are above all interested in the economics and practicalities of provision. It should come as no surprise that people who work in schools of planning, architecture, urban geography and design *are* interested in pursuing these questions. For them, the discursive framing of urban and infrastructure figures as an enabling device: it permits significant debates to be had about how and where roads, power and water supplies are built, about urban governance, and about the consequences and future implications of such arrangements for issues of resource consumption, access, equity and justice (Bulkeley, Broto et al. 2012).

But if these are *not* the questions we want to follow then the urban focus, as mobilised above, is something to be overcome, re-worked or worked around. As Brenner observes, other concepts and terms are required to show how forms of accumulation and spatial regulation "cascade along intercontinental transportation corridors; large-scale infrastructural, telecommunications, and energy networks" (Brenner 2013: 88). When abstracted from the city and reinserted in the borderless world of business, infrastructures reappear as opportunities for financial gain, for transnational innovation, and for increasing market share. With these questions in mind, different features of the city pop into view: if you look down as you walk along many streets in Europe you will discover that the manufacture of simple elements like manhole covers is dominated by a handful of companies.[2] To give another example, conceptualising infrastructures as sites of regional political cohesion (Hård and Misa 2008) or of international tension and conflict (Collier and Kemoklidze 2014) calls for a repertoire of non-urban terms borrowed from neighbouring fields like politics and international relations. Further methodological challenges arise when looking at infrastructures not as networks in their own right, but as material arrangements that are integral to the conduct of daily life and that are enmeshed in the ongoing dynamics of social practice – that is, to doing things like travelling to work, lighting and heating homes, keeping food cool and so forth. What terms and spatial scales should we work with in conceptualising and understanding infrastructures-in-use?

Following Brenner, I agree that an emphasis on social practice calls for new ways of conceptualising the "*patterning* of modern sociospatial landscapes" (Brenner 2013: 95). But what does this involve, and where should we begin? The rest of this chapter attempts to address these questions, first commenting on the relation between infrastructures and practices in general terms and then working through a series of more specific questions with reference to a selection of empirical cases and examples.

Infrastructures and social practices

At one level, the relation between infrastructures and the range of social practices enacted in contemporary society is direct and simple. Road and rail systems matter in how and when people travel and for where they go; mains water is an integral element of daily showering and washing with an automatic machine; electric power is significant for all manner of activities including forms of entertainment, communication, lighting, cooking and so forth. The snag is that observations like these tell us little about *how* infrastructures constitute and are in turn outcomes of daily life.

Fortunately, the literature on social practice theory provides some clues. Within this literature, social practices are generally defined as entities that exist across space and time (Giddens 1984). A practice is thus something that persists beyond any one moment of enactment; it is socially shared, and it involves the active integration of a range of constitutive elements including meanings, forms of competence and materials. From this point of view, practices have recognisable 'lives' of their own: they circulate and travel, and they emerge, endure and disappear. In addition, some of these dynamic processes relate to the ways in which practices interact with each other and to how bundles or complexes of practice hang together (Shove, Pantzar et al. 2012).

Thinking about how infrastructures relate to the lives of practices, so defined, is clearly not a matter of understanding instances of individual user behaviour, or of quantifying demand for the resources and services that infrastructures deliver. Instead, and as illustrated below, understanding the various interconnections involved calls for a more subtle account of the historical co-evolution and spatial distribution of infrastructures, resources and the devices and appliances on which different practices depend (Watson 2012).

In approaching this topic there are several generic features to bear in mind. One is that infrastructures typically enable many practices at once. This is important for the ways in which infrastructural histories have been and might be organised. Although there are numerous excellent accounts of infrastructures-in-the-making, these singular narratives are rarely matched by parallel analyses of the necessarily *multiple* transitions in practice that are also entailed.

To elaborate, the electrification of Western society depended on a whole lot more than the reliable provision of electric power. It also required the electrification of a multitude of social practices, each differently transformed by the introduction of some new device, and by the consequent reconfiguration of competence and meaning. As Cowan reports, electric washing machines made a real difference to the cultural significance of doing the laundry and to the skills involved (Cowan 1983). Electric cookers also transformed methods and habits of food preparation, but not in the same way. To complicate matters, a number of social practices, and a number of material elements/appliances, rely on the co-existence of several infrastructures at once. Whether multiple systems are involved or not, the key point is that accounting for infrastructures-in-use calls for a detailed and differentiated understanding of how diverse social practices are reconfigured.

A second, also generic feature is that the relation between social practices and infrastructures is almost always mediated by 'appliances', by which I mean things like computers, refrigerators, cars, bikes, showerheads etc. This is important for how infrastructures and practices influence each other and for how such relationships might be revealed.

The fact that relations between infrastructures and practices are multiply mediated means that there is (consequently) scope for substitution, adaptation and dislocation. The 'same' practices can be enacted in much the same way despite sometimes significant differences in how resources and services are provided. As a result, changes in infrastructural arrangements (e.g. from centralised to decentralised power supply, or from local to more distant forms of water treatment etc.) are unlikely to show up *in* practice unless they affect the 'proper' functioning of one or more mediating devices. By the same token, individual practices and mediating devices/appliances may well change, sometimes radically so, but with little or no tangible impact on the infrastructures on which they multiply and variously depend. Among other things, this means that analyses of the relation between infrastructures and practices need to consider the *mediating* function of appliances – broadly defined. As material elements of practice, appliances are implicated in the stability of what people do, and in how that changes (Shove, Pantzar et al. 2012). They also constitute what we might think of as the sensitive tips of the infrastructures to which they connect: defining both the extent and the patterning of demand, and the institutional and material configuration of supply.

A third feature, again related to the mediated character of infrastructure-practice relations, is that there are likely to be different but co-existing forms of adaptation on the one hand, and obduracy on the other. In other words, infrastructures and complexes of practice can change at different rates and in different directions. For example, canal systems built for freight are now important sites of leisure. There have certainly been changes, including major programmes of restoration, but the point is that the basic hardware, including material arrangements of locks, bridges, towpaths etc., is capable of sustaining really very different forms and patterns of use.

In combination, these comments indicate that understanding infrastructures-in-use depends on understanding the ongoing dynamics of different practices, and the range of material elements through which infrastructural relations are mediated and made. These ideas provide a basis from which to consider methods of addressing three more specific questions: how do infrastructures and practices co-evolve? How do infrastructures enable the circulation of practices, and how do infrastructures relate to changing complexes of practice? The challenge of analysing and representing these different aspects calls for some methodological ingenuity and for creative thinking about how to define infrastructure-in-use and identify relevant processes of change. It is obvious that conventional methods, such as tracking the various actors with a stake in managing and governing networks of one kind or another, will not do, so what other approaches might we take?

The next three sections explore different ways of responding to the questions outlined above. The first strategy is to pick on a specific location – a bend in a road – and consider the co-evolution of infrastructure and practice from that one spot. The second is to focus on a process: to ask how food is cooled and take that as a way into the analysis of infrastructure-practice relations. The third is to take stock of how an 'unchanging' infrastructure (the UK road network) accommodates overlapping and rival practices.

These exercises and examples should not be interpreted as historically significant case studies in their own right, or as means of illustrating what I take to be generic trends and processes. Instead, each is designed and written with the aim of articulating particular features of the mutually constitutive relation between infrastructures and social practices. As becomes obvious, in putting these accounts together I have foregrounded aspects of materiality, and I have done so at the expense of other crucial media through which infrastructures and practices interact.[3] This is evidently not the only way to go, but it is a way that allows me to capture and conceptualise the spatial and temporal dynamics of infrastructures-in-use.

How do infrastructures and practices co-evolve?

This question led me to a section of road some 70 kilometres south of London. Other stretches would do as well, but the piece I have selected is especially well documented. At the point where the current A3 passes underneath a village called Hindhead and behind an impressively steep valley called the 'Devil's Punch Bowl', there are in fact three co-existing versions of the same road. The first version, which was the original Portsmouth Road, was re-routed by the Kingston-upon-Thames to Sheet Bridge Turnpike Company in 1826 to provide a gentler incline for the many horses and wagons that used it. The original steeper route (the old A3) remained as a rough pedestrian track. As cars replaced horses the sharp bend around the head of the valley became notoriously dangerous: for a time the A3 had 40% more traffic accidents than other roads of its grade. There were numerous attempts to solve this problem, including cutting the road deeper into the hillside (in 1968), but in 2011 what was by then the last remaining stretch of single carriageway was finally abandoned. Some 20,000 to 30,000 vehicles a day[4] now speed along the 'new' A3 and through the Hindhead tunnel. The tunnel, which is by all accounts a triumph of modern engineering (Stafford 2013)[5] is also an outcome of more than thirty years of transport planning, and a final answer to years of local and national controversy. At a cost of £386 million for just 1.83 kilometres, it is the second most expensive piece of road (per metre) ever constructed in Britain.

To summarise, the Portsmouth Road (the first A3) was paved with stone, and when it was first abandoned it became a path for walking, not for horse-drawn traffic or cars. The second A3 was re-routed for horses, surfaced with tarmac, upgraded and then abandoned. Having been covered with earth from the tunnel excavation it is now turning back to heath land. Meanwhile, the new A3 runs underground, an expensive route designed to reduce accidents, ease congestion and provide a fast dual carriageway from London through to Portsmouth.

All three forms certainly counts as infrastructure, but how do they relate to changes in social practice? One answer is that the three A3s are outcomes of successive, generic transitions from walking on foot to the use of horse-drawn carts (hence the need for a gentler incline), and from a regime of wagons (Sturt 1923) to a fast-moving system of automobility. These developments are inseparable from innovations in the production and ownership of cars, the diffusion of driving skills and changing patterns both of work and leisure. More abstractly, patterns of mobility constitute and are occasioned by the changing organisation of space and time: the A3 today is thus part of an historically and geographically specific patterning of origins and destinations, including forms of commuting and leisure, in this case involving week-ends spent at the seaside.

More immediately, the A3 simultaneously enables and depends upon connections to other roads and places. It is used for local and long-distance journeys and is not only or singularly a road to Portsmouth. At the same time, it is not 'any' road: the details of the route reflect the movement of people and goods between London and Portsmouth, a major harbour and naval base on England's south coast. Furthermore, the specificities of the local geology matter hugely. Less obvious, but also important, the A3 in use depends on other 'local' infrastructures including the London-Portsmouth railway which opened in 1858.

Looking down from the head of the Devil's Punch Bowl, we can see the different incarnations of the A3 and its presently co-existing forms, and we can see that these simultaneously figure as: a) elements in certain localised performances of generic practices (walking, driving a wagon and horses, travelling by car); b) elements of a much more extensive road network that facilitates multiple forms of living, working and moving around; and c) parts of infrastructure that are constantly over-written by changing configurations of daily life. From this point of view, the A3 (or more precisely, the three A3s) provides some insight into the persistently dynamic relation between infrastructures-as-material-arrangements and infrastructures-in-use.

How do infrastructures enable practices to circulate?

I now turn from a discussion of the co-existence and successive transformation of infrastructures and practices, as revealed from the vantage point of one location, to an exercise which is primarily concerned with processes of diffusion and circulation. I also turn from roads to networks of electric power, focusing on the relation between electrification, the availability of freezers and refrigerators, global systems of food provision and changes in cooking and eating.

Infrastructures typically link dispersed people and practices. In the case of electricity, the scale of interconnection varies: relatively de-centralised systems exist alongside highly integrated grids. However, the general pattern is one in which the output from one or more power stations is *distributed* to multiple, disparately located consumers. Although distribution is integral to these arrangements, there is relatively little academic literature on the central role of wire, miles of which were involved in making even simple city-centric systems of power supply. This is surprising in that the economics of distribution, themselves closely related to issues of density and topography, proved crucial for the directions and rates at which electrification proceeded, and for the very different calculations involved in taking electricity to rural areas (Harrison 2013).

In practical terms, the challenges of developing high and low voltage transmission systems called for innovations in engineering and for significant investment in labour and specialist expertise, not to mention raw materials. Dealing with the hazards of lightning strikes, frost and ice, thermal differential and sagging cables involved the development methods of extending the distance between sites of production and consumption. These forms of spatial extension are integral, not incidental, to the making of a functioning infrastructure, and to the project of establishing a network the tentacles of which now reach far and wide.

Wiring, and in particular the wiring involved in long-distance 'point to point' transmission, figures in Hughes' account of how distant power sources (hydropower) were conveyed to urban centres of consumption (Hughes 1993 [1983]). Nye also writes about how the first industrial connections established a network of cables that facilitated further extension to domestic consumers (Nye 1992). However, neither foreground the importance of distribution for the circulation of power, and neither follow wires as they cross thresholds, threading their way through walls and ceiling spaces and into the fabric of the home. Despite being vital for systems of electric power and for the daily operation of the grid, homes and workplaces are hardly ever thought of as part of the infrastructure itself (Patterson (2008) is a notable exception). Although largely unnoticed by those who analyse the 'big' infrastructures of the day, innovations in this area include the provision of an increasing number of power sockets[6] and extension cables, each connected to an also increasing number of appliances.

The question of how this extensive diffusion of power translates into the practices of daily life is complicated, not least because electricity (mediated by a multitude of appliances) has many uses. Rather than considering this topic in general, I will focus on freezing and refrigeration. Electrically-powered fridges and freezers represented a step change in the production and distribution of food, and in habits of cooking and eating. As such, they illustrate the part that infrastructures play in separating sites of consumption and production, in the same move enabling the circulation and diffusion of new practices.

Previously important methods of keeping food cool, including the use of ice or larders, were neither sufficiently consistent nor sufficiently widespread to enable what is now described as the "cold chain". The cold chain depends on an unbroken series of spaces, from sites of production, through to transportation, retailing and the home itself that are maintained at set temperatures. The global food industry, which is now thoroughly reliant on the cold chain, is also critically dependent on the very widespread, very reliable diffusion of both electrical power and fridges and freezers.

Accounts of how refrigerators were first developed (Cowan 1985), and of their changing role in the home (Shove and Southerton 2000) swing between narratives of technological innovation and innovation in the organisation and management of food provisioning. Initially used to 'beat the seasons' and preserve gluts of home-grown food, refrigerators and freezers acquired new roles alongside the commercial use of cold storage and the supply of ready-frozen food via supermarkets (Hand and Shove 2007). In Friedberg's words, "the very rhythms and geographies of daily life changed … as the cold chain reached from the markets and shops into consumers' kitchens" (Freidberg 2015: 8). As the cold chains of food provisioning have extended, so the 'need' for this particular appliance and for the electricity to power it have grown. One consequence is that fridges have risen to third position in the list of priorities for the poorest of electrified households (after lighting and a television) (McNeil and Letschert 2009). Fridges do not, of themselves, change practices of eating and cooking, but in constituting and contributing to wider systems of provision, they have led to the emergence and diffusion of what Garnett refers to as forms of refrigerator dependence (Garnett 2007: 78).

As represented here, the spatially distributed character of the electricity infrastructure enabled transformations in the spatially distributed character of food provisioning, facilitating the emergence and circulation of new fridge-related practices (bulk buying, ready-meals, fast food, eating seasonal products all year round etc.). In making freezing possible around the world, the widespread availability of electric power (up to and including the wiring of the home) has arguably enabled the development and persistence of systems of food provision in which consumers are again distanced from sites of production. In doing this, it has enabled the circulation not only of food products, but also of tastes, habits and skills.

This example highlights the extent to which infrastructures sustain each other, impacting on commercial and domestic practice not in isolation but in combination. As Freidberg notes, changes in methods of food supply depended on electric power, on the widespread diffusion of refrigerators *and* on expanding transportation networks (Freidberg 2015). As with the three A3s, infrastructures and practices co-evolve, but in this case on a global scale. In effect, the distribution of electric power enables the circulation of other goods and appliances, doing so in ways that reconfigure geographies of consumption and production. One practical consequence, illustrated here through an account of the interweaving of power lines, household wiring, cold chains, and the fridge itself, is that the 'need' for electricity becomes embedded not just in one location, but everywhere.

How do infrastructures relate to changing complexes of practice?

Thus far I have underlined the extent to which practices and infrastructures depend on each other, whether at one location or across many. For example, I have suggested that systemic transitions in mobility and in the social arrangements that underpin the movement of things and people in space and time depended on, and occasioned, the sequential 'upgrading' of the A3. I have also underlined the co-dependence of freezer- and refrigerator-reliant complexes of shopping, cooking and eating and related networks of electricity and frozen space. In both cases it is important to appreciate the two-way nature of these connections: there would be no need for the A3 tunnel if the road was only rarely used; likewise, the ongoing reproduction of the frozen-food chain absolutely depends on millions of people buying and cooking frozen foods. However, my third example complicates this account.

As already mentioned, infrastructures routinely sustain many practices at once. Because of this multiple anchoring, and because of the heavy weight of materiality that is often involved, infrastructures frequently outlive at least some of the practices around which they were designed. Similarly, the mere provision of relevant infrastructure does not guarantee that specific practices will take hold or endure. To illustrate these ideas I consider the changing relation between cycling and driving.

The development of cycling as a normal means of transport laid the foundations for many of the elements on which the coming system of automobility depended, including relatively smooth well-surfaced roads (Geels 2005). While the A3 has changed course, the much more common situation is one in which routes remain the same and roads are repeatedly repaired and re-surfaced. From this point of view, they constitute relatively enduring, relatively obdurate forms of material infrastructure. However, the same cannot be said of roads-in-use. In the UK in the 1930s, "approximately one fifth of men cycled to work, and around one tenth of women" (Pooley and Turnbull 2000: 19). According to the UK Cabinet Office, "the *annual distance* cycled fell from 23 billion kilometres (13% modal share) to 5 billion kilometres (1% modal share)" between 1952 and 1970 (Cabinet Office Strategy Unit 2009: 15). As these few figures indicate, the road infrastructure is not as strongly tied to the practice of cycling as it once was (Shove 2012).

One obvious explanation has to do with the car (Watson 2012). Insofar as cycling and driving have elements in common (e.g. the road), improvements in this shared infrastructure promise to enhance the development of both. On the other hand, these common elements may prove to be crucial sites of competition: as roads become busier (i.e. filled with cars), cyclists are quite literally edged aside. From this point of view, the changing character of the road infrastructure-in-use is a consequence both of the development and diffusion of automobility and the decline in cycling.

Many of the necessary elements of cycling endured, but during the 1960s and 1970s it seems that the links which held cycling in place as a normal means of transport for the many, and not just for a few, broke apart. This trend was not confined to the UK. Pucher and Buehler note that levels of cycling dropped massively between the 1950s and 1970s across much of Europe. To quote: "From 1950 to 1975, the bike share of trips declined by roughly two-thirds in a sample of Dutch, Danish and German cities" (Pucher and Buehler 2008: 502). However, not all the figures fell as far as they did in the UK, and in some countries the bike share of trips has since risen by as much as 20%. As a result, rates of cycling within Europe currently differ by a factor of ten – ranging from an average of 0.2 kilometres per day in the UK to 2.5 kilometres in the Netherlands (Pucher and Buehler 2008: 497).

This is odd in that cars are no less established in the Netherlands than they are in the UK. In thinking about where the difference lies, de la Bruheze suggests that although levels of cycling also fell spectacularly in the Netherlands, the idea of cycling as normal retained its grip on the public consciousness. Dutch transport planners consequently hung onto the idea that roads should be for bikes as well as for cars, and acted accordingly (de la Bruheze 2000). This was apparently not the case in the UK, where rates of cycling dropped so low that for a time cyclists' needs were not provided for at all: indeed, some policies were designed to drive them from the road. The point here is that changing *practices* are of consequence for infrastructures-in-use, and that changes in how infrastructures are used are, in turn, of consequence for future planning and investment in infrastructures-as-structures.

That said, infrastructures and practices do not correspond, one to one. Likewise, local forms of infrastructural provision do not determine the longer-term evolution of practices like cycling. It is consequently possible to find bits of infrastructure that were designed with cyclists in mind, but that are no longer, or perhaps have never been, used. Stevenage, a new town developed not far from London in the late 1950s, has twenty-six miles of segregated cycleway built to allow residents to pedal back and forth between home, work or school. Some of these routes are still in use, but none carry the volume of cycles for which they were designed. In this case the local infrastructure is in place, but for reasons that have to do with changing patterns of mobility and with trends that extend well beyond Stevenage itself, the practice is missing.

As this example shows, and as has already been mentioned, infrastructures and the multiple practices they enable co-exist in ways that are complicated, contested and sometimes in conflict. Second, and as indicated above, it is possible to find different combinations of obduracy and innovation. These include instances in which practices persist in the absence of suitable infrastructures, and in which infrastructures outlast or fail to engender the practices for which they were constructed. Such otherwise puzzling situations make sense provided we think of infrastructures not as physical, geographically bounded entities, but as elements in more extensive and also more dynamic transitions in practice. From this point of view, roads constitute material elements that are both part of the practices they sustain, and at the same time are partly constituted by those same practices (Shove, Pantzar et al. 2012).

Discussion

The strategy of analysing infrastructures-in-use as opposed to infrastructures-in-the-making has allowed me to catch sight of the co-existing scales across which infrastructures act. Among other things, this method positions infrastructures as systems of *connection* – variously providing links between sites of production and consumption, between one location (London) and another (Portsmouth), and between co-existing bundles and complexes of social practice (shopping, eating, cooking).

I have worked with a selection of examples, each chosen to illustrate a different facet of infrastructure-practice relations, but in none of these has the city been an obvious starting point. Instead, all three suggest that road infrastructures, cold chains and networks of power are usefully and appropriately conceptualised as material arrangements that enable the emergence and diffusion of practices that are socially and spatially extended. The idea that infrastructures (and parts of them) are involved in reproducing what Schatzki terms 'large' social phenomena makes a lot of sense (Schatzki 2011). Ironically, it also helps make sense of how urban environments change. It does so in that city planners and authorities 'want' infrastructures not for their own sakes, but because they enable citizens to reproduce practices and ways of life that are valued and desired and that exist *elsewhere.*

When viewed as material entities, infrastructures have definite, though sometimes distributed, coordinates. They are possible to map, locate and find. By contrast, the spatial ranges and registers that apply to infrastructures-in-use are inherently fluid, being defined by the always shifting outlines of specific social practices and/or constellations of practices. This is complicated by the mediated character of the relationship between infrastructure and practice. Since electricity is (now) needed for refrigeration the geographical distribution of power is important, but electricity alone does not make food cold: for this a fridge is also required. The diffusion of appliances, and of home wiring, is therefore also part of the story. In addition, places where fridges and power are present constitute links in cold chains through which consumers, producers and retailers are actually or potentially connected around the world. On top of this we might also consider the circulation and changing contours of knowledge (how do skills in handling frozen food develop and change?) and of diet: what does the fridge contain, and to which culinary traditions does it contribute?

To reiterate, infrastructures-in-use have no firm spatial coordinates, but are instead determined by the layering and overlaying of the various elements of which specific practices are made – and at this point it is important to remember that infrastructures typically enable many practices at once.

Despite this diversity, despite the fact that any two locations are likely to be served by co-existing infrastructures of very different vintage, and despite real inequalities of access, European cities have much in common. One way or another, electricity, mains water, regular waste collection, sewerage systems, telephones, parking spaces and internet connections are increasingly available as standard. This standardisation (together with standardisation in home appliances) is such that practices of urban living appear to be converging regardless of the networked or post-networked arrangements on which they depend, and regardless of the hugely varied combinations of commercial, regulatory and state interests that lie behind them. Hård and Misa view these trends with horror, arguing that previously distinct urban identities are being eroded by a "menace of creeping homogenisation" (Hård and Misa 2008: 5). Others take a more positive view, but either way there is agreement that behind the scenes, common standards of infrastructural provision constitute new forms of spatial and cultural interconnection.

Discussions of this kind routinely suppose that in one way or another, infrastructures enable forms of social and practical convergence. Another option is to think about how past and present complexes of social practice shape the infrastructures on which they also depend. From this point of view, ambitions such as taking a daily shower have impelled people to upgrade electric circuits, add pumps, rip out cupboards and do whatever is necessary to make showering possible. Similar work goes on at a macro scale: grids are reinforced, motorways are widened and tunnels built in order to cater for an array of underspecified, overlapping and interlinked complexes of practice. Forecasts of future demand are routinely organised around units of resource (litres of water, barrels of oil) or of service (passenger kilometres travelled), and as such they skate over the different dynamics of practice on which summary estimates implicitly depend. It is nonetheless the case that at every scale, infrastructures (as forms of hardware) are made and mended in anticipation of multiple future practices, not all of which transpire. This means that as grounded, materialised entities, infrastructures are also full of imagination, embodying and carrying ideas and visions of welfare, well-being and profit[7] that circulate and extend far beyond the spots in which they are built.

In conclusion, the intellectual challenges around which this book is organised are consistent with a tradition which deals with urban infrastructures-in-the making and in-the-reconfiguration. In this context, it is appropriate to compare dominant and alternative models of infrastructure supply and to consider these processes within and as part of a corresponding dynamic of "urban change". Up to a point, it is possible to include patterns of use and demand within this frame. However, if we are to really get to grips with the reconfiguration of infrastructures-in-use, the city is *not* an especially meaningful point of reference.

As described above, the forms of socio-spatial patterning that arise from the changing outlines of infrastructures-in-use and from the circulation and disappearance of appliances and practices are complicated, fluid and exceptionally hard to define, but they are clearly not urban. At first sight, this leaves us with something of a conundrum: urban infrastructures – those which are made in the city and by the city, and which define and enable city life – are *non-urban* when viewed as infrastructures-in-use.

The way out is to realise that the contours of cities, infrastructures and social practices do not correspond. The next step is to develop methods of characterising and describing networks and forms of connection and circulation that link infrastructures and practices, that extend beyond the city, but that also make cities what they are. In this chapter I hope to have made a first move in that direction.

Acknowledgements

Thanks to Frank Trentmann and Olivier Coutard for their comments and advice.

This work was undertaken as part of the DEMAND Research Centre (www.demand.ac.uk), supported by the Engineering and Physical Sciences Research Council [grant number EP/K011723/1] as part of the RCUK Energy Programme and by EDF as part of the R&D ECLEER programme.

Notes

1 Questions of consumption and anticipated use are important for designers and planners, and for methods of predicting and providing for future demand. However, techniques of anticipating and providing for future demand typically depend upon methods of aggregating, averaging and modelling generic trends – for instance, in traffic flow or energy consumption. The part that infrastructures themselves play in creating and sustaining the very practices on which demand depends is routinely overlooked.

2 For example, Saint Gobain-PAM produces manhole covers that are exported around the world. http://www.pamcovers.com/ (last accessed 20 September 2015).

3 For example, infrastructures can be seen as carriers of moral and political ideas; they often reproduce and embody forms of knowledge (both expert and lay), and they have various organising and 'disciplining' effects on those who interact with them.

4 https://www.whatdotheyknow.com/request/traffic_volume_pre_and_post_hind and http://www.roadtraffic-technology.com/projects/a3hindhead/ (both last accessed 20 September 2015).

5 It was awarded the status of 'International Tunnelling Project of the Year' in 2010. http://www.roadtraffic-technology.com/projects/a3hindhead/ (last accessed 20 September 2015).

6 The UK's Electrical Contractors' Association currently recommends 37 double power sockets for a small home. http://www.eca.co.uk/search/?q=increasing%20power%20sockets&Module=FileLibrary (last accessed 20 September 2015).

7 Infrastructure projects generally have double lives: existing and being designed to work as vehicles of financial accumulation and profit *and* as roads, sewerage systems, power stations etc.

References

Brenner, N. (2013) "Theses on urbanization." *Public Culture* **25**(1): 85–114.

Bulkeley, H., V. Broto et al. (2012) *Cities and Low Carbon Transitions*. London: Routledge.

Cabinet Office Strategy Unit. (2009). *Urban Transport Analysis*. From http://webarchive.nationalarchives.gov.uk/20091204001009/http:/www.cabinetoffice.gov.uk/media/308292/urbantransportanalysis.pdf (last accessed 20.09.15).

Collier, S. and N. Kemoklidze (2014) "Pipes and Wires. In N.J. Thrift, A. Tickell, S. Woolgar and W. Rupp (Eds.), *Globalization in Practice*. Oxford: Oxford University Press.

Coutard, O., R.E. Hanley et al. (2005) *Sustaining urban networks: the social diffusion of large technical systems*. London ; New York, Routledge.

Cowan, R.S. (1983) *More Work for Mother: The ironies of household technology from the open hearth to the microwave*. New York: Basic Books.

Cowan, R.S. (1985) How the refrigerator got its hum. In M.D. and J. Wajcman (Eds.), *The Social Shaping Of Technology: How The Refrigerator Got Its Hum*. Milton Keynes: Open University Press, 201–218.

de la Bruheze, A. (2000) "Bicycle use in twentieth century Western Europe: the comparison of nine cities." http://www.velomondial.net/velomondiall2000/PDF/BRUHEZE.PDF (last accessed 20.09.15).

Edgerton, D. (1999) "From innovation to use: Ten eclectic theses on the historiography of technology." *History and Technology* **16**(2): 111–136.

Freidberg, S. (2015) "Moral economies and the cold chain." *Historical Research* **88**(239): 125–137.

Garnett, T. (2007) "Food refrigeration: what is the contribution to greenhouse gas emissions and how might emissions be reduced?" *Food Climate Research Network Working Paper*, from http://fcrn.org.uk/sites/default/files/FCRN%20refrigeration%20paper%20final.pdf (last accessed 20 September 2015).

Geels, F. (2005) The dynamics of transitions in socio-technical systems: A multi-level analysis of the transition pathway from horse-drawn carriages to automobiles (1860–1930). *Technology Analysis & Strategic Management* **17**(4): 445–476

Giddens, A. (1984) *The Constitution of Society*. Cambridge: Polity Press.

Graham, S. and S. Marvin (2001) *Splintering urbanism: networked infrastructures, technological mobilities and the urban condition*. London: Routledge.

Guy, S., S. Marvin, et al. (Eds.) (2001) *Urban infrastructure in transition: networks, buildings and plans*. London, Earthscan.

Hand, M. and E. Shove (2007) "Condensing Practices: Ways of living with a freezer." *Journal of Consumer Culture* **7**(1): 79–104.

Hård, M. and T.J. Misa (2008) *Urban machinery: inside modern European cities*. Cambridge, MA: MIT Press.

Harrison, C. (2013) "'Accomplished by methods which are indefensible': Electric utilities, finance, and the natural barriers to accumulation." *Geoforum* **49**: 173–183.

Hodson, M. and S. Marvin (2009) "Cities mediating technological transitions: understanding visions, intermediation and consequences." *Technology Analysis & Strategic Management* **21**(4): 515–534.

Hughes, T. (1993 [1983]) *Networks of Power: Electrification in Western Society, 1880–1930*. Baltimore: Johns Hopkins University.

McNeil, M.A. and V.E. Letschert (2009) "Modeling diffusion of electrical appliances in the residential sector." *Energy and Buildings* **42**(6): 783–790.

Nye, D.E. (1992) *Electrifying America: Social Meanings of a New Technology, 1880–1940*. Cambridge, MA: MIT Press.

Patterson, W. (2008) Managing Energy Wrong. Chatham House/SPRU University of Sussex, Energy, Environment and Resource Governance working paper http://www.waltpatterson.org/mewfinal.pdf (last accessed: 20.09.15).

Pooley, C. and J. Turnbull (2000) Modal choice and modal change: the journey to work in Britain since 1890. *Journal of Transport Geography* **8**: 11–24.

Pucher, J. and R. Buehler (2008) Making Cycling Irresistable: Lessons from The Netherlands, Denmark and Germany. *Transport Reviews* **28**(4): 495–528.

Rosen, C. (1986) "Infrastructure improvement in nineteenth-century cities." *Journal of Urban History* **12**(3): 211–256.

Schatzki, T.R. (2011) *Where the Action Is – On Large Social Phenomena Such as Sociotechnical Regimes*. Sustainable Practices Research Group Working paper 1., University of Manchester.

Schott, D. (2010) Gas and Electricity in the Urban Environment. In M. Hård and T. Misa (Eds.), *Urban Machinery: Inside modern European cities. Cambridge, MA: MIT Press,* 165–186.

Shove, E. (2012) "The shadowy side of innovation: unmaking and sustainability." *Technology Analysis & Strategic Management* **24**(4): 363–375.

Shove, E., M. Pantzar et al. (2012) *The Dynamics of Social Practice: Everyday life and how it changes*. London: Sage.

Shove, E. and D. Southerton (2000) "Defrosting the freezer: From novelty to convenience (a narrative of normalization)." *Journal of Material Culture*, **5**(3): 301–319.

Stafford, J. (2013) "Briefing: Participation, consensus and adjudication in designing the A3 Hindhead tunnel, UK." *Proceedings of the ICE-Engineering Sustainability* **166**(2): 57–60.

Sturt, G. (1923) *The Wheelwright's Shop*. Cambridge: Cambridge University Press.

Watson, M. (2012). "How theories of practice can inform transition to a decarbonised transport system." *Journal of Transport Geography* **24**: 488–496.

13 Coda

Jonathan Rutherford and Olivier Coutard

The chapters in this book have offered a diverse and compelling look at the forms, meanings and implications of 'post-networked' urbanism – an urbanism of infrastructure that no longer assumes, either ontologically or epistemologically, some kind of 'material' or 'ideal' convergence of socio-technical systems around a networked configuration, model or paradigm. Working from in-depth empirical investigation across very different urban contexts, they have explored the shifting social signification of infrastructure and the transformative changes it does and may well bring about. The collection makes a substantial contribution to current debates about the nature and repercussions of urban socio-technical change.

In recent years, even since the summer 2012 seminar from which many of the contributions to this book were drawn, it could be said that infrastructure has begun to move into the mainstream in urban studies. Whereas for most of the last three decades and more it was largely understudied, pushed to the margins, even plain ignored, it has now come to the surface, exposed and opened up as a crucial arena for understanding the urban in all its practical and theoretical diversity, from localised conflicts to long-distance planetary urbanisms.

Of course, this is partly a reflection of what might be said to be an increasing visibility of infrastructure and technology in the city and in urban practice. Discourses, strategies, policies and actions in the urban domain – whether they mobilise notions and aims around 'sustainability', 'transition', 'resilience' or 'smartness' – are aware of the potential benefits of adapting, transforming and enrolling technological systems, flows and mobilities into wider urban development.

However, it is also part of a broader epistemic turn towards exploring intersections between STS and the urban, uncovering the wide-ranging and contingent socio-technical components upon and through which urban change takes place, while critically and reflexively analysing how (or indeed whether) the urban offers new scales and sites for understanding the knowledge production, material and technological development, and 'heterogeneous engineering' operated by an ever broader realm of actors who together constitute the 'seamless web' of society and technology.

The flourishing work around infrastructure in the recent literature begs a number of questions. What is this infrastructure that has been, or is being, 'unblackboxed'? How do new perspectives change our view and understanding of it? How can we continue to extend and expose its conceptual and empirical worth while ensuring that it keeps its analytical acuity? We are excited by the many paths and perspectives that have been pursued or opened up recently, while conscious of a need to be wary of not overstretching the explanatory purchase of concepts of or around infrastructure (which, we argue, demands some engagement with material forms, technical arrangements and reconfigured territorialities). The chapters in this book have hopefully pointed to some fruitful ways forward in this delicate balancing act, stimulating further study of what appear to be on the surface 'mundane' urban objects, but which quickly reveal themselves to be captivating 'mediators' of all kinds of relations, flows, symbolisms, powers and politics.

To this end, part of the work of this book has been to try and move debate and discussion around urban infrastructure into new fertile (and invigorating) theoretical and empirical territory. Long-held frames of reference such as the 'networked city' continue to be relevant but are no longer, on their own, sufficient to capture either the diversity of ways in which infrastructure – as a loose arrangement of different systems/networks of provision, service configurations, rules and norms, finance, material and resource flows, practices, etc – is being configured and reconfigured in practice across the global South and North, or the sheer vivacity of recent conceptual reflections which contribute to rethinking infrastructure in/as theory.

This will be a continuing discussion across geographies, disciplines and perspectives. By way of, or in lieu of, conclusion to this collection, we want to expand briefly on three elements which have emerged from the chapters here and which seem to be pertinent with regard to the questions posed above and a future agenda for urban infrastructure studies.

Infrastructure as alternative

Looking beyond the networked city, as the chapters here have shown, means seeking out the sites, actors, practices and configurations that digress from the infrastructural norm, model or status quo. This is a significant shift in perspective – to stories about other ways of seeing – when a rather homogeneous style of socio-technical system organised around an expanding large centralised network form has dominated imaginaries, representations, practices and analyses for more than a century. Rather than focusing only on the recomposition of this form, and thus on infrastructure as continuous regime, the authors in this book have preferred other, new or renewed ways of doing, using and conceiving of 'post-networked' infrastructure and urbanism. The analyses uncover the practices, processes and politics of infrastructure in the margins, shadows or interstices of 'usual' urban development and 'habitual' urban lives, or show how an apparently inevitable model, configuration and/or pattern must be, and has been, displaced or even jettisoned by actual everyday infrastructure uses and routines (Furlong, Jaglin).

The chapters have not, however, converged on one perspective or approach to these alternative infrastructure configurations. Instead they have shown the variety of ways of seeing and analysing these. While infrastructure as alternative in practice means hybridity and heterogeneity, the authors here embrace and engage with these in different ways, focusing on: hybrid topologies combining mixes of local technical configurations, materials, expertise, practices and regulations (Monstadt and Schramm, Jaglin); local actors doing things differently to regime practices (Webb); experimenting with new ideas or other ways of doing (Guy and Karvonen); and revisiting and putting to work old configurations (Shove, Barles et al.). In methodological terms, this means either tracking the various local components and their interrelations within socio-technical assemblages (Furlong), situating local actions and capacities within larger political economic contexts but read in different ways (Graham et al., Cavé, Beveridge and Naumann), or abandoning network topologies for life support ecologies of atmosphere, domes and encapsulation (Marvin). As ever, though, this does mean a collective concern for excavating the practices and politics hidden within technology and through which urban socio-technical systems are (re)produced, operate, evolve and are contested.

Infrastructure as alternative is as fundamentally ambivalent in its forms and outcomes as 'networked urbanism' has been, connecting up spaces and groups in some places at some times and disconnecting others elsewhere or on other occasions. Infrastructure can thus, as we know, be put to work for very different social and political projects which need to be made visible. Even redistributive alternatives can quickly become part of dominant narratives. The authors here are aware of the inherent risks of emerging systems being captured by particular interests, and objectives becoming reoriented to ensure their reproduction. Others are critical from the outset about the likely forms and implications of 'post-networked' infrastructure. Marvin's chapter extends the analysis of splintering urbanism from premium network enclaves to a concern for selective control and management of the atmospheric qualities of volumetric enclosure which are rebundling particular ecologies into eco-securitised domic urbanisms. Graham et al.'s focus on conflicts and the reproduction of social inequalities around water provision in Mumbai lays bare the logics and practices sustaining disconnection of the urban poor for the benefit of specific elite social groups. There is contestation here not so much over network nodes and topologies, but over extended urban space: who and what is 'inside' and 'outside' particular infrastructure configurations is no longer just about access to basic services but about inclusion in or exclusion from political projects which are rebundling urban 'commons' or even the technical systems of life support for valued users.

At the same time, the papers on European cities trace the emergence of alternatives to dominant infrastructure regimes, thus offering a potentially more empowering or progressive view of who has agency or the capacity to change things. By delving into the rocky processes and complex relations through which infrastructure alternatives come to be, whether within the UK centralised liberalised national energy landscape (Webb, Guy and Karvonen) or within contexts dominated by powerful transnational operators (Beveridge and Naumann) or pressures to abandon alternative systems (Barles et al.), the authors lay out compelling narratives of socio-technical possibility capable of contesting hegemonic interests and arrangements.

Infrastructure as methodology/epistemology

Seeking out alternative infrastructure projects and arrangements means reflexive, expanded engagement with ways and means of knowing and doing research on infrastructure. We cannot argue that socio-technical arrangements are being fundamentally transformed, and then continue to use only the same tools, methods and theories as before. If the object is changing, then our ways of conceiving of and knowing the object must change too. The book has begun to outline at least three implications of this observation.

First, infrastructure cannot be seen or conceived as just (the planning and management of) water, energy, waste, transport and communications networks and systems. This traditional sectoral focus has either dilated or collapsed (depending on the situation) into something more resembling a broad infrastructure arena, zone, sphere or territory, or a set of socio-technical spaces, currently under excavation, wherein the study of the fixed deployment of systems within a set area gives way to tracking the circulation, appropriation and 'inhabiting' of artefacts, devices, techniques, regulations and practices which, as a whole, make up a politically charged infrastructure underpinning urban functioning, life support and/or social reproduction in particular ways.

Even if questions of access and affordability to the most basic and fundamental services and fluids will long be of the utmost importance, and even if each has its distinctive materialities and flows, there are many ways in which these 'sectors' are becoming blurred, combined and intertwined, as low carbon, urban sustainability, eco-security and smart urbanism objectives, among others, take precedence in some city strategies, and as urban actors combine and rebundle aims to adjust and adapt to constraints on budgets and finance, on knowledge and expertise, and on their specific powers and capacities in relation to others.

The chapters have highlighted crucial links to housing, land rights and tenure, and a whole package of urban living and urban functioning, that interconnects essential services, the systems through which these are provided, all the socio-technical components that mediate between resource extraction and consumption, and the social practices which both reproduce and are reproduced through this intricate assemblage.

So it is less about infrastructural form (and reform) in different sectors than seeking out, capturing and analysing the ways in which infrastructure as assemblage (of which technical configurations and service forms are just one part) is put to work for different political projects and intentions, the variety of claims made around infrastructure circulations, how these translate into appropriations, rebundlings and enclosures, and the uneven, often unequal, outcomes produced or mediated by infrastructure in terms of access to and use of broader 'rights' to the urban, to metabolic resources and flows, to democratic participation and to renewal.

Second, infrastructure agency is not just the remit of the usual suspects of official policy actors and utility companies, although following the off-the-record activities and outcomes of the work of these actors is crucial in exposing the limits to (their) existing ways of doing. The chapters have followed a much wider array of actors involved in transforming urban socio-technical systems, including technicians and engineers, activists and community group members, individuals, households and users of all kinds. They have also collectively highlighted something of the active role that materiality, technicity and non-humans play in this transformation. Urban infrastructure is full of life and vitality, some of which comes from its 'inhabiting' by humans, and some of which can be said to reflect the activity of more-than-human actors within evolving collective assemblages. We have followed through the chapters an incredible variety of infrastructural materials, artefacts, calculations and regulations which give themselves to the varied and flexible interpretation, appropriation and deployment of actors. Whether it is the devices for rainwater harvesting in Quibdó or the technical equations and calculations for configuring heat systems in the UK, these elements may be seen to be contingently adjusting and shape-shifting to the requirements and demands of actors, at least as much as the other way round. 'Lived materiality' (Graham et al.) is coproduced by and in assemblages.

Following these actors and agencies, and their evolving participation in collective infrastructure assemblages, produces new ways of knowing infrastructure. Enrolling into narratives the realm of local and not so local, formal and informal, human and more-than-human actors and perspectives, as well as the myriad array of material and imagined sites of displacement, disruption and deviation of given pathways of socio-technical change, is a key way of envisaging alternatives and new possibilities.

Third, tracking urban infrastructure means going beyond 'the city'. Infrastructure is so much bigger than the city, enveloping resource flows, material circulations and political agencies which cut across scales and distances. It is not for nothing that infrastructure is one of the key arenas through which urban studies scholars are re-conceiving the nature and signification of the urban. None of the chapters here have studied urban infrastructure reconfiguration within a bounded, pre-conceived territory of a single city (whatever that might be). Reconfigurations are intrinsically relational and translocal, maybe even planetary, but also dependent on extraction from, exchange with, and extension into non-urban, i.e. rural, spaces. They are also inherently differentiated between distinctive urban contexts across the global North and South. It almost goes without saying that the ways in which infrastructure is configured/reconfigured in Berlin and Paris bear only partial relation to how this is done in Hanoi or Quibdó, and, similarly, that our epistemologies of infrastructure in each case are and must be distinctive. Nevertheless, a key task now is to continue to develop global comparative urbanisms of infrastructure which make translocal similarities and differences manifest, bring situated analysis and knowledge to a more general theoretical level, and begin to form a globally relevant politics of infrastructure that can offer other pathways and more hopeful ways of seeing than the further embedding of depressingly unequal neoliberal configurations.

Infrastructure as meaningful socio-technical change

A third observation concerns, as ever, the nature, degree and prospects of the kinds of urban infrastructural change that the chapters in the book have analysed. Some authors see potentially transformative shifts in a variety of sites and events, while others focus on continuities, modest developments, and even inertias in urban socio-technical systems. This has been, and will long continue to be, the source of debate and (productive) tension within STS and urban studies. However, as we argued at the end of the introduction, there is an urgent need to unsettle and repoliticise pre-conceived notions of the changes infrastructure can contribute to. Infrastructure has been shown here to be a fragile, but fertile, zone of socio-technical controversy, suggesting that it is a key battleground for the direction of social, political and territorial change. The fundamental question behind the analyses in all the chapters remains the kinds of inclusive urban futures and urban societies that we can, and indeed should, envisage and work towards, whether as practitioners, researchers and/or citizens. This is clearly a political issue, and one that demonstrates that the politics of infrastructure will continue to be a crucial zone of negotiation, engagement and contestation for years to come.

Perhaps the key point is that we may see infrastructure as an immanent alternative, always coming or being brought into being, as a fluid process before it becomes a 'fixed' result, rather than through the lens of a pre-existing or pre-conceived hegemonic paradigm or model (or regime) under pressure to shift with the times. Viewing it this way means we see both socio-technical change and the political status and value of infrastructure in a different light. Small-scale, incremental change can thus be discerned to be meaningful and potentially transformative, and it can come to matter. Indeed, in many of the chapters, whether in Berlin, Aberdeen or Coimbatore, infrastructure constitutes a progressive site of political struggle for alternative futures which offers something different to 'business as usual', or to the dominant regime within which it is increasingly difficult to imagine path-breaking change. From these moments, and across these sites, infrastructure takes on new momentum, building up and forging new pathways to potentially more inclusive global urban socio-technical futures.

Index

FSC
www.fsc.org
MIX
Papier | Fördert
gute Waldnutzung
FSC® C083411

Kolibri 360 GmbH
Ferdinand-Jühlke-Straße 7
99095 Erfurt, Deutschland
produktsicherheit@kolibri360.de